BIOREMEDIATION OF AQUATIC
AND
TERRESTRIAL ECOSYSTEMS

BIOREMEDIATION OF AQUATIC AND TERRESTRIAL ECOSYSTEMS

Editors

Milton Fingerman
Rachakonda Nagabhushanam
Department of Ecology and Evolutionary Biology
Tulane University
New Orleans, Louisiana 70118
USA

CRC Press
Taylor & Francis Group
Boca Raton London New York

CRC Press is an imprint of the
Taylor & Francis Group, an **informa** business

A SCIENCE PUBLISHERS BOOK

First published 2005 by Science Publishers Inc.

Published 2019 by CRC Press
Taylor & Francis Group
6000 Broken Sound Parkway NW, Suite 300
Boca Raton, FL 33487-2742

© 2005, Copyright Reserved
CRC Press is an imprint of Taylor & Francis Group, an Informa business

First issued in paperback 2019

No claim to original U.S. Government works

ISBN 13: 978-0-367-45422-7 (pbk)
ISBN 13: 978-1-57808-364-0 (hbk)

Visit the Taylor & Francis Web site at
http://www.taylorandfrancis.com

and the CRC Press Web site at
http://www.crcpress.com

Library of Congress Cataloging-in-Publication Data

Bioremediation of aquatic and terrestrial ecosystem /
    editors, Milton Fingerman, Rachakonda Nagabhushanam.
        p. cm.
    Includes bibliographical references.
    ISBN 1-57808-364-8
        1. Bioremediation. I. Fingerman, Milton,1928-
        II. Nagabhushanam, Rachakonda.

TD192.5B55735 2005
628.5--dc22

Preface

Bioremediation, the use of microorganisms, by virtue of their bioconcentrating and metabolic properties, to degrade, sequester, or remove environmental contaminants, has about a 45-year history. Such uses of microorganisms for this purpose now involve freshwater, marine, and terrestrial environments. Bioremediation is a multidisciplinary area of knowledge and expertise that involves basic and applied science. Microbiologists, chemists, toxicologists, environmental engineers, molecular biologists, and ecologists have made major contributions to this subject.

The use of microorganisms to clean up polluted areas is increasingly drawing attention because of the high likelihood that such bioremediation efforts will indeed attain the effectiveness in the environment that laboratory investigations have indicated would be the case. Among the current broad array of research efforts in bioremediation are some directed toward identifying organisms that possess the ability to degrade specific pollutants. With such organisms, which have already been identified, studies are being conducted to identify the mechanisms whereby heavy metals are concentrated and sequestered. There are also ongoing efforts to tailor microorganisms through genetic engineering for specific cleanup activities. Herein, specifically, are chapters, among others, that are devoted to petroleum spill bioremediation, bioremediation of heavy metals, the use of genetically engineered microorganisms in bioremediation, the use of microbial surfactants for soil remediation, and phytoremediation using constructed treatment wetlands. A broad-based approach to bioremediation of aquatic and terrestrial habitats, as exemplified by the chapters herein, is required because of the wide variety of contaminants that are now present in these ecosystems.

This volume, which presents the most recent information on bioremediation, was written by a highly talented group of scientists who are not only able to communicate very effectively through their writing, but are also responsible for many of the advances that are described herein. We, the editors, have been most fortunate in attracting a highly talented, internationally respected group of investigators to serve as authors. We intentionally set out to present a truly international scope to this volume.

Consequently, appropriate authors from several countries were sought, and to everyone's benefit, our invitations to contribute were accepted.

We take pleasure in thanking the authors for their cooperation and excellent contributions, and for keeping to the publication schedule. The efforts of these individuals made our task much less difficult than it might have been. Also, we especially wish to thank our wives, Maria Esperanza Fingerman and Rachakonda Sarojini, for their constant and undiminishing encouragement and support during the production of this volume. We trust that you, the readers, will agree with us that the efforts of the authors of the chapters in this volume will serve collectively to provide a major thrust toward a better understanding of environmental bioremediation and what must be done to improve the health of our planet.

Milton Fingerman
Rachakonda Nagabhushanam

Contents

The Contributors

Ziad Deeb Al-Ghzawi
Department of Civil Engineering
College of Engineering
Jordan University of Science and Technology
Irbid-22110, Jordan

Yves Andrès
Ecole des Mines de Nantes
GEPEA UMR CNRS 6144
BP 20722, 4 rue Alfred Kastler
44307 Nantes cedex 03, France

Nick Christofi
Pollution Research Unit
School of Life Sciences
Napier University
10 Colinton Road
Edinburgh, EH10 5DT
Scotland, United Kingdom

Steve Comfort
School of Natural Resources
University of Nebraska
Lincoln, Nebraska 68583-0915, USA

Maia Fleming-Singer
Ecological Engineering Group
Department of Civil and Environmental Engineering
University of California
Berkeley, California 94720, USA

David J. Glass
D. Glass Associates, Inc., and
Applied PhytoGenetics, Inc.
124 Bird Street
Needham, Massachusetts 02492, USA

Alex J. Horne
Ecological Engineering Group
Department of Civil and Environmental Engineering
University of California
Berkeley, California 94720, USA

K. Inoue
Biotechnology Research Center
The University of Tokyo
1-1-1 Yayoi, Bunkyo-ku
Tokyo 1 13-8657, Japan

Irena Ivshina
Alkanotrophic Bacteria Laboratory
Institute of Ecology and Genetics of Microorganisms
Russian Academy of Sciences
13 Golev Street
Perm 614081, Russian Federation

Pierre Le Cloirec
Ecole des Mines de Nantes
GEPEA UMR CNRS 6144
BP 20722, 4 rue Alfred Kastler
44307 Nantes cedex 03, France

H. Nojiri
Biotechnology Research Center
The University of Tokyo
1-1-1 Yayoi, Bunkyo-ku
Tokyo 113-8657, Japan

T. Omori
Department of Industrial Chemistry
Shibaura Institute of Technology
3-9-14 Shibaura, Minato-ku
Tokyo 108-8548, Japan

Hanadi S. Rifai
Department of Civil and Environmental Engineering
University of Houston
4800 Calhoun Road
Houston, Texas 77204-4003, USA

Ismail M. K. Saadoun
Department of Applied Biological Sciences
College of Arts and Sciences
Jordan University of Science and Technology
Irbid-22110, Jordan

Lisa C. Strong
Department of Biochemistry,
Molecular Biology and Biophysics and
Biotechnology Institute
University of Minnesota
St. Paul, Minnesota 55108, USA

Albert D. Venosa
U.S. Environmental Protection Agency
26 W. Martin Luther King Drive
Cincinnati, Ohio 45268, USA

Lawrence P. Wackett
Department of Biochemistry,
Molecular Biology and Biophysics and
Biotechnology Institute
University of Minnesota
St. Paul, Minnesota 55108, USA

J. Widada
Laboratory of Soil and Environmental Microbiology
Department of Soil Science
Faculty of Agriculture
Gadjah Mada University
Bulaksumur, Yogyakarta 55281, Indonesia

David B. Wilson
Department of Molecular Biology and Genetics
458 Biotechnology Building
Cornell University
Ithaca, New York 14853, USA

Xueqing Zhu
Department of Civil and Environmental Engineering
University of Cincinnati
Cincinnati, Ohio 45221, USA

Molecular Techniques of Xenobiotic-Degrading Bacteria and Their Catabolic Genes in Bioremediation

K. Inoue[1], J. Widada[2], T. Omori[3] and H. Nojiri[1]

[1]Biotechnology Research Center, The University of Tokyo, 1-1-1 Yayoi, Bunkyo-ku, Tokyo 113-8657, Japan
[2]Laboratory of Soil and Environmental Microbiology, Department of Soil Science, Faculty of Agriculture, Gadjah Mada University, Bulaksumur, Yogyakarta 55281, Indonesia
[3]Department of Industrial Chemistry, Shibaura Institute of Technology, 3-9-14 Shibaura, Minato-ku, Tokyo 108-8548, Japan

Introduction

The pollution of soil and water with xenobiotics is a problem of increasing magnitude (Moriarty 1988). *In situ* clean-up may include bioremediation (Madsen 1991, Madsen *et al.* 1991), which can be defined as: (1) a method of monitoring the natural progress of degradation to ensure that the contaminant decreases with sampling time (bioattenuation), (2) the intentional stimulation of resident xenobiotic-degrading bacteria by electron acceptors, water, nutrient addition, or electron donors (biostimulation), or (3) the addition of laboratory-grown bacteria that have appropriate degradative abilities (bioaugmentation).

Molecular approaches are now being used to characterize the nucleic acids of microorganisms contained in the microbial community from environmental samples (Fig. 1). The major benefit of these molecular approaches is the ability to study microbial communities without culturing of bacteria and fungi, whereas analyses using incubation in the laboratory (classic microbiology) are indirect and produce artificial changes in the microbial community structure and metabolic activity. In addition, direct molecular methods preserve the *in situ* metabolic status and microbial community composition, because samples are frozen immediately after acquisition. Also, direct extraction of nucleic acids from environmental samples can be used for the very large proportion of microorganisms (90.0-

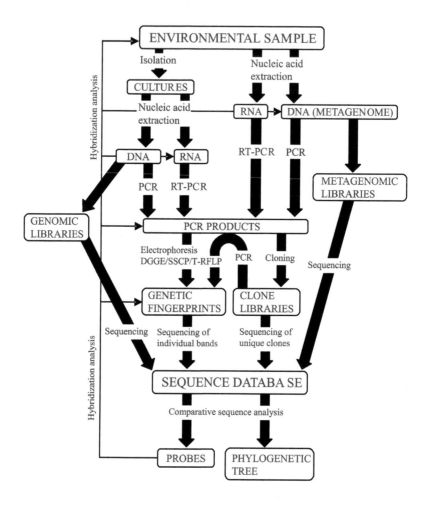

Figure. 1. Molecular approaches for detection and identification of xenobiotic-degrading bacteria and their catabolic genes from environmental samples (adapted from Muyzer and Smalla 1998, Widada *et al.* 2002c).

99.9%) that are not readily cultured in the laboratory, but that may be responsible for the majority of the biodegradative activity of interest (Brockman 1995). When combined with classic microbiological methods, these molecular biological methods will provide us with a more comprehensive interpretation of the *in situ* microbial community and its response to both engineered bioremediation and natural attenuation processes (Brockman 1995).

In this review chapter we summarize recent developments in molecular-biology-based techniques of xenobiotic-degrading bacteria and their catabolic genes in bioremediation.

In situ analysis of the microbial community and activity in bioremediation

DNA-based methods

A probe DNA may detect genes or gene sequences in total DNA isolated and purified from environmental samples by a variety of methods. DNA hybridization techniques, using labeled DNA as a specific probe, have been used in the past for identification of specific microorganisms in environmental samples (Atlas 1992, Sayler and Layton 1990). Although these techniques are still useful for monitoring a specific genome in nature, they have some limitations. Colony hybridization can only be used for detection of culturable cells, and slot blot and Southern blot hybridization methods are not adequately sensitive for the detection when the number of cells is small. On the other hand, greater sensitivity of detection, without reliance on cultivation, can be obtained using PCR (Jansson 1995).

One of the earliest studies on the use of direct hybridization techniques for monitoring xenobiotic degraders monitored the TOL (for toluene degradation) and NAH (for naphthalene degradation) plasmids in soil microcosms (Sayler *et al.* 1985). Colonies were hybridized with entire plasmids as probes to quantify the cells containing these catabolic plasmids. A positive correlation was observed between plasmid concentrations and the rates of mineralization. Exposure to aromatic substrates caused an increase in plasmid levels (Sayler *et al.* 1985). A similar technique has been reported recently for monitoring the *xylE* and *ndoB* genes involved in creosote degradation in soil microcosms (Hosein *et al.* 1997). Standard Southern blot hybridization has been used to monitor bacterial populations of naphthalene-degraders in seeded microcosms induced with salicylate (Ogunseitan *et al.* 1991). In this study, probes specific for the *nah* operon were used to determine the naphthalene-degradation potential of the microbial population. Dot-blot hybridizations with isolated polychlorinated biphenyl (PCB) catabolic genes have been used to measure the level of PCB-degrading organisms in soil microbial communities (Walia *et al.* 1990).

Molecular probing has been used in conjunction with traditional most-probable-number (MPN) techniques in several studies. A combination of MPN and colony hybridization was used to monitor the microbial community of a flow-through lake microcosm seeded with a chlorobenzoate-degrading *Alcaligenes* strain (Fulthorpe and Wyndham 1989). This study revealed a correlation between the size and activity of a specific catabolic population during exposure to various concentrations of 3-chlorobenzoate. In another study, Southern hybridization with *tfdA* and *tfdB* gene probes was used to measure the 2,4-dichlorophenoxyacetic acid (2,4-D)-degrading populations in field soils (Holben *et al.* 1992). It was shown that amendment of the soil with 2,4-D increased the level of hybridization and that these changes agreed with the results of MPN analyses.

RNA-based methods

One disadvantage of DNA-based methods is that they do not distinguish between living and dead organisms, which limits their use for monitoring purposes. The mRNA level may provide a valuable estimate of gene expression and/or cell viability under different environmental conditions (Fleming *et al.* 1993). Retrieved mRNA transcripts can be used for comparing the expression level of individual members of gene families in the environment. Thus, when properly applied to field samples, mRNA-based methods may be useful in determining the relationships between the environmental conditions prevailing in a microbial habitat and particular *in situ* activities of native microorganisms (Wilson *et al.* 1999). Extraction of RNA instead of DNA, followed by reverse-transcription-PCR (RT-PCR), gives a picture of the metabolically active microorganisms in the system (Nogales *et al.* 1999, Weller and Ward 1989). RT-PCR adds an additional twist to the PCR technique. Before PCR amplification, the DNA in a sample is destroyed with DNase. Reverse transcriptase and random primers (usually hexamers) are added to the reaction mixture, and the RNA in the sample - including both mRNA and rRNA - is transcribed into DNA. PCR is then used to amplify the specific sequences of interest. RT-PCR gives us the ability to detect and quantify the expression of individual structural genes. In a recent study, the fate of phenol-degrading *Pseudomonas* was monitored in bioaugmented sequencing batch reactors fed with synthetic petrochemical wastewater by using PCR amplification of the *dmpN* gene (Selvaratnam *et al.* 1995, 1997). In addition, RT-PCR was used to measure the level of transcription of the *dmpN* gene. Thus, not only was the presence of organisms capable of phenol degradation detected, but the specific catabolic activity of interest was also measured. A positive correlation was observed between the level of transcription, phenol degradation, and periods of aeration. In a similar study, transcription of the *tfdB* genes was measured by RT-PCR in activated-sludge bioreactors augmented with a 3-chlorobenzoate-degrading *Pseudomonas* (Selvaratnam *et al.* 1997), and the expression of a chlorocatechol 1,2-dioxygenase gene (*tcbC*) in river sediment was measured by RT-PCR (Meckenstock *et al.* 1998). Similarly, with this approach Wilson *et al.* (1999) isolated and characterized *in situ* transcribed mRNA from groundwater microorganisms catabolizing naphthalene at a coal-tar-waste-contaminated site using degenerate primer sets. They found two major groups related to the dioxygenase genes *ndoB* and *dntAc*, previously cloned from *Pseudomonas putida* NCIB 9816-4 and *Burkholderia* sp. strain DNT, respectively. Furthermore, the sequencing of the cloned RT-PCR amplification product of 16S rRNA generated from total RNA extracts has been used to identify presumptive metabolically active members of a bacterial community in soil highly polluted with PCB (Nogales *et al.* 1999).

Differential display (DD), an RNA-based technique that is widely used almost exclusively for eukaryotic gene expression, has been recently optimized to assess bacterial rRNA diversity (Yakimov *et al.* 2001). Double-stranded cDNAs of rRNAs were synthesized without a forward primer, digested with endonuclease, and ligated with a double-stranded adapter. The fragments obtained were then amplified using an adapter-specific extended primer and a 16S rDNA universal primer pair, and displayed by electrophoresis on a polyacrylamide gel (Yakimov *et al.* 2001). In addition, the DD technique has been optimized and used to directly clone actively expressed genes from soil-extracted RNA (Fleming *et al.* 1998). Using this approach, Fleming *et al.* (2001) successfully cloned a novel salicylate-inducible naphthalene dioxygenase from *Burkholderia cepacia* (Fleming *et al.* 1998), and identified the bacterial members of a 2,4,5-trinitrophenoxyacetic acid-degrading consortium.

Nucleic acid extraction and purification methods for environmental samples

Nucleic acid isolation from an environmental sample is the most important step in examining the microbial community and catabolic gene diversity. Procedures for DNA isolation from soil and sediment were first developed in the 1980s, and can be divided into two general categories: (1) direct cell lysis followed by DNA purification steps, and (2) bacterial isolation followed by cell lysis and DNA purification. Since then, these methods have been continually modified and improved. The methods for fractionation of bacteria as a preliminary step (Bakken and Lindahl 1995, Torsvik *et al.* 1995) and for direct extraction (Saano *et al.* 1995, Trevors and van Elsas 1995) have recently been compiled. In general, DNA isolation methods are moving from the use of large samples and laborious purification procedures towards the processing of small samples in microcentrifuge tubes

(Dijkmans *et al.* 1993, More *et al.* 1994). In addition, methods for efficient bacterial cell lysis have been evaluated and improved (Zhou *et al.* 1996, Gabor *et al.* 2003). Bead-mill homogenization has been shown to lyse a higher percentage of cells (without excessive DNA fragmentation) than freeze-thaw lysis although 'soft lysis' by freezing and thawing is useful for obtaining high molecular weight DNA (Erb and Wagner-Dobler 1993, Miller *et al.* 1999). The efficiency of cell lysis and DNA extraction varies with sample type and DNA extraction procedure (Erb and Wagner-Dobler 1993, Zhou *et al.* 1996, Frostegard *et al.* 1999, Miller *et al.* 1999). Therefore, in order to obtain accurate and reproducible results, the variation in the efficiency of cell lysis and DNA extraction must be taken into account. Co-extraction with standard DNA has been used to overcome the bias in extraction of DNA from Baltic Sea sediment samples (Moller and Jansson 1997). In contrast to extraction of DNA, extraction of mRNA from environmental samples is quite difficult and is further hampered by the very short half-lives of prokaryotic mRNA.

An ideal procedure for recovering nucleic acids from environmental samples has recently been summarized by Hurt *et al.* (2001). They state that an ideal procedure should meet several criteria: (1) the nucleic acid recovery efficiency should be high and not biased so that the final nucleic acids are representative of the total nucleic acids within the naturally occurring microbial community; (2) the RNA and DNA fragments should be as large as possible so that molecular studies, such as community gene library construction and gene cloning, can be carried out; (3) the RNA and DNA should be of sufficient purity for reliable enzyme digestion, hybridization, reverse transcription, and PCR amplification; (4) the RNA and DNA should be extracted simultaneously from the same sample so that direct comparative studies can be performed (this will also be particularly important for analyzing samples of small size); (5) the extraction and purification protocol should be kept simple as much as possible so that the whole recovery process is rapid and inexpensive; and (6) the extraction and purification protocol should be robust and reliable, as demonstrated with many diverse environmental samples. However, none of the previously mentioned nucleic acid extraction methods have been evaluated and optimized based on all the above important criteria.

Genetic fingerprinting techniques

Genetic fingerprinting techniques provide a pattern or profile of the genetic diversity in a microbial community. Recently, several fingerprinting techniques have been developed and used in microbial ecology studies such as bioremediation.

The separation of, or detection of small differences in, specific DNA sequences can give important information about the community structure and the diversity of microbes containing a critical gene. Generally, these techniques are coupled to a PCR reaction to amplify sequences that are not abundant. PCR-amplified products can be examined by using techniques that detect single substitutions in the nucleotide sequence (Schneegurt-Mark and Kulpa-Chaler 1998). These techniques are important in separating and identifying PCR-amplified products that might have the same size but slightly different nucleotide sequences. For example, the amplified portions of *nahAc* genes from a mixed microbial population might be of similar size when amplified with a particular set of *nahAc*-specific degenerate primers, but have small differences within the PCR-amplified products at the nucleotide level. One way of detecting these differences is to digest the PCR-amplified product with restriction endonucleases and examine the pattern of restriction fragments. The PCR-amplified product can be end-labeled or uniformly labeled for this technique.

In one study, natural sediments were tested for the presence of *nahAc* gene sequences by using PCR (Herrick *et al.* 1993). Polymorphisms in this gene sequence were detected by restricting the PCR-amplified products. In another study, PCR amplification of *bphC* genes by using total DNA extracted from natural soils as template allowed further investigation of the PCB degradation pathway (Erb and Wagner-Dobler 1993). No restriction polymorphisms were observed in the PCR-amplified products, suggesting limited biodiversity in this PCB-degrading population. Contaminated soils gave positive results, whereas pristine lake sediments did not contain appreciable amounts of the *bphC* gene.

Matrix-assisted laser desorption/ionization time-of-flight mass spectrophotometry (MALDI-TOF-MS) has been developed as a rapid and sensitive method for analyzing the restriction fragments of PCR-amplified products (Taranenko *et al.* 2002). A mass spectrum can be obtained in less than 1 min.

Another advanced method, terminal restriction fragment length polymorphism (T-RFLP) analysis, measures the size polymorphism of terminal restriction fragments from a PCR-amplified marker. It combines at least three technologies, including comparative genomics/RFLP, PCR, and electrophoresis. Comparative genomics provides the necessary insight to allow design of primers for amplification of the target product, and PCR amplifies the signal from a high background of unrelated markers. Subsequent digestion with selected restriction endonucleases produces terminal fragments appropriate for sizing on high resolution (±1-base) sequencing gels. The latter step is conveniently performed on automated systems such as polyacrylamide gel or capillary electrophoresis systems that provide digital output. The use of a fluorescently tagged primer limits the analysis to only the terminal fragments of the digestion. Because size markers bearing a different fluorophore from the samples can be included in every lane, the sizing is extremely accurate (Marsh 1999).

Denaturing gradient gel electrophoresis (DGGE) and its cousin TGGE (thermal-GGE) is a method by which fragments of DNA of the same length but different sequence can be resolved electrophoretically (Muyzer and Smalla 1998, Muyzer 1999). Separation is based on the decreased electrophoretic mobility of a partially melted double-stranded DNA molecule in polyacrylamide gels containing a linear gradient of a denaturing reagent (a mixture of formamide and urea) or a linear temperature gradient (Muyzer *et al.* 1993). As the duplex DNA fragments are subjected to electrophoresis, partial melting occurs at denaturant concentrations specific for various nucleotide sequences. An excellent study by Watanabe and coworkers (Watanabe *et al.* 1998) used a combination of molecular-biological and microbiological methods to detect and characterize the dominant phenol-degrading bacteria in activated sludge. TGGE analysis of PCR products of 16S rDNA and of the gene encoding phenol hydroxylase (LmPH) showed a few dominant bacterial populations after a 20-day incubation with phenol as a carbon source. Comparison of sequences of different bacterial isolates and excised TGGE bands revealed two dominant bacterial strains responsible for the phenol degradation (Watanabe *et al.* 1998).

Watts *et al.* (2001) recently analyzed PCB-dechlorinating communi-ties in enrichment cultures using three different molecular screening techniques, namely, amplified ribosomal DNA restriction analysis (ARDRA), DGGE, and T-RFLP. They found that the methods have different biases, which were apparent from discrepancies in the relative clone frequencies (ARDRA), band intensities (DGGE) or peak heights (T-RFLP) from the same enrichment culture. However, all of these methods were useful for qualitative analysis and could identify the same organisms (Watts *et al.* 2001). Overall, in community fingerprinting and preliminary identification, DGGE proved to be the most rapid and effective tool for monitoring microorganisms within a highly enriched culture. T-RLFP results corroborated DGGE fingerprint analysis, but the identification of the bacteria detected required the additional step of creating a gene library. ARDRA provided an in-depth analysis of the community and this technique detected slight intra-species sequence variation in 16S rDNA (Watts *et al.* 2001).

Another such approach takes advantage of sequence-dependent conformational differences between re-annealed single-stranded products (SSCP), which also result in changes in electrophoretic mobility; DNA fragments are separated on a sequencing gel under non-denaturing conditions based on their secondary structures (Schiwieger and Tebbe 1998).

Recently, a method using denaturing high performance liquid chromatography (DHPLC) was developed that can detect single base-pair mutations within a specific sequence (Taliani *et al.* 2001). This is a rapid, sensitive and accurate method of detecting sequence variation, but has not yet been used for analyzing the diversity of specific sequences from environmental samples. DHPLC could be a useful, rapid and sensitive method for ecological studies in bioremediation.

Discovery of novel catabolic genes involved in xenobiotic degradation

There are two different approaches to investigate the diversity of catabolic genes in environmental samples: culture-dependent and culture-independent methods. In culture-dependent methods, bacteria are isolated from environmental samples with culture medium. Nucleic acid is then extracted from the bacterial culture. By contrast, culture-independent methods employ direct extraction of nucleic acids from environmental samples (Lloyd-Jones *et al.* 1999, Okuta *et al.* 1998, Watanabe *et al.* 1998). The description of catabolic gene diversity by culture-independent molecular biological methods often involves the amplification of DNA or cDNA from RNA extracted from environmental samples by PCR, and the subsequent analysis of the diversity of amplified molecules (community fingerprinting). Alternatively, the amplified products may be cloned and sequenced to identify and enumerate bacterial species present in the sample.

To date, more than 300 catabolic genes involved in catabolism of aromatic compounds have been cloned and identified from culturable bacteria. Several approaches, such as shotgun cloning by using indigo formation (Ensley *et al.* 1983, Goyal and Zylstra 1996), clearing zone formation (de Souza *et al.* 1995), or meta-cleavage activity (Sato *et al.* 1997) as screening methods for cloning; applying proteomics (two dimensional gel electrophoresis analysis) of xenobiotic-inducible proteins to obtain genetic information (Khan *et al.* 2001), transposon mutagenesis to obtain a defective mutant (Foght and Westlake 1996), transposon mutagenesis using a transposon-fused reporter gene (Bastiaens *et al.* 2001), applying a degenerate primer to generate a probe (Saito *et al.* 2000), and applying a short probe from a homologous gene (Moser and Stahl 2001), have been used to discover catabolic genes for aromatic compounds from various bacteria.

The emergence of methods using PCR to amplify catabolic sequences directly from environmental DNA samples now appears to offer an alternative technique to discover novel catabolic genes in nature. Most research focusing on analysis of the diversity of the catabolic genes in environmental samples has employed PCR amplification using a degenerate primer set (a primer set prepared from consensus or unique DNA sequence), and the separation of the resultant PCR products either by cloning or by gel electrophoresis (Allison *et al.* 1998, Hedlund *et al.* 1999, Lloyd-Jones *et al.* 1999, Watanabe *et al.* 1998, Wilson *et al.* 1999, Bakermans and Madsen 2002). To confirm that the proper gene has been PCR-amplified, it is necessary to sequence the product, after which the resultant information can be used to reveal the diversity of the corresponding gene(s). Over the last few years, these molecular techniques have been systematically applied to the study of the diversity of aromatic-compound-degrading genes in environmental samples (Table 1).

Application of a degenerate primer set to isolate functional catabolic genes directly from environmental samples has been reported (Okuta *et al.* 1998). Fragments of catechol 2,3-dioxygenase (C23O) genes were isolated from environmental samples by PCR with degenerate primers, and the gene fragments were inserted into the corresponding region of the *nahH* gene, the structural gene for C23O encoded by the catabolic plasmid NAH7, to reconstruct functional hybrid genes reflecting the diversity in the natural gene pool. In this approach, the only information necessary is knowledge of the conserved amino acid sequences in the protein family from which the degenerate primers should be designed. This method is generally applicable, and may be useful in establishing a divergent hybrid gene library for any gene family (Okuta *et al.* 1998).

When degenerate primers cannot be used for amplification of DNA or RNA targets, PCR has limited application for investigating novel catabolic genes from culture collections or from environmental samples. Dennis and Zylstra (1998) developed a new strategy for rapid analysis of genes for Gram-negative bacteria. They constructed a minitransposon containing an origin of replication in an *Escherichia coli* cell. These artificially derived transposons are called plasposons (Dennis and Zylstra 1998). Once a desired mutant has been constructed by transposition, the region around the insertion point can be rapidly cloned and sequenced. Mutagenesis with these plasposons can be used as an alternative tool for investigating novel catabolic genes from culture collections, although such approaches cannot be taken for environmental samples. The *in vitro* transposon mutagenesis by plasposon containing a reporter gene without a promoter will provide an alternative technique to search for desired xenobiotic-inducible promoters from environmental DNA samples.

Table 1. Molecular approaches for investigating the diversity and identification of catabolic genes involved in degradation of xenobiotics. RT, Reverse transcription; PCR, polymerase chain reaction; DGGE, denaturing gradient gel electrophoresis; RHD, ring hydroxylating dioxygenase; PAH, polycyclic aromatic hydrocarbon.

Target gene	Molecular approach	Source	Reference
nahAc	RT-PCR with degenerate primers	Groundwater (culture-independent)	Wilson *et al.* 1999
phnAc, nahAc, Jones and glutathione -S-transferase	PCR with several primers	Soil samples (culture-independent)	Lloyd- *et al.* 1999
Phenol hydroxylase (LmPH)	PCR-DGGE with degenerate primers	Activated sludge (culture-independent)	Watanabe *et al.* 1998
RHD	PCR with degenerate primers	Prestine- and aromatic hydrocarbon-contami-nated soils (culture-independent)	Yeates *et al.* 2000
PAH Jones dioxygenase	PCR with several primers	PAH soil bacteria (culture-dependent)	Lloyd- *et al.* 1999
nahAc	PCR with degenerate primers	Marine sediment bacteria (culture-dependent)	Allison *et al.* 1998
nahAc	PCR with degenerate primers	Marine sediment bacteria (culture-dependent)	Hedlund *et al.* 1999
nahAc	PCR with degenerate primers	Coal-tar-waste contaminated aquifer waters(culture-independent)	Bakermans *et al.* 2002
NahR	PCR with degenerate primers	Coal tar waste-contaminated site (culture-independent)	Park *et al.* 2002
Nah	PCR with degenerate primers	Soil bacteria (culture-dependent)	Hamann *et al.* 1999
TfdC	PCR with degenerate primers	Soil bacteria (culture-dependent)	Cavalca *et al.* 1999
PAH dioxygen-ase and catechol dioxygenase	PCR with degenerate primers	Wastewater and soil bacteria (culture-dependent)	Meyer *et al.* 1999
phnAc, nahAc and PAH dioxygenase	PCR with several degenerate primers	River water, sediment, and soil bacteria (culture-dependent)	Widada *et al.* 2002a

(contd.)

Table 1. (contd.)

Target gene	Molecular approach	Source	Reference
RHD	PCR-DGGE with degenerate primers	*Rhodococcus* sp. strain RHA1 (culture-dependent)	Kitagawa *et al.* 2001
dszABC	PCR-DGGE with several primers	Sulfurous-oil-containing soils (culture-independent)	Duarte *et al.* 2001

Monitoring of bioaugmented microorganisms in bioremediation

Because different methods for enumeration of microorganisms in environmental samples sometimes provide different results, the method used must be chosen in accordance with the purpose of the study. Not all detection methods provide quantitative data; some only indicate the presence of an organism and others only detect cells in a particular physiological state (Jansson and Prosser 1997). Several molecular approaches have been developed to detect and quantify specific microorganisms (Table 2).

Quantification by PCR/RT-PCR

PCR is now often used for sensitive detection of specific DNA in environmental samples. Sensitivity can be enhanced by combining PCR with DNA probes, by running two rounds of amplification using nested primers (Moller *et al.* 1994), or by using real-time detection systems (Widada *et al.* 2001). Detection limits vary for PCR amplification, but usually between 102 and 103 cells/g soil can routinely be detected by PCR amplification of specific DNA segments (Fleming *et al.* 1994b, Moller *et al.* 1994). Despite its sensitivity, until recently it has been difficult to use PCR quantitatively to calculate the number of organisms (gene copies) present in a sample. Three techniques have now been developed for quantification of DNA by PCR, namely: MPN-PCR, replicative limiting dilution-PCR (RLD-PCR), and competitive PCR (cPCR) (Chandler 1998).

MPN-PCR is carried out by running multiple PCR reactions of samples that have been serially diluted, and amplifying each dilution in triplicate. The number of positive reactions is compared with the published MPN tables for an estimation of the number of target DNA copies in the sample (Picard *et al.* 1996). In MPN-PCR, DNA extracts are serially diluted before PCR amplification and limits can be set on the number of genes in the sample by reference to known control dilutions.

Table 2. Molecular approaches for detection and quantification of specific microorganisms in environmental samples (adapted from Jansson and Prosser 1997). CPCR, Competitive PCR; MPN-PCR, most probable number PCR; RLD-PCR, replicative limiting dilution PCR.

Identification method	Detection and quantification method	Cell type monitored
Fluorescent tags on	Microscopy	Primary active cells
rRNA probes	Flow cytometry	
lux or *luc* gene	Luminometry/scintillation counting	Active cells
	Cell extract luminescence	Total cells with translated luciferase protein
	Luminescent colonies	Culturable luminescent cells
gfp gene	Fluorescent colonies	
	Microscopy	Culturable fluorescent cells
	Flow cytometry	Total cells, including starved
Specific DNA sequence	cPCR MPN-PCR, RLD-PCR	Total DNA (living and dead cell and free DNA)
	Slot/dot blot hybridization	Culturable cells
	Colony hybridization	
Specific mRNA transcript	Competitive RT-PCR Slot/dot blot hybridization	Catabolic activity of cells
Other marker genes (e.g., *lacZY, gusA, xylE,* and antibiotic resistance genes)	Plate counts colony hybridization	Culturable marked cells and indigenous cells with marker phenotype
	Quantitative PCR Slot/dot blot hybridization	Total DNA (living and dead cells and free DNA)

RLD-PCR, an alternate quantitative PCR for environmental application, is based on RLD analysis and the pragmatic tradeoffs between analytical sensitivity and practical utility (Chandler 1998). This method has been used to detect and quantify specific biodegradative genes in aromatic-compound-contaminated soil. The catabolic genes *cdo, nahAc,* and *alkB* were used as target genes (Chandler 1998).

Quantitative cPCR is based on the incorporation of an internal standard in each PCR reaction. The internal standard (or competitor DNA) should be as similar to the target DNA as possible and be amplified with the same primer set, yet still be distinguishable from the target, for example, by size (Diviacco et al. 1992). A standard curve is constructed using a constant series of competitor DNA added to a dilution series of target DNA. The ratio of PCR-amplified DNA yield is then plotted versus initial target DNA concentration. This standard curve can be used for calculation of unknown target DNA concentrations in environmental samples. The competitive standard is added to the sample tube at the same concentration as used for preparation of the standard curve (Diviacco et al. 1992). Since both competitor and target DNAs are subjected to the same conditions that might inhibit the performance of DNA polymerase (such as humic acid or salt contaminants), the resulting PCR product ratio is still valid for interpolation of target copy number for the standard curve. Recently, Alvarez et al. (2000) have developed a simulation model for cPCR, which takes into account the decay in efficiency as a linear function of product yield. Their simulation data suggested that differences in amplification efficiency between target and standard templates induced biases in quantitative cPCR. Quantitative cPCR can only be used when both efficiencies are equal (Alvarez et al. 2000).

In bioremediation, quantitative PCR has been used to monitor and to determine the concentration of some catabolic genes from bioaugmented bacteria in environmental samples (Table 3). Recently, quantitative competitive RT-PCR has been used to quantify the mRNA of the tcbC of Pseudomonas sp. strain P51 (Meckenstock et al. 1998).

Molecular marker gene systems

In many laboratory biodegradation studies, bacterial cells that are metabolically capable of degrading/mineralizing a pollutant are added to contaminated environmental samples to determine the potential biodegradation of target compound(s). Assessment of the environmental impact and risk associated with the environmental release of augmented bacteria requires knowledge of their survival, persistence, activity, and dispersion within the environment. Detection methods that take advantage of unique and identifiable molecular markers are useful for enumerating and assessing the fate of microorganisms in bioremediation (Prosser 1994). The application of molecular techniques has provided much greater precision through the introduction of specific marker genes. Some of the requirements for marker systems include the ability to allow unambiguous identification of the marked strain within a large indigenous microbial

Table 3. PCR detection and quantification of introduced bacteria in bioremediation of xenobiotics.

Bacteria	Target gene	Detection and quantification method	Reference
Desulfitobacterium frappieri strain PCP-1 (pentachlorophenol-degrader)	16 rRNA	Nested PCR	Levesque *et al.* 1997
Mycobacterium chlorophenolicum strain PCP-1 (pentachlorophenol-degrader)	16 rRNA	MPN-PCR	van Elsas *et al.* 1997
Sphingomonas chlorophenolica (pentachlorophenol-degrader)	16rRNA	Competitive PCR	van Elsas *et al.* 1998
Pseudomonas sp. strain B13 (chloroaromatic-degrader)	16 rRNA	Competitive PCR	Leser *et al.* 1995
Pseudomonas putida strain mx (toluene-degrader)	*xylE*	Competitive PCR	Hallier-Soulier *et al.* 1996
P. putida strain G7 (naphthalene-degrader)	*nahAc*	PCR-Southern blot	Herrick *et al.* 1993
P. putida strain mt2 (toluene-degrader)	*xylM*	Multiplex PCR-Southern blot	Knaebel and Crawford 1995
P. putida ATCC 11172 (phenol-degrader)	*dmpN*	PCR and RT-PCR	Selvaratnam *et al.* 1995, 1997
Pseudomonas sp. strain P51 (trichlorobenzene-degrader)	*tbcAa, tbcC*	PCR	Tchelet *et al.* 1999
Pseudomonas sp. strain P51 (trichlorobenzene-degrader)	*tbcC*	Competitive RT-PCR	Meckenstock *et al.* 1998
Psuedomonas resinovorans strain CA10 (carbazole- and dibenzo-*p*-dioxin-degrader)	*carAa*	Real-time competitive PCR	Widada *et al.* 2001, 2002b

community, its stable maintenance in the host cell, and adequate expression for detection (Lindow 1995).

Antibiotic resistance genes, such as the *nptII* gene encoding resistance to kanamycin, were the first genes to be employed as markers. Although they are still in use, these phenotypic marker genes are generally falling out of favor because of the small risk of contributing to the undesirable spread of antibiotic resistance in nature (Lindow 1995).

Genes encoding metabolic enzymes have also been used as non-selective markers. These include *xylE* (encoding catechol 2,3-oxygenase), *lacZY* (encoding galactosidase and lactose permease) and *gusA* (encoding glucuronidase). The *xylE* gene product can be detected by the formation of a yellow catabolite (2-hydroxymuconic semialdehyde) from catechol. The enzymes encoded by *lacZ* and *gusA* cleave the uncolored substrates 5-bromo-4-chloro-3-indolyl--D-galactopyranoside (X-gal) and 5-bromo-4-chloro-3-indolyl--D-glucuronide cyclohexyl ammonium salt (X-gluc), res-pectively, producing blue products. Some advantages and disadvantages of these phenotypic markers have recently been discussed (Jansson 1995). For example, one useful application of *xylE* is the specific detection of intact or viable cells, because catechol 2,3-oxygenase is inactivated by oxygen and rapidly destroyed outside the cell (Prosser 1994).

Two disadvantages of the above mentioned marker genes are the potentially high background of marker enzyme activity in the indigenous microbial population and the requirement for growth and cultivation in the detection methods. DNA hybridization is another potentially useful method for detecting these phenotypic marker genes as long as background levels are sufficiently low. Both *lacZ* and *gusA* have limited application in soil, however, because of their presence in the indigenous microbiota.

The *gfp* gene, encoding the green fluorescent protein (GFP) from the jellyfish *Aequorea victoria* is an attractive marker system with which to monitor bacterial cells in the environment. An advantage of the application of the *gfp* gene over that of other marker genes is the fact that the detection of fluorescence from GFP is independent of substrate or energy reserves (Tombolini *et al.* 1997). Since the *gfp* gene is eukaryotic in origin, it was first necessary to develop an optimized construct for expression of *gfp* in bacteria (Unge *et al.* 1999). Another reason that *gfp* is becoming so popular is that single cells tagged with *gfp* can easily be visualized by epifluorescence microscopy (Tombolini *et al.* 1997). In addition, fluorescent cells may be rapidly enumerated by flow cytometry (Ropp *et al.* 1995). The flow cytometer measures parameters related to size, shape and fluorescence of individual cells (Tombolini *et al.* 1997).

Another promising marker of cellular metabolic activity is bacterial or eukaryotic luciferase. Bacterial luciferase catalyzes the following reaction: $RCHO + FMNH_2 + O_2$. $RCOOH + FMN + H_2O + light$ (490 nm), where R is a long chain aldehyde (e.g., n-decanal). Due to the requirement of reducing equivalent ($FMNH_2$), the bioluminescence output is directly related to the metabolic activity of the cells (Unge *et al.* 1999). The marker systems

Table 4. The application of marker genes and methods used to detect introduced bacteria in bioremediation of xenobiotics.

Marker gene	Microorganism	Detection method	References
lux or lac	Pseudomonas cepacia (2,4-D-degrader)	Non-selective plating, selective plating and autophotography	Masson et al. 1993
lux or lac	Pseudomonas aeruginosa (biosurfactant-producer)	Non-selective plating, selective plating, charge-coupled device (CCD)-enhanced detection, PCR and Southern blotting	Fleming et al. 1994b
lux	P. aeruginosa (biosurfactant-producer)	Bioluminescent-MPN (microplate assay), luminometry and CCD-enhanced detection	Fleming et al. 1994a
lux	Alcaligenes eutrophus strain H850 (PCB-degrader)	Selective plating and bioluminescence	van Dyke et al. 1996
gfp	Ralstonia eutropha strain H850 (PCB-degrader)	Selective plating	Irwin Abbey et al. 2003
lac	Sphingomonas wittichii strain RW1 (dibenzo-p-dioxin- and dibenzo-furan-degrader)	Non-selective plating and selective plating	Megharaj et al. 1997
gfp or lux	Pseudomonas sp. strain UG14Gr (phenanthrene-degrader)	Non-selective plating, selective plating and CCD-enhanced detection	Errampalli et al. 1998
gfp	Moraxella sp. (p-nitrophenol-degrader)	Non-selective plating and selective plating	Tresse et al. 1998
xyl	S. wittichii strain RW1 (dibenzo-p-dioxin- and dibenzofuran-degrader)	Selective plating	Halden et al. 1999
gfp	P. resinovorans CA10 (carbazole- and dibenzo-p-dioxin-degrader)	Selective plating	Widada et al. 2002b
gfp or luc	Arthobacter chlorophenolicus A6 (4-chlorophenol-degrader)	Selective plating, luminometry, and flow cytometry	Elvang et al. 2001

mentioned above for monitoring of augmented bacteria in bioremediation have been broadly applied (Table 4).

Recent development of methods increasing specificity of detection

A new approach that permits culture-independent identification of microorganisms responding to specified stimuli has been developed (Borneman 1999). This approach was illustrated by the examination of microorganisms that respond to various nutrient supplements added to environmental samples. A thymidine nucleotide analog, bromode-oxyuridine (BrdU), and specified stimuli were added to environmental samples and incubated for several days. DNA was then extracted from an environmental sample, and the newly synthesized DNA was isolated by immunocapture of the BrdU-labelled DNA. Comparison of the microbial community structures obtained from total environmental sample DNA and the BrdU-labelled fraction showed significantly different banding patterns between the nutrient supplement treatments, although traditional total DNA analysis revealed no notable differences (Borneman 1999). Similar to BrdU strategy, stable isotope probing (SIP) is an elegant method for identifying the microorganisms involved in a particular function within a complex environmental sample (Radajewski et al. 2000). After enrichment of environmental samples with ^{13}C-labeled substrate, the bacteria that can use the substrate incorporate ^{13}C into their DNA, making it denser than normal DNA containing ^{12}C. SIP has been used for labeling and separating DNA and RNA (Radajewski et al. 2003). Density gradient centrifugation cleanly separates the labeled from unlabeled nucleic acids. These approaches provide new strategies to permit identification of DNA from a stimulus- or substrate-responsive organism in environmental samples. Application of such approaches in bioremediation by using the desired xenobiotic as a substrate or stimulus added to an environmental sample may provide a robust strategy for discovering novel catabolic genes involved in xenobiotic degradation.

Bacteria belonging to the newly recognized phylogenetic groups are widely distributed in various environments (Dojka et al. 1998, Hugenholtz et al. 1998). The 16S rDNA sequences of these groups are very diverse and include mismatches to the bacterial universal primer designed from conserved regions in bacterial 16S rDNA sequences (Dojka et al. 1998, von Wintzingerode et al. 2000). Mismatches between PCR primer and a template greatly reduce the efficiency of amplification (von Wintzingerode et al. 1997). To overcome such problems, Watanabe et al. (2001) designed new universal primers by introducing inosine residues at positions where mismatches were frequently found. Using the improved primers, they could detect the phylotypes affiliated with *Verrucomicrobia* and candidate division OP11, which had not been detected by PCR-DGGE with conventional universal primers (Watanabe *et al.* 2001).

The number of bands in a DGGE gel does not always accurately reflect the number of corresponding species within the microbial community; one organism may produce more than one DGGE band because of multiple, heterogeneous rRNA operons (Cilia *et al.* 1996). Microbial community pattern analysis using 16S rRNA gene-based PCR-DGGE is significantly limited by this inherent heterogeneity (Dahllöf *et al.* 2000). As an alternative to 16S rRNA gene sequences in community analysis, Dahllöf *et al.* (2000) employed the gene for the β subunit of RNA polymerase (*rpoB*), which appears to exist in only one copy in bacteria. This approach proved more accurate compared with 16S rRNA gene-based PCR-DGGE for a mixture of bacteria isolated from red algae.

Recently, DNA microarrays have been developed and introduced for analyzing microbes and their activity in environmental samples (Cho and Tiedje 2002, Small *et al.* 2001, Wu *et al.* 2001). These are particularly powerful tools because of the large number of hybridizations that can be performed simultaneously on glass slides: over 100,000 spots per cm^2 can be accommodated (Kuipers *et al.* 1999). As with conventional dot blot hybridization, sample nucleic acids can be spotted onto the carrier material or reverse hybridization can be performed using immobilized probes. If PCR is involved, specific primers can be used to amplify partial or whole rRNA genes of the microorganisms of interest. Small *et al.* (2001) recently developed and validated a simple microarray method for the direct detection of intact 16S rRNA from unpurified soil extracts. In addition, it has been reported that DNA array technology is also a potential method for assessing the functional diversity and distribution of selected genes in the environment (Cho and Tiedje 2002, Wu *et al.* 2001).

The vast majority of environmental microorganisms have yet to be cultured. Consequently, a major proportion of the genetic diversity within nature resides in the uncultured organisms (Stokes *et al.* 2001). Isolation of these genes is limited by lack of sequence information, and PCR amplification techniques can be employed for the amplification of only partial genes. Thus a strategy to recover complete open reading frames from environmental DNA samples has been developed (Stokes *et al.* 2001). PCR assays targeted to the 59-base element family of recombination sites that flank gene cassettes associated with integrons were designed. Using such assays, diverse gene cassettes could be amplified from the vast majority of the environmental DNA samples tested. These gene cassettes contained a complete open reading frame, the majority of which were associated with ribosome binding sites. Such a strategy applied together with the BrdU or SIP strategy (Borneman 1999, Radajewski *et al.* 2000, Schloss and Handelsman 2003) should provide a robust method for discovering catabolic gene cassettes from environmental samples.

It is becoming increasingly apparent that the best solution for monitoring an introduced microorganism in the environment is to use either several markers simultaneously or multiple detection methods. Sometimes single markers or certain combinations of markers are not selective enough, such as *lacZY* used either alone or together with antibiotic selection. Even so, the use of antibiotic selection, in combination with bioluminescence, has been found to be very effective and useful for selection of low numbers of tagged cells (Jansson and Prosser 1997). A dual-marker system was developed for simultaneous quantification of bacteria and their activity by the *luxAB* and *gfp* gene products, respectively. Generally, the bioluminescence phenotype of the *luxAB* biomarker is dependent on cellular energy status. Since cellular metabolism requires energy, bioluminescence output is directly related to the metabolic activity of the cells. In contrast, the fluorescence of GFP has no energy requirement. Therefore, by combining these two biomarkers, total cell number and metabolic activity of a specific marked cell population could be monitored simultaneously (Unge *et al.* 1999).

The specificity of detection can be increased by detecting marker DNA in total DNA isolated and purified from an environmental sample by a variety of molecular-biology-based methods, such as gene probing, DNA hybridization, and quantitative PCR (Jansson 1995, Jansson and Prosser 1997).

Recently, we developed a rapid, sensitive, and accurate quantification method for the copy number of specific DNA in environmental samples by combining the fluorogenic probe assay, cPCR and co-extraction with internal standard cells (Widada *et al.* 2001). The internal standard DNA was modified by replacement of a 20-bp-long region responsible for binding a specific probe in fluorogenic PCR (TaqMan; Applied Biosystems, Foster City, Calif.). The resultant DNA fragment was similar to the corresponding region of the intact target gene in terms of G+C content. When used as a competitor in the PCR reaction, the internal standard DNA was distinguishable from the target gene by two specific fluorogenic probes with different fluorescence labels, and was automatically detected in a single tube using the ABI7700 sequence detection system (Applied Biosystems). By using an internal standard designed for cPCR, we found that the amplification efficiency of target and standard templates was quite similar and independent of the number of PCR cycles (Widada *et al.* 2001). The internal standard cell was used to minimize the variations in the efficiency of cell lysis and DNA extraction between the samples. A mini-transposon was used to introduce competitor DNA into the genome of a non-target bacterium in the same genus, and the resultant transformant was used as an internal standard cell. After adding a known amount of the internal standard cells to soil samples, we extracted the total DNA (co-extraction). Using this method, the copy number of the target gene in environmental samples can be quantified rapidly and accurately (Widada *et al.* 2001).

Conclusions

Molecular-biology-based techniques in bioremediation are being increasingly used, and have provided useful information for improving bioremediation strategies and assessing the impact of bioremediation treatments on ecosystems. Several recent developments in molecular techniques also provide rapid, sensitive, and accurate methods of analyzing bacteria and their catabolic genes in the environment. In addition, these molecular techniques have been used for designing active biological containment systems to prevent the potentially undesirable spread of released microorganisms, mainly genetically engineered microorganisms. However, a thorough understanding of the limitations of these techniques is essential to prevent researchers from being led astray by their results.

Acknowledgement

We are indebted to Prof. David E. Crowley of the University of California, Riverside, for kindly providing suggestions and discussions. This work was partly supported by the Program for Promotion of Basic Research Activities for Innovative Biosciences (PROBRAIN) in Japan.

REFERENCES

Allison, D.G., B. Ruiz, C. San-Jose, A. Jaspe, and P. Gilbert. 1998. Analysis of biofilm polymers of *Pseudomonas fluorescens* B52 attached to glass and stainless steel coupons. *Abstracts of the General Meeting of the American Society for Microbiology*, Atlanta, Georgia, 98: 325.

Alvarez, M.J., A.M. Depino, O.L. Podhajcer, and F.J. Pitossi. 2000. Bias in estimations of DNA content by competitive polymerase chain reaction. *Anal. Biochem.* 28: 87-94.

Atlas, M. 1992. Molecular methods for environmental monitoring and containment of genetically engineered microorganisms. *Biodegradation* 3: 137-146.

Bakermans, C., and E.L. Madsen. 2002. Diversity of 16S rDNA and naphthalene dioxygenase genes from coal-tar-waste-contaminated aquifer waters. *Microb. Ecol.* 44: 95-106.

Bakken, L.R., and V. Lindahl. 1995. Recovery of bacterial cells from soil. Pages 9-27 in *Nucleic Acids in the Environment: Methods and Applications*, J.T. Trevors and J.D. van Elsas, eds., Springer, Berlin, Heidelberg, New York.

Bastiaens, L., D. Springael, W. Dejonghe, P. Wattiau, H. Verachtert, and L. Diels. 2001. A transcriptional *luxAB* reporter fusion responding to fluorine in *Sphingomonas* sp. LB126 and its initial characterizations for whole-cell bioreporter purposes. *Res. Microbiol.* 152: 849-859.

Borneman, J. 1999. Culture-independent identification of microorganisms that respond to specified stimuli. *Appl. Environ. Microbiol.* 65: 3398-3400.

Brockman, F.J. 1995. Nucleic-acid-based methods for monitoring the performance of in situ bioremediation. *Mol. Ecol.* 4: 567-578.

Cavalca, L., A. Hartmann, N. Rouard, and G. Soulas. 1999. Diversity of *tfdC* genes: distribution and polymorphism among 2,4-dichlorohenoxyacetic acid degrading soil bacteria. *FEMS Microbiol. Ecol.* 29: 45-58.

Chandler, D.P. 1998. Redefining relativity: quantitative PCR at low template concentrations for industrial and environmental microbiology. *J. Ind. Microbiol. Biotechnol.* 21: 128-140.

Cho, J.-C., and J.M. Tiedje. 2002. Quantitative detection of microbial genes by using DNA microarrays. *Appl. Environ. Microbiol.* 68: 1425-1430.

Cilia, V., B. Lafay, and R. Christen. 1996. Sequence heterogeneities among 16S RNA sequences, and their effect on phylogenetic analyses at species level. *Mol. Biol. Evol.* 13: 415-461.

Dahllöf, I., H. Baillie, and S. Kjelleberg. 2000. *rpoB*-Based microbial community analysis avoids limitations inherent in 16S rRNA gene intraspecies heterogeneity. *Appl. Environ. Microbiol.* 66: 3376-3380.

Dennis, J.J., and G.J. Zylstra. 1998. Plasposon: modular self-cloning mini-transposon derivatives for the rapid genetic analysis of Gram-negative bacterial genomes. *Appl. Environ. Microbiol.* 64: 2710-2715.

de Souza, M.L., L.P. Wackett, K.L. Boundy-Mills, R.T. Mandelbaum, and M.J. Sadowsky. 1995. Cloning, characterization, and expression of a gene region from *Pseudomonas* sp. strain ADP involved in the dechlorination of atrazine. *Appl. Environ. Microbiol.* 61: 3373-3378.

Dijkmans, R., A. Jagers, S. Kreps, J.M. Collard, and M. Mergeay. 1993. Rapid method for purification of soil DNA for hybridization and PCR analysis. *Microb. Releases* 2: 29-34.

Diviacco, S., P. Norio, L. Zentilin, S. Menzo, M. Clementi, G. Biamonti, S. Riva, A. Falaschi, and M. Giacca. 1992. A novel procedure for quantitative polymerase chain reaction by coamplification of competitive templates. *Gene* 122: 313-320.

Dojka, M.A., P. Hugenholtz, S.K. Haack, and N.R. Pace. 1998. Microbial diversity in a hydrocarbon- and chlorinated-solvent-contaminated aquifer under-going intrinsic bioremediation. *Appl. Environ. Microbiol.* 64: 3869-3877.

Duarte, G.F., A.S. Rosado, L. Seldin, W. de Araujo, and J.D. van Elsas, 2001. Analysis of bacterial community structure in sulfurous-oil-containing soils and detection of species carrying dibenzothiophene desulfurization (*dsz*) genes. *Appl. Environ. Microbiol.* 67: 1052-1062.

Elvang, A.M., K. Westerberg, C. Jernberg, and J.K. Jansson. 2001. Use of green fluorescent protein and luciferase biomarkers to monitor survival and activity of *Arthrobacter chlorophenolicus* A6 cells during degradation of 4-chlorophenol in soil. *Environ. Microbiol.* 3: 32-42.

Ensley, B.D., B.J. Ratzkin, T.D. Osslund, M.J. Simon, L.P. Wackett, and D.T. Gibson. 1983. Expression of naphthalene oxidation genes in *Escherichia coli* results in biosynthesis of indigo. *Science* 222: 167-169.

Erb, R.W., and I. Wagner-Dobler. 1993. Detection of polychlorinated biphenyl degradation genes in polluted sediments by directed DNA extraction and polymerase chain reaction. *Appl. Environ. Microbiol.* 59: 4065-4073.

Errampalli, D., H. Okamura, H. Lee, J.T. Trevors, and J.T. van Elsas. 1998. Green fluorescent protein as a marker to monitor survival of phenanthrene-mineralizing *Pseudomonas* sp. UG14Gr in creosote-contaminated soil. *FEMS Microbiol. Ecol.* 26: 181-191.

Fleming, C.A., H. Lee, and J.T. Trevors. 1994a. Bioluminescent most-probable-number method to enumerate *lux*-marked *Pseudomonas aeruginosa* UG2Lr in soil. *Appl. Environ. Microbiol.* 60: 3458-3461.

Fleming, C.A., K.T. Leung, H. Lee, J.T. Trevors, and C.W. Greer. 1994b. Survival of *lux-lac*-marked biosurfactant-producing *Pseudomonas aeruginosa* UG2L in soil monitored by nonselective plating and PCR. *Appl. Environ. Microbiol.* 60: 1606-1613.

Fleming, J.T., J. Sanseverino, and G.S. Sayler. 1993. Quantitative relationship between naphthalene catabolic gene frequency and expression in predicting PAH degradation in soil at town gas manufacturing sites. *Environ. Sci. Technol.* 27: 1068-1074.

Fleming, J.T., W.-H. Yao, and G.S. Sayler. 1998. Optimization of differential display of prokaryotic mRNA: application to pure culture and soil microcosms. *Appl. Environ. Microbiol.* 64: 3698-3706.

Fleming, J.T., A.C. Nagel, J. Rice, and G.S. Sayler. 2001. Differential display of prokaryote messenger RNA and application to soil microbial communities. Pages 191-205 in *Environmental Molecular Microbiology: Protocol and Applications*, P.A. Rochelle, ed., Horizon Press, Norfolk, England.

Foght, J.M., and D.W.S. Westlake. 1996. Transposon and spontaneous deletion mutants of plasmid-borne genes encoding polycyclic aromatic hydrocarbon degradation by a strain of *Pseudomonas fluorescens*. *Biodegradation* 7: 353-366.

Frostegard, A., S. Courtois, V. Ramisse, S. Clerc, D. Bernillon, F. Le Gall, P. Jeannin, X. Nesme, and P. Simonet. 1999. Quantification of bias related to the extraction of DNA directly from soils. *Appl. Environ. Microbiol.* 65: 5409-5420.

Fulthorpe, R.R., and R.C. Wyndham. 1989. Survival and activity of a 3-chlorobenzoate-catabolic genotype in a natural system. *Appl. Environ. Microbiol.* 55: 1584-1590.

Gabor, E.M., E.J. de Vries, and D.B. Janssen. 2003. Efficient recovery of environmental DNA for expression cloning by indirect extraction methods. *FEMS Microbiol. Ecol.* 44: 153-163.

Goyal, A.K., and G.J. Zylstra. 1996. Molecular cloning of novel genes for polycyclic aromatic hydrocarbon degradation from *Comamonas testosteroni* GZ39. *Appl. Environ. Microbiol.* 62: 230-236.

Halden, R.U., B.G. Halden, and D.F. Dwyer. 1999. Removal of dibenzofuran, dibenzo-*p*-dioxin, and 2-chlorodibenzo-*p*-dioxin from soils inoculated with *Sphingomonas* sp. strain RW1. *Appl. Environ. Microbiol.* 65: 2246-2249.

Hallier-Soulier, S., V. Ducrocq, N. Mazure, and N. Truffaut. 1996. Detection and quantification of degradative genes in soils contaminated by toluene. *FEMS Microbiol. Ecol.* 20: 121-133.

Hamann, C., J. Hegemann, and A. Hildebrandt. 1999. Detection of polycyclic aromatic hydrocarbon degradation genes in different soil bacteria by polymerase chain reaction and DNA hybridization. *FEMS Microbiol. Lett.* 173: 255-263.

Hedlund, B.P., A.D. Geiselbrecht, J.B. Timothy, and J.T. Staley. 1999. Polycyclic aromatic hydrocarbon degradation by a new marine bacterium, *Neptunomonas napthovorans,* sp. nov. *Appl. Environ. Microbiol.* 65: 251-259.

Herrick, J.B., E.L. Madsen, C.A. Batt, and W.C. Ghiorse. 1993. Polymerase chain reaction amplification of naphthalene-catabolic and 16S rRNA gene sequences from indigenous sediment bacteria. *Appl. Environ. Microbiol.* 59: 687-694.

Holben, W.E., B.M. Schroeter, V.G.M. Calabrese, R.H. Olsen, J.K. Kukor, U.D. Biederbeck, A.E. Smith, and J.M. Tiedje. 1992. Gene probe analysis of soil microbial populations selected by amendment with 2,4-dichlorophenoxyacetic acid. *Appl. Environ. Microbiol.* 58: 3941-3948.

Hosein, S.G., D. Millette, B.J. Butler, and C.W. Greer. 1997. Catabolic gene probe analysis of an aquifer microbial community degrading creosote-related polycyclic aromatic and heterocyclic compounds. *Microb. Ecol.* 34: 81-89.

Hugenholtz, P., B.M. Goebel, and N.R. Pace. 1998. Impact of culture-independent studies on the emerging phylogenetic view of bacterial diversity. *J. Bacteriol.* 180: 4765-4774.

Hurt, R.A., X. Qiu, L. Wu, Y. Roh, A.V. Palumbo, J.M. Tiedje, and J. Zhou. 2001. Simultaneous recovery of RNA and DNA from soils and sediments. *Appl. Environ. Microbiol.* 67: 4495-4503.

Irwin Abbey, A.-M., L.A. Beaudette, H. Lee, and J.T. Trevor. 2003. Polychlorinated biphenyl (PCB) degradation and persistence of a *gfp*-marked *Ralstonia eutropha* H850 in PCB-contaminated soil. *Appl. Microbiol. Biotechnol.* 63: 222-230.

Jansson, J.K. 1995. Tracking genetically engineered microorganisms in nature. *Curr. Opin. Biotechnol.* 6: 275-283.

Jansson, J.K., and J.I. Prosser. 1997. Quantification of the presence and activity of specific microorganisms in nature. *Mol. Biotechnol.* 7: 103-120.

Khan, A.A., R.-F. Wang, W.-W. Cao, D.R. Doerge, D. Wennerstrom, and C.E. Cerniglia. 2001. Molecular cloning, nucleotide sequence and expression of genes encoding a polycyclic aromatic ring dioxygenase from *Mycobacterium* sp. strain PYR-1. *Appl. Environ. Microbiol.* 67: 3577-3585.

Kitagawa, W., A. Suzuki, T. Hoaki, E. Masai, and M. Fukuda. 2001. Multiplicity of aromatic ring hydroxylation dioxygenase genes in a strong PCB degrader, *Rhodococcus* sp. strain RHA1 demonstrated by denaturing gel electrophoresis. *Biosci. Biotechnol. Biochem.* 65: 1907-1911.

Knaebel, D.B., and R.L. Crawford. 1995. Extraction and purification of microbial DNA from petroleum-contaminated soils and detection of low numbers of toluene, octane and pesticide degraders by multiplex polymerase chain reaction and Southern analysis. *Mol. Ecol.* 4: 579-591.

Kuipers, O.P., A. de Jong, S. Holsappel, S. Bron, J. Kok, and L.W. Hamoen. 1999. DNA-microarrays and food-biotechnology. *Antonie van Leeuwenhoek* 76: 353-355.

Leser, T.D., M. Boye, and N.B. Hendriksen. 1995. Survival and activity of *Pseudomonas* sp. strain B13(FR1) in a marine microcosm determined by quantitative PCR and an rRNA-targeting probe and its effect on the indigenous bacterioplankton. *Appl. Environ. Microbiol.* 61: 1201-1207.

Levesque, M.J., S. La-Boissiere, J.C. Thomas, R. Beaudet, and R. Villemur. 1997. Rapid method for detecting *Desulfitobacterium frappiri* strain PCP-1 in soil by the polymerase chain reaction. *Appl. Microbiol. Biotechnol.* 47: 719-725.

Lindow, S.E. 1995. The use of genes in the study of microbial ecology. *Mol. Ecol.* 4: 555-566.

Lloyd-Jones, G., A.D. Laurie, D.W.F. Hunter, and R. Fraser. 1999. Analysis of catabolic genes for naphthalene and phenanthrene degradation in contaminated New Zealand soils. *FEMS Microbiol. Ecol.* 29: 69-79.

Madsen, E.L. 1991. Determining in situ biodegradation: facts and challenges. *Environ. Sci. Technol.* 25: 1663-1673.

Madsen, E.L., J.L. Sinclair, and W.C. Ghiorse. 1991. In situ biodegradation: microbiological patterns in a contaminated aquifer. *Science* 252: 830-833.

Marsh, T.L. 1999. Terminal restriction fragment length polymorphism (T-RFLP): an emerging method for characterizing diversity among homologous populations of amplification products. *Curr. Opin. Microbiol.* 2: 323-327.

Masson, L., Y. Comeau, R. Brousseau, R. Samson, and C. Greer. 1993. Construction and application of chromosomally integrated *lac-lux* gene markers to monitor the fate of a 2,4-dichlorophenoxyacetic acid-degrading bacterium in contaminated soils. *Microb. Releases* 1: 209-216.

Meckenstock, R., P. Steinle, J.R. van der Meer, and M. Snozzi. 1998. Quantification of bacterial mRNA involved in degradation of 1,2,4-trichlorobenzene by *Pseudomonas* sp. strain P51 from liquid culture and from river sediment by reverse transcriptase PCR (RT/PCR). *FEMS Microbiol. Lett.* 167: 123-129.

Megharaj, M., R.-M. Wittich, R. Blasco, D.H. Pieper, and K.N. Timmis. 1997. Superior survival and degradation of dibenzo-*p*-dioxin and dibenzofuran in soil by soil-adapted *Sphingomonas* sp. strain RW1. *Appl. Microbiol. Biotechnol.* 48: 109-114.

Meyer, S., R. Moser, A. Neef, U. Stahl, and P. Kämpfer. 1999. Differential detection of key enzymes of polyaromatic-hydrocarbon-degrading bacteria using PCR and gene probes. *Microbiology* 145: 7131-1741.

Miller, D.N., J.E. Bryant, E.L. Madsen, and W.C. Ghiorse. 1999. Evaluation and optimization of DNA extraction and purification procedures for soil and sediment samples. *Appl. Environ. Microbiol.* 65: 4715-4724.

Moller, A., and J.K. Jansson. 1997. Quantification of genetically tagged cyanobacteria in Baltic Sea sediment by competitive PCR. *Biotechniques* 22: 512-518.

Moller, A., K. Gustafsson, and J.K. Jansson. 1994. Specific monitoring by PCR amplification and bioluminescence of firefly luciferase gene-tagged bacteria added to environmental samples. *FEMS Microbiol. Ecol.* 15: 193-206.

More, M.I., J.B. Herrick, M.C. Silva, W.C. Ghiorse, and E.L. Madsen. 1994. Quantitative cell lysis of indigenous microorganisms and rapid extraction of microbial DNA from sediment. *Appl. Environ. Microbiol.* 60: 1572-1580.

Moriarty, F. 1988. *Exotoxicology: The Study of Pollutants in Ecosystems*, 2nd edn. Academic Press, New York.

Moser, R., and U. Stahl. 2001. Insights into the genetic diversity of initial dioxygenases from PAH-degrading bacteria. *Appl. Microbiol. Biotechnol.* 55: 609-618.

Muyzer, G. 1999. DGGE/TGGE: a method for identifying genes from natural ecosystems. *Curr. Opin. Microbiol.* 2: 317-322.

Muyzer, G., and K. Smalla. 1998. Application of denaturing gradient gel electrophoresis (DGGE) and temperature gradient gel electrophoresis (TGGE) in microbial ecology. *Antonie van Leeuwenhoek* 73: 127-141.

Muyzer, G., E.C. de Waal, and A.G. Uitterlinden. 1993. Profiling of complex microbial populations by denaturing gradient gel electrophoresis analysis of polymerase chain reaction-amplified genes encoding for 16S rRNA. *Appl. Environ. Microbiol.* 59: 695-700.

Nogales, B., E.R.B. Moore, W.-R. Abraham, and K.N. Timmis. 1999. Identification of the metabolically active members of a bacterial community in a polychlorinated biphenyl-polluted moorland soil. *Environ. Microbiol.* 1: 199-212.

Ogunseitan, O.A., I.L. Delgado, Y.L. Tsai, and B.H. Olson. 1991. Effect of 2-hydroxybenzoate on the maintenance of naphthalene-degrading pseudomonads in seeded and unseeded soil. *Appl. Environ. Microbiol.* 57: 2873-2879.

Okuta, A., K. Ohnishi, and S. Harayama. 1998. PCR isolation of catechol 2,3-dioxygenase gene fragments from environmental samples and their assembly into functional genes. *Gene* 212: 221-228.

Park, W., P. Padmanabhan, S. Padmanabhan, G.J. Zylstra, and E.L. Madsen. 2002. *nahR*, encoding a LysR-type transcriptional regulator, is highly conserved among naphthalene-degrading bacteria isolated from coal tar waste-contaminated site and in extracted community DNA. *Microbiology* 148: 2319-2329.

Picard, C., X. Nesme, and P. Sinomet. 1996. Detection and enumeration of soil bacteria using the MPN-PCR technique. Pages 1-9 in *Molecular Ecology Manual*, Vol. 2, J.D. van Elsas, ed., Kluwer, Dordrecht.

Prosser, J.I. 1994. Molecular marker systems for detection of genetically engineered microorganisms in the environment. *Microbiology* 140: 5-17.

Radajewski, S., P. Ineson, N.R. Parekh, and J.C. Murrell. 2000. Stable-isotope probing as a tool in microbial ecology. *Nature* 403: 646-649.

Radajewski, S., I.R. McDonald, and J.C. Murrell. 2003. Stable-isotope probing of nucleic acids: a window to the function of uncultured microorganisms. *Curr. Opin. Biotechnol.* 14: 296-302.

Ropp, J.D., C.J. Donahue, D. Wolfgang-Kimball, J.J. Hooley, J.Y.W. Chin, R.A. Hoffman, R.A. Cuthbertson, and K.D. Bauer. 1995. Aequorea green fluorescent protein and analysis by flow cytometry. *Cytometry* 21: 309-317.

Saano, A., E. Tas, S. Piippola, K. Lindström, and J.D. van Elsas. 1995. Extraction and analysis of microbial DNA from soil. Pages 49-67 in *Nucleic Acids in the Environment: Methods and Applications*. J.T. Trevors and J.D. van Elsas, eds., Springer, Berlin, Heidelberg, New York.

Saito, A., T. Iwabuchi, and S. Harayama. 2000. A novel phenanthrene dioxygenase from *Nocardiodes* sp. strain KP7: expression in *Escherichia coli*. *J. Bacteriol.* 182: 2134-2141.

Sato, S., J.-W. Nam, K. Kasuga, H. Nojiri, H. Yamane, and T. Omori. 1997. Identification and characterization of the gene encoding carbazole 1,9a-dioxygenase in *Pseudomonas* sp. strain CA10. *J. Bacteriol.* 179: 4850-4858.

Sayler, G.S., and A.C. Layton. 1990. Environmental application of nucleic acid hybridization. *Annu. Rev. Microbiol.* 44: 625-648.

Sayler, G.S., M.S. Shields, E.T. Tedford, A. Breen, and S.W. Hooper. 1985. Application of DNA-DNA colony hybridization to the detection of catabolic genotypes in environmental samples. *Appl. Environ. Microbiol.* 49: 1295-1303.

Schiwieger, F., and C.C. Tebbe. 1998. A new approach to utilize PCR-single-strand-conformation polymorphism for 16S rRNA gene-based microbial community analysis. *Appl. Environ. Microbiol.* 64: 4870-4876.

Schloss, P.D., and J. Handelsman. 2003. Biotechnological prospects from metagenomics. *Curr Opin Biotechnol* 14: 303-310.

Schneegurt-Mark, A., and F. Kulpa-Charler, Jr. 1998. The application of molecular techniques in environmental biotechnology for monitoring microbial systems. *Biotechnol. Appl. Biochem.* 27: 73-79.

Selvaratnam, S., B.A. Schoedel, B.L. McFarland, and C.F. Kulpa. 1995. Application of reverse transcriptase PCR for monitoring expression of the catabolic *dmpN* gene in a phenol-degrading sequencing batch reactor. *Appl. Environ. Microbiol.* 61: 3981-3985.

Selvaratnam, S., B.A. Schoedel, B.L. McFarland, and C.F. Kulpa. 1997. Application of the polymerase chain reaction (PCR) and reverse transcriptase/PCR for determining the fate of phenol-degrading *Pseudomonas putida* ATCC 11172 in a bioaugmented sequencing batch reactor. *Appl. Microbiol. Biotechnol.* 47: 236-240.

Small, J., D.R. Call, F.J. Brockman, T.M. Straub, and D.P. Chandler. 2001. Direct detection of 16S rRNA in soil extracts by using oligonucleotide microarrays. *Appl. Environ. Microbiol.* 67: 4708-4716.

Stokes, H.W., A.J. Holmes, B.S. Nield, M.P. Holley, K.M.H. Nevalainen, B.C. Mabbutt, and M.R. Gillings. 2001. Gene cassette PCR: sequence-independent recovery of entire genes from environmental DNA. *Appl. Environ. Microbiol.* 67: 5240-5246.

Taliani, M.R., S.C. Roberts, B.A. Dukek, R.K. Pruthi, W.L. Nichols, and J.A. Heit. 2001. Sensitivity and specificity of denaturing high-pressure liquid chromatography for unknown protein C gene mutations. *Genet. Test.* 5: 39-44.

Taranenko, N.I., R. Hurt, J. Zhou, N.R. Isola, H. Huang, S.H. Lee, and C.H. Chen. 2002. Laser desorption mass spectrometry for microbial DNA analysis. *J. Microbiol. Methods* 48: 101-106.

Tchelet, R., R. Meckenstock, P. Steinle, and J.R. van der Meer. 1999. Population dynamics of an introduced bacterium degrading chlorinated benzenes in a soil column and in sewage sludge. *Biodegradation* 10: 113-125.

Tombolini, R., A. Unge, M.E. Davey, F.J. de Bruijn, and J.K. Jansson. 1997. Flow cytometric and microscopic analysis of GFP-tagged *Pseudomonas fluorescens* bacteria. *FEMS Microbiol. Ecol.* 22: 17-28.

Torsvik, V., F.L. Daae, and J. Goksoyr. 1995. Extraction, purification and analysis of DNA from soil bacterial. Pages 29-48 in *Nucleic Acids in the Environment: Methods and Applications,* J.T. Trevors and J.D. van Elsas, eds., Springer, Berlin, Heidelberg, New York.

Tresse, O., D. Errampalli, M. Kostrzynska, K.T. Leung, H. Lee, J.T. Trevors, and J.D. van Elsas. 1998. Green fluorescent protein as a visual marker in a *p*-nitrophenol degrading *Moraxella* sp. *FEMS Microbiol. Lett.* 164: 187-193.

Trevors, J.T., and J.D. van Elsas. 1995. Introduction to nucleic acids in the environment: methods and applications. Pages 1-7 in *Nucleic Acids in the*

Environment: Methods and Applications, J.T. Trevor and J.D. van Elsas, eds., Springer, Berlin, Heidelberg, New York.

Unge, A., R. Tombolini, L. Mølbak, and J.K. Jansson. 1999. Simultaneous monitoring of cell number and metabolic activity of specific bacterial populations with a dual *gfp-luxAB* marker system. *Appl. Environ. Microbiol.* 65: 813-821.

van Dyke, M.I., H. Lee, and J.T. Trevors. 1996. Survival of *luxAB*-marked Alcaligenes *eutrophus* H850 in PCB-contaminated soil and sediment. *J. Chem. Technol. Biotechnol.* 65: 115-122.

van Elsas, J.D., V. Mantynen, and A.C. Wolters. 1997. Soil DNA extraction and assessment of the fate of *Mycobacterium chlorophenolicum* strain PCP-1 in different soils by 16S ribosomal RNA gene sequence based most-probable-number PCR and immunofluorescence. *Biol. Fertil. Soil* 24: 188-195.

van Elsas, J.D., A. Rosado, A.C. Moore, and V. Karlson. 1998. Quantitative detection of *Sphingomonas chlorophenoliza* in soil via competitive polymerase chain reaction. *J. Appl. Microbiol.* 85: 463-471.

von Wintzingerode, F., U.B. Gobel, and E. Stackkenbrandt. 1997. Determination of microbial diversity in environmental samples: pitfalls of PCR-based rRNA analysis. *FEMS Microbiol. Rev.* 21: 213-229.

von Wintzingerode, F., O. Landt, A. Ehrlich, and U.B. Gobel. 2000. Peptide nucleic acid-mediated PCR clamping as a useful supplement in the determination of microbial diversity. *Appl. Environ. Microbiol.* 66: 549-557.

Walia, S., A. Khan, and N. Rosenthal. 1990. Construction and applications of DNA probes for detection of polychlorinated biphenyl-degrading genotypes in toxic organic-contaminated soil environments. *Appl. Environ. Microbiol.* 56: 254-259.

Watanabe, K., M. Teramoto, H. Futamata, and S. Harayama. 1998. Molecular detection, isolation, and physiological characterization of functionally dominant phenol-degrading bacteria in activated sludge. *Appl. Environ. Microbiol.* 64: 4396-4402.

Watanabe, K., Y. Kodama, and S. Harayama. 2001. Design and evaluation of PCR primers to amplify bacterial 16S ribosomal DNA fragments used for community fingerprinting. *J. Microbiol. Methods* 44: 253-262.

Watts, J.E.M., Q. Wu, S.B. Schreier, H.D. May, and K.R. Sowers. 2001. Comparative analysis of polychlorinated biphenyl-dechlorinating communities in enrichment cultures using three different molecular screening techniques. *Environ. Microbiol.* 3: 710-719.

Weller, R., and D.M. Ward. 1989. Selective recovery of 16S ribosomal RNA sequences from natural microbial communities in the form of complementary DNA. *Appl. Environ. Microbiol.* 55: 1818-1822.

Widada, J., H. Nojiri, K. Kasuga, T. Yoshida, H. Habe, and T. Omori. 2001. Quantification of carbazole 1,9a-dioxygenase gene by real-time competitive PCR combined with co-extraction of internal standards. *FEMS Microbiol. Lett.* 202: 51-57.

Widada, J., H. Nojiri, K. Kasuga, T. Yoshida, H. Habe, and T. Omori. 2002a. Molecular detection and diversity of polycyclic aromatic hydrocarbon-degrading bacteria isolated from geographically diverse sites. *Appl. Microbiol. Biotechnol.* 58: 202-209.

Widada, J., H. Nojiri, T. Yoshida, H. Habe, and T. Omori. 2002b. Enhanced degradation of carbazole and 2,3-dichlorodibenzo-*p*-dioxin in soils by *Pseudomonas resinovorans* strain CA10. *Chemosphere* 49: 485-491.

Wilson, M.S., C. Bakerman, and E.L. Madsen. 1999. In situ, real-time catabolic gene expression: extraction and characterization of naphthalene dioxygenase mRNA transcripts from groundwater. *Appl. Environ. Microbiol.* 65: 80-87.

Widada, J., H. Nojiri, and T. Omori. 2002c. Recent developments in molecular techniques for identification and monitoring of xenobiotic-degrading bacteria and their catabolic genes in bioremediation. *Appl. Microbiol. Biotechnol.* 60: 45-59.

Wu, L., D.K. Thompson, G. Li, R.A. Hurt, J.M. Tiedje, and J. Zhou. 2001. Development and evaluation of functional gene arrays for detection of selected genes in the environment. *Appl. Environ. Microbiol.* 67: 5780-5790.

Yakimov, M.M., L. Giuliano, K.N. Timmis, and P.N. Golyshin. 2001. Upstream-independent ribosomal RNA amplification analysis (URA): a new approach to characterizing the diversity of natural microbial communities. *Environ. Microbiol.* 3: 662-666.

Yeates, C., A.J. Holmes, and M.R. Gillings. 2000. Novel forms of ring-hydroxylating dioxygenases are widespread in pristine and contaminated soils. *Environ. Microbiol.* 2: 644-653.

Zhou, J.-Z., M.A. Bruns, and J.M. Tiedje. 1996. DNA recovery from soils of diverse composition. *Appl. Environ. Microbiol.* 62: 316-322.

Genetic Engineering of Bacteria and Their Potential for Bioremediation

David B. Wilson

Department of Molecular Biology and Genetics, Cornell University,
458 Biotechnology Building, Ithaca, NY 14853, USA

Introduction

Genetic engineering of bacteria to improve their ability to degrade contaminants in the environment was the subject of the first patent for a living organism issued to Dr. Chakrabarty, who constructed an organism to degrade petroleum (Chakrabarty *et al.* 1978). However, these organisms were never used in bioremediation, partially because of regulatory constraints. This pattern of extensive research leading to the development of many potentially useful microorganisms that are not used because of strict regulations, continues today. In many cases, natural organisms have been isolated that can degrade manmade pollutants and these can be used with fewer tests, so that even when genetically modified organisms with higher activity have been developed, natural organisms are more likely to be used. However, there are still problems with bioremediation by non-modified organisms, so it is not always used.

A recent mini-review of the use of genetically engineered bacteria for bioremediation remains hopeful that this approach will ultimately be used (de Lorenzo 2001) and this area was thoroughly reviewed in 2000 (Pieper and Reineke 2000). Genetically modified organisms have been developed to degrade or modify many different compounds including carbozole, a petroleum component that inhibits catalysts used in refining (Riddle *et al.* 2003), pesticides (Qiao *et al.* 2003), explosives (Duque *et al.* 1993), aromatic compounds (Lorenzo *et al.* 2003, Watanabe *et al.* 2003), sulfur containing compounds (Noda *et al.* 2003), dioxins (Saiki *et al.* 2003) and heavy metals (Chen and Wilson 1997).

Bioremediation of Radioactive Sites

A major effort is being made by the U.S. Department of Energy (DOE) to develop radiation resistant bacteria to remediate radioactive sites contaminated during the production of nuclear weapons. *Deinococcus radiodurons* is a mesophilic radiation resistant bacterium, whose genome has been sequenced (Makarova *et al.* 2001) by the DOE Joint Genome Institute, while *D. geothermalis* is a moderately thermophilic radiation resistant bacterium that can grow at 55°C. Derivatives of *D. radiodurons* have been constructed that contain the *mer* operon for Hg^{++} resistance (Brim *et al.* 2000) or the *Pseudomonas tol* operon for degrading toluene (Lange *et al.* 1998). In a recent paper, *D. geothermalis* was transformed with plasmids isolated from *D. radiodurons* and a mercury resistant strain was produced (Brim *et al.* 2003). The combination of radiation, heavy metals, organic pollutants and high temperature present at some of these sites clearly provides a major opportunity for genetically modified organisms, as natural organisms that can function in remediating them are extremely unlikely to be found.

Bioremediation of Heavy Metals

A number of bacteria have been genetically engineered to remove a specific heavy metal from contaminated water by overexpressing a heavy metal binding protein, such as metallothionein, along with a specific metal transport system. This was first done with a Hg^{++} transport system (Chen and Wilson 1997) and the organisms that were constructed removed 99.8% of the Hg^{++} from water passed through induced cells in a hollow fiber reactor from both distilled water and a sample of polluted water containing many other ions (Chen *et al.* 1998, Deng and Wilson 2001), even in the absence of a carbon source. Organisms capable of removing Ni^{++}, Cd^{++} and Cu^{++} have also been constructed and characterized (Krishnaswamy and Wilson 2000, Zagorski and Wilson 2004).

It is not likely that naturally occurring bacteria will be found that specifically take up a single heavy metal, as this would not benefit the organism. Furthermore, induced organisms that contain large amounts of the metallothionein fusion protein cannot grow, although they still possess the ability to accumulate the heavy metal, so that these organisms provide little potential to escape and cause environmental problems. In theory, it should be possible to remove and separate several heavy metals from contaminated water by using multiple reactors in series, each containing an organism specific for a given heavy metal. The amounts of heavy metals found in bacteria that are saturated with metal are large enough so that it would be possible to recycle each metal from metal saturated cells. Calculations show that Hg^{++} should make up about 40% of the ash from mercury saturated cells. An enzyme that codes for phytochelatin synthesis in *Escherichia coli* was overexpressed and it was shown that the modified bacteria accumulate more heavy metals than WT cells (Sauge-Merle *et al.* 2003). However, these cells do not express metal transport genes and appear only to concentrate Cd^{++}, Cu^{++} and As^{++}. Furthermore, the maximum amount of metal found, 7 µmoles/gram, is lower than seen with some other methods. The use of organisms containing the *mer* operon for mercury resistance in mercury bioremediation was reviewed recently (Nascimento and Chartone-Souza 2003). One problem with organisms containing the complete *mer* operon is that mercury ions are converted to mercury, which remains in the environment.

Bioremediation of Chlorinated Compounds

There have been significant advances in the identification of bacteria that can degrade chlorinated hydrocarbons such as tetrachlorothene (PCE), 1,1,1-trichlorothene (TCA), polychlorinated biphenyls (PCBs), which are major environmental contaminants because of their widespread use and persistence, and the degradation of chlorophenols was recently reviewed (Solyanikova and Golovleva 2004). The genome of *Dehalococcoides ethenogenes* has been sequenced by the DOE Joint Genome Institute. This organism can completely degrade PCE to CO_2, whereas most organisms produce vinyl chloride, a toxic substance, so that *D. ethenogenes* is an excellent organism for bioremediation of PCE (Fennell *et al.* 2004).

PCB degradation is complex as there are many different forms and it has been shown that orthochlorinated PCBs inhibit and inactivate a key enzyme in the degradation pathway, dehydroxybiphenyl oxygenase (Dai *et al.* 2002). The first enzyme in the pathway of PCB degradation is biphenyl dioxygenase and DNA shuffling has been used to produce modified enzymes that have higher activity on highly resistant PCBs including 2,6-dichlorobiphenyl, which is very resistant to degradation by natural organisms (Barriault *et al.* 2002). The shuffled genes were expressed in *E. coli* and the best strain degraded a broad range of PCBs from 6 to 10 times faster than strains containing the parent gene. Recombinant organisms with improved ability to degrade TCE have also been constructed (Maeda *et al.* 2001). The use of modified organisms to degrade chlorinated compounds was the subject of a recent review (Furukawa 2003).

Another important pollutant, pentachlorophenol (PCP), is slowly degraded by *Sphingobium chlorophenolicum*, but only at low concentrations. Genome shuffling, which is carried out by generating a set of mutant strains that have improved activity and then carrying out multiple rounds of protoplast fusion, allowed the construction of strains that could grow in the presence of 6 mM PCP, ten times higher than the starting strain, and the new strains can completely degrade 3 mM PCP, while the WT strain can only degrade 0.3 mM PCP (Dai and Copley 2004).

A major contaminant in farm soils is atrazine, a chlorinated herbicide. A successful field trial was reported in which killed recombinant *E. coli* overproducing atrazine chlorohydrolase were applied to soil along with inorganic phosphate (Strong *et al.* 2000). In the plots receiving only the killed bacteria (0.5% w/w), atrazine was 52% lower after eight weeks, while in plots receiving the bacteria and phosphate, atrazine was 77% lower. In the control plots or ones receiving only phosphate, there was no degradation of atrazine. A natural organism able to degrade atrazine at 250 ppm was isolated recently (Singh *et al.* 2004).

2,4,5-T is a chlorinated aromatic compound that is used as a herbicide and was extensively used as a defoliant in the Vietnam war. A strain of *Pseudomonas cepacia* was isolated from a chemostat, fed with a low concentration of a carbon source and a high concentration of 2,4,5-T, that could use it as a sole carbon and energy source (Ogawa *et al.* 2003).

TecA is a tetrachlorobenzene dioxygenase from *Ralsonla* sp. PS12, which can react with many chlorinated benzenes and toluene. Its substrate specificity is determined by its α-subunit, as is true for several oxygenases. Using sequence alignments, five substitutions were identified in two residues that were likely to be important for substrate specificity (Pollmann *et al.* 2003). Site directed mutations were made containing each of the changes and caused some changes in product formation, but all the mutations reduced the activity.

Real-time PCR was used to monitor the population of a genetically engineered strain of *P. putida* that could degrade 2-chlorobenzoate. This strain also contained a gene for green fluorescent protein so that the population determined by PCR could be compared to that determined by direct culturing of fluorescent bacteria and the growth curves measured by the two methods were very similar. This method was tested in three different soils and in each case the rate of 2-chlorobenzoate degradation matched the level of the modified bacteria in the culture (Wang *et al.* 2004).

Organophosphate Bioremediation

Parathion is a powerful organophosphorous insecticide that is very toxic. A dual species consortium was constructed by cloning the gene for parathion hydrolase into *E. coli* and the operon for p-nitrophenol degradation, a product of parathion hydrolysis, into *Pseudomonas putida* (Gilbert *et al.*, 2003). The mixed culture was shown to degrade 6 mg parathion/g dry weight of cells/h with a Km of 47 mg/L. These two strains could form a mixed biofilm, but it was not tested for its ability to degrade parathion. Another group engineered a strain of *Moraxella*, which can grow on dinitrophenol to degrade parathion and other organophosphorous pesticides by expressing organophosphorous hydrolase (OPH) on the surface of the engineered cells (Shimazu *et al.* 2001). These cells degraded 0.4 mM paradoxin within 40 minutes, although p-nitrophenol degradation was much slower. The rate of paradoxin degradation at 30°C was 9 μmol/h/mg dry weight, while the PNP degradation rate was 0.6 μmol/h/mg.

This same group constructed a recombinant *E. coli* strain that expressed both OPH and a cellulose binding domain on the outer membrane outer surface. This strain could be immobilized on cellulose and the immobilized cells completely degraded 0.25 mM paradoxin in an hour (Wang *et al.* 2002). The immobilized cells were stable for 45 days, while a cell suspension lost more than 50% of its activity over the same period. A cotton fabric coated with immobilized cells had a degradation rate of 6.7 μM/min/0.24 gram at 25°C. Another group has used a genetically engineered enzyme to degrade organophosphate compounds (Qiao *et al.* 2003).

Phytoremediation

Phytoremediation of water soluble, volatile organic compounds often results in the release of the compounds into the atmosphere. By colonizing a plant with recombinant endophytic bacteria that could degrade toluene, its release was cut to less than 50% of that of control plants or plants with unmodified bacteria (Barac *et al.* 2004). A surprising finding was that a related strain of bacteria, which was selected to degrade toluene but was not endophytic, gave higher cell numbers inside the plant, inhibiting plant growth, but the presence of the native toluene degrading bacteria did not reduce toluene release. The plants containing the recombinant bacteria degraded more toluene than any of the other plants.

A biological system to prevent long-term survival of rhizoremediating bacteria in the soil, in the absence of the pollutant being degraded, was developed (Ronchel and Ramos 2001). The *Pseudomonas putida asd* gene was deleted in a strain and a plasmid that contained the *lacI* gene regulated by the Pm promoter along with a Plac promoter linked to *gef*, which encodes a lethal porin protein, was introduced. When inducers of Pm are present (modified benzoates), the cells survive, as porin synthesis is repressed and the essential compounds required by the *asd* mutant strain are produced from the benzoate compounds. This strain survived in the rhizosphere, as well as WT cells in the presence of pollutant, but disappeared in less than 20 days in its absence, where as WT cells lasted much longer (Ronchel and Ramos 2001).

A recombinant strain of *Rhizobium* was constructed that expressed carbozole 1,9a-dioxygenase. This strain colonized the roots of siratrol (a legume) and caused significant degradation of dibenzofuran, a very insoluble dioxin (48% in 3 days) (Saiki *et al.* 2003). The bacteria were able to colonize this plant in all non-sterile soils tested, except wet paddy soils (Saiki *et al.* 2003).

Aromatic Hydrocarbon Bioremediation

An organism was constructed that actively degrades styrene and also contains a gene containment system to reduce lateral transfer of the styrene degrading genes to other hosts (Lorenzo *et al.* 2003). *Pseudomonas putida* F1 was transformed with both the pWWO *tol* plasmid and a styrene plasmid to produce a strain that could degrade mixtures of styrene, toluene and xylene. In further work, a mini transposon cassette was prepared, which contained the ColE$_3$ gene, and it was integrated into the genome of bacteria that contain an E$_3$ resistance gene. This cassette was integrated into *P. putida* kt24421CS and the resulting strain could grow on styrene (Lorenzo *et al.* 2003).

A very interesting approach is to produce bacteria that convert waste chemicals to useful chemicals. Modified oxygenases have been created that convert arenes (polycyclic compounds) into novel products (Shindo *et al.* 2000) such as 4-hydroxyfluorene and 10-hydroxyphen anthridine (Ronchel and Ramos 2001). Finally, *P. putida* was modified so that it was unable to metabolize medium chain length alcohols such as decanol. The modified strain was shown to degrade phenol at the same rate as the wildtype strain. However, the modified strain could be used in a two-phase partitioning bioreactor with decanol as the solvent and gave rapid phenol degradation without degradation of decanol (Vrionis *et al.* 2002).

REFERENCES

Barac, T., S. Taghavi, B. Borremans, A. Provoost, L. Oeyen, J.V. Colpaert, J. Vangronsveld, and D. van der Lelie. 2004. Engineered endophytic bacteria improve phytoremediation of water-soluble, volatile, organic pollutants. *Nat. Biotechnol.* 22: 583-588.

Barriault, D., M.M. Plante, and M. Sylvestre. 2002. Family shuffling of a targeted bphA region to engineer biphenyl dioxygenase. *J. Bacteriol.* 184: 3794-3800.

Brim, H., S.C. McFarlan, J.K. Fredrickson, K.W. Minton, M. Zhai L.P. Wackett, and M.J. Daly. 2000. Engineering *Deinococcus radiodurans* for metal remediation in radioactive mixed waste environments. *Nat. Biotechnol.* 18: 85-90.

Brim, H., A. Venkateswaran, H.M. Kostandarithes, J.K. Fredrickson, and M.J. Daly. 2003. Engineering *Deinococcus geothermalis* for bioremediation of high-temperature radioactive waste environments. *Appl. Environ. Microbiol.* 69: 4575-4582.

Chakrabarty, A.M., D.A. Friello, and L.H. Bopp. 1978. Transposition of plasmid DNA segments specifying hydrocarbon degradation and their expression in various microorganisms. *Proc. Natl. Acad. Sci. USA* 75: 3109-3112.

Chen, S., E. Kim, M.L. Shuler, and D.B. Wilson. 1998. Hg^{2+} removal by genetically engineered *Escherichia coli* in a hollow fiber bioreactor. *Biotechnol. Prog.* 14: 667-671.

Chen, S. and D.B. Wilson. 1997. Construction and characterization of *Escherichia coli* genetically engineered for Hg^{2+} bioremediation. *Appl. Environ. Microbiol.* 63: 2442-2445.

Dai, M. and S.D. Copley. 2004. Genome shuffling improves degradation of the anthropogenic pesticide pentachlorophenol by *Sphingobium chlorophenolicum* ATCC 39723. *Appl. Environ. Microbiol.* 70: 2391-2397.

Dai, S., F.H. Vaillancourt, H. Maaroufi, N.M. Drouin, D.B. Neau, V. Snieckus, J.T. Bolin, and L.D. Eltis. 2002. Identification and analysis of a bottleneck in PCB biodegradation. *Nat. Struct. Biol.* 9: 934-939.

de Lorenzo, V. 2001. Cleaning up behind us. The potential of genetically modified bacteria to break down toxic pollutants in the environment. *EMBO Rep.* 2: 357-359.

Deng, X. and D.B. Wilson. 2001. Bioaccumulation of mercury from wastewater by genetically engineered *Escherichia coli*. *Appl. Microbiol. Biotech.* 56: 276-279.

Duque, E., A. Haidour, F. Godoy, and J.L. Ramos. 1993. Construction of a Pseudomonas hybrid strain that mineralizes 2,4,6-trinitrotoluene. *J. Bacteriol.* 175: 2278-2283.

Fennell, D.E., I. Nijenhuis, S.F. Wilson, S.H. Zinder, and MM. Haggblom. 2004. *Dehalococcoides ethenogenes* strain 195 reductively dechlorinates diverse chlorinated aromatic pollutants. *Environ. Sci. Technol.* 38: 2075-2081.

Furukawa, K. 2003. Related Articles, 'Super bugs' for bioremediation. *Trends Biotechnol.* 21: 187-90.

Gilbert, E.S., A.W. Walker, and J.D. Keasling. 2003. A constructed microbial consortium for biodegradation of the organophosphorus insecticide parathion. *Appl. Microbiol. Biotechnol.* 61: 77-81.

Krishnaswamy, R, and D.B. Wilson. 2000. Construction and characterization of *Escherichia coli* genetically engineered for Ni(II) bioaccumulation. *Appl. Environ. Microbiol.* 66: 5383-5386.

Lange C.C., L.P. Wackett, K.W. Minton, and M.J. Daly. 1998. Engineering a recombinant *Deinococcus radiodurans* for organopollutant degradation in radioactive mixed waste environments. *Nat. Biotechnol.* 16: 929-33.

Lorenzo, P., S. Alonso, A. Velasco, E. Diaz, J.L. Garcia, and J. Perera. 2003. Design of catabolic cassettes for styrene biodegradation. *Antonie van Leeuwenhoek* 84: 17-24.

Maeda, T., Y. Takahashi, H. Suenaga, A. Suyama, M. Goto, and K. Furukawa. 2001. Functional analyses of Bph-Tod hybrid dioxygenase, which exhibits high degradation activity toward trichloroethylene. *J. Biol. Chem.* 276: 29833-29838.

Makarova, K.S., L. Aravind, Y.I. Wolf, R.L. Tatusov, K.W. Minton, E.V. Koonin, and M.J. Daly. 2001. Genome of the extremely radiation-resistant bacterium *Deinococcus radiodurans* viewed from the perspective of comparative genomics. *Microbiol. Mol. Biol. Rev.* 65: 44-79.

Nascimento, A.M., and E. Chartone-Souza. 2003. Operon *mer:* bacterial resistance to mercury and potential for bioremediation of contaminated environments. *Genet. Mol. Res.* 2: 92-101.

Noda, K., K. Watanabe, and K. Maruhashi. 2003. Recombinant *Pseudomonas putida* carrying both the dsz and hcu genes can desulfurize dibenzothiophene in n-tetradecane. *Biotechnol. Lett.* 25: 1147-1150.

Ogawa, N., K. Miyashita, and A.M. Chakrabarty. 2003. Microbial genes and enzymes in the degradation of chlorinated compounds. *Chem. Rec.* 3: 158-71.

Pieper, D.H., and W. Reineke. 2000. Engineering bacteria for bioremediation. *Curr. Opin. Biotechnol.* 11: 262-270.

Pollmann, K., V. Wray, H.J. Hecht, and D.H. Pieper. 2003. Rational engineering of the regioselectivity of TecA tetrachlorobenzene dioxygenase for the transformation of chlorinated toluenes. *Microbiology* 149: 903-913.

Qiao, ChL., J. Huang, X. Li, B.C. Shen, and J.L. Zhang. 2003. Bioremediation of organophosphate pollutants by a genetically-engineered enzyme. *Bull. Environ. Contam. Toxicol.* 70: 455-61.

Qiao, ChL., YCh. Yan, H.Y. Shang, X.T. Zhou and Y. Zhang. 2003. Biodegradation of pesticides by immobilized recombinant *Escherichia coli. Bull. Environ. Contam. Toxicol.* 71:455-61.

Riddle, R.R., P.R. Gibbs, R.C. Willson, and M.J. Benedik. 2003. Recombinant carbazole-degrading strains for enhanced petroleum processing. *J. Ind. Microbiol. Biotechnol.* 30: 6-12.

Ronchel, M.C., and J.L. Ramos. 2001. Dual system to reinforce biological containment of recombinant bacteria designed for rhizoremediation. *Appl. Environ. Microbiol.* 67: 2649-2656.

Saiki, Y., H. Habe, T. Yuuki, M. Ikeda, T. Yoshida, H. Nojiri, and T. Omori. 2003. Rhizoremediation of dioxin-like compounds by a recombinant *Rhizobium tropici* strain expressing carbazole 1,9a-dioxygenase constitutively. *Biosci. Biotechnol. Biochem.* 67: 1144-1148.

Sauge-Merle, S., S. Cuine, P. Carrier, C. Lecomte-Pradines, D.T. Luu, and G. Peltier. 2003. Enhanced toxic metal accumulation in engineered bacterial cells expressing *Arabidopsis thaliana* phytochelatin synthase. *Appl. Environ. Microbiol.* 69: 490-494.

Shimazu, M., A. Mulchandani, and W. Chen. 2001. Simultaneous degradation of organophosphorus pesticides and p-nitrophenol by a genetically engineered *Moraxella* sp. with surface-expressed organophosphorus hydrolase. *Biotechnol. Bioeng.* 76: 318-324.

Shindo, K., Y. Ohnishi, H.K. Chun, H. Takahashi, M. Hayashi, A. Saito, K. Iguchi, K. Furukawa, S. Harayama, S. Horinouchi, and N. Misawa. 2000. Oxygenation reactions of various tricyclic fused aromatic compounds using *Escherichia coli* and *Streptomyces lividans* transformants carrying several arene dioxygenase genes. *Biosci. Biotechnol. Biochem.* 65: 2472-2481.

Singh, P., C.R. Suri, and S.S. Cameotra. 2004. Isolation of a member of Acinetobacter species involved in atrazine degradation. *Biochem. Biophys. Res. Commun.* 317: 697-702.

Solyanikova, I.P., L.A. Golovleva. 2004. Bacterial degradation of chlorophenols: pathways, biochemical, and genetic aspects. *J. Environ. Sci. Health B.* 39: 333-351.

Strong, L.C., H. McTavish, M.J. Sadowsky, and L.P. Wackett. 2000. Field-scale remediation of atrazine-contaminated soil using recombinant *Escherichia coli* expressing atrazine chlorohydrolase. *Environ. Microbiol.* 2: 91-98.

Vrionis, H.A., A.M. Kropinski, and A.J. Daugulis. 2002. Enhancement of a two-phase partitioning bioreactor system by modification of the microbial catalyst: demonstration of concept. *Biotechnol. Bioeng.* 79: 587-594.

Wang, A.A., A. Mulchandani, and W. Chen. 2002. Specific adhesion to cellulose and hydrolysis of organophosphate nerve agents by a genetically engineered *Escherichia coli* strain with a surface-expressed cellulose-binding domain and organophosphorus hydrolase. *Appl. Environ. Microbiol.* 68: 1684-1689.

Wang, G., T.J. Gentry, G. Grass, K. Josephson, C. Rensing, and I.L. Pepper. 2004. Real-time PCR quantification of a green fluorescent protein-labeled, genetically engineered *Pseudomonas putida* strain during 2-chlorobenzoate degradation in soil. *FEMS Microbiol. Lett.* 233: 307-314.

Watanabe, K., K. Noda, J. Konishi, and K. Maruhashi. 2003. Desulfurization of 2,4,6,8-tetraethyl dibenzothiophene by recombinant Mycobacterium sp. strain MR65. *Biotechnol. Lett.* 25: 1451-1456.

Zagorski, N., and D.B. Wilson. 2004. Characterization and comparison of metal accumulation in two *Escherichia coli* strains expressing either CopA or MntA, heavy metal-transporting bacterial P-type adenosine triphosphatases. *Appl. Biochem. Biotechnol.* 117: 33-48.

Commercial Use of Genetically Modified Organisms (GMOs) in Bioremediation and Phytoremediation

David J. Glass

D. Glass Associates, Inc. and Applied PhytoGenetics, Inc., 124 Bird Street, Needham, MA 02492 USA

Introduction

Ever since the advent of recombinant DNA and other genetic engineering technologies in the late 1970s, and the growth of the biotechnology industry beginning shortly thereafter, it has been widely assumed that these biotechnologies would be used for environmentally-beneficial purposes, including the clean-up of contaminated soils and waters. Many observers have expected that genetically modified organisms (GMOs) would quickly find broad applicability in remediation of hazardous chemicals from the environment, and these expectations persisted even as the uses of biology for clean-up began to extend to plants, as the phytoremediation industry arose in the 1990s. However, as of this writing, genetically engineered microorganisms have not yet been used in commercial site remediation, with few if any current plans for such uses, and transgenic plants are only beginning to find applicability in commercial phytoremediation projects. Why is this so?

Although there are many compelling reasons to consider the use of advanced biotechnology to improve on naturally occurring plants and microbes for use in remediation, there are many more reasons why this has not yet come to pass. Many of these reasons have their origins in the regulatory and public controversies that surrounded uses of GMOs for agricultural purposes in the 1980s and which to some degree still exist. Other reasons are more particular to the economic and other realities of the remediation business, and to the economics of conducting advanced biological research.

This article will describe the potential need for engineered organisms in commercial remediation; summarize some of the ways that academic and industrial research groups are considering modifying naturally occurring organisms for this purpose; discuss where these efforts stand and how close to commercial markets they are; and examine the prospects for the use of GMOs in commercial remediation. It is beyond the scope of this article to exhaustively or comprehensively review R&D efforts in academic and commercial laboratories to modify microorganisms and plants for use in hazardous waste remediation, but there are several such reviews recently published (e.g., Wilson this volume, and other references cited below). Instead, we will focus on a discussion of the reasons one might plausibly wish to use GMOs in commercial remediation, and an analysis of the feasibility of seeing such organisms used commercially.

Overview: What Barriers do GMOs Face in the Remediation Market?

In assessing the possible role that GMOs may play in commercial remediation, it is first useful to consider the existing market for remediation products and services, in particular several aspects of this market most relevant to introduction of GMOs. Unlike other fields of commerce where GMOs and their products have been adopted, in some cases enthusiastically, by the marketplace, the unique nature of the environmental industry has placed obstacles and challenges in the way of the introduction of innovative products and technologies such as GMOs, and there are unusually powerful economic, technical and regulatory factors that affect the ability of new technologies to enter commercial markets.

Although a relatively young industry, dating back only to the 1970s, the U.S. hazardous waste remediation business has been dominated throughout its history by a very conservative approach to technology. The vast majority of contaminated sites have been remediating using the traditional techniques of disposal (i.e., landfilling) and containment, even though regulatory and other governmental initiatives over the past two decades or more have promoted a shift to "treatment technologies" using more cutting-edge methodology. The U.S. Environmental Protection Agency (EPA) defines two major categories of treatment technology: "established technologies", primarily including incineration and solidification/stabilization for soil remediation and pump-and-treat for groundwater remediation; and "innovative technologies" such as bioremediation, phytoremediation, soil vapor extraction and others. The major difference between the two is that the EPA considers cost and performance data to be available for "established" technologies, but not for "innovative" techniques (U.S. EPA 1999). In spite of efforts to promote treatment technologies, including innovative technologies, traditional methods still dominate much of the nation's remediation, while the better understood physical or chemical treatment methods have the lion's share of the treatment market (U.S. EPA 1997), leaving only a small share to newer techniques like the biological remediation technologies.

GMOs designed for remediation will be entering a market (at least in the U.S.) that is mature, slow-growing, and fragmented among a very large number of providers and a large number of competing technologies. In addition, it is a service industry rather than a products-based industry, and together these factors create a very small market niche for engineered plants and microbes to compete.

The U.S. remediation market, which exhibited explosive growth in the 1970s and 1980s, has in recent years become a mature, conservative market that has seen flat or even negative growth for much of the past decade and a half. The overall U.S. remediation market is perhaps U.S. $6-8 billion per year, depending on which products and services are included in the estimate, but this market has declined or remained steady throughout the 1990s and the early years of the present decade (Glass 2000, Environmental Business Journal 2003). The U.S. has the largest remediation market in the world, but markets outside the U.S., while smaller, exhibit faster, stronger growth. The current world remediation market is about U.S. $20-25 billion per year.

As innovative technologies, both bioremediation and phytoremediation command only small shares of this overall market. We have previously estimated that altogether, the two dozen or so different innovative remedial technologies used in the U.S. make up no more than 30-50% of the total remediation market, or approximately $2-4 billion per year (Glass 2000). Bioremediation is the better established of the two biological technologies considered in this chapter, and we estimate the current U.S. bioremediation market to be U.S. $600 million, a level it has taken most of bioremediation's twenty-year history to reach. Phytoremediation is a newer technology which has attracted a lot of attention but which has been slow to penetrate the market, and the current (2004) market for phytoremediation is probably no more than U.S. $100-150 million, somewhat lower than our previous estimates (Glass 1999).

The bulk of the bioremediation market consists of services rather than sales of microorganisms (see below). It is important, when considering the market potential for GMOs, to realize how little of the U.S. bioremediation market is attributable to sales of isolated microbial cultures. One early (1990) estimate of the U.S. market for packaged microbial cultures was U.S. $30-50 million, but a 1994 estimate put the 1993 market for microbes at only U.S. $6-7 million. Consensus figures published in the 1990s placed the market at U.S. $25-55 million, and we estimate that the market for microbial remediation products in the early years of the current decade was probably about U.S. $30-50 million, or perhaps a little higher. This constitutes less than 10% of the bioremediation market, although this share may have risen in recent years due to recent product introductions. Although likely not documented in any publication, the same is true in the phytoremediation market: it is almost certain that only a small percentage of phytoremediation revenues is attributable to sales of plants and trees, with the majority of revenues being devoted to the service component of any remediation job (e.g., site preparation, planting, maintenance and monitoring).

As noted above, several features of the remediation market will affect the adoption and widespread use of GMOs in bioremediation or phytoremediation (see Glass 1999 for a longer discussion of many of these issues). As mentioned, the U.S. remediation industry has historically been quite cautious and conservative with regard to adopting innovative technologies. Many site owners, consultants and regulators are more comfortable choosing technologies and methods with which they are familiar, and which have a long track record of success and thus a greater predictability. Site owners are often unwilling to fund "research", and will therefore not be willing to consider the use of a possibly experimental method at a site under their control. For example, Dümmer and Bjornstad (2004) refer to the "incredible inertia" of the U.S. Department of Energy's (DOE's) institutional framework for remediation, saying that it causes new technologies to be "less than fully attractive to locals". It should be noted, however, that newer markets elsewhere in the world have seemed somewhat more willing to use innovative technologies, particularly once they had begun to be demonstrated in the U.S.

A corollary to this is that the regulations themselves often favor existing technologies: under several applicable federal and state regulatory programs in the U.S., endpoint concentrations for certain contaminants have been established as the levels achievable using "best demonstrated available technology", under circumstances where the best technology is an established one such as incineration. These regulations may apply even in remediation scenarios where less-stringent endpoints would be acceptable in view of the proposed end-use of the site. In those cases where an innovative technology is incapable of delivering the "6 logs" clean-up standard achievable by incineration, but where the innovative technology could nevertheless clean the site to an otherwise-sufficient degree, the innovative technology is often unlikely to be chosen as the remedial option.

In addition, the economics of the remediation business work against the desire to introduce new technologies. In mature markets like the U.S., where there are numerous technologies, traditional and innovative, competing for market share, remediation has become a commodity business, with a large number of vendors in the market competing on price and often offering "me too" technologies that are not proprietary and which can often not be distinguished from other available methods on the basis of performance. This creates a market with very small profit margins, and with so many vendors offering competing services of many kinds, it is hard for a new company to achieve a significant market share.

A recent report by the EPA (U.S. EPA 2000a) identified 42 barriers to the introduction of innovative treatment technologies that had been consistently cited by the authors of ten different reports and documents since 1995. Included among these were institutional barriers, regulatory and legislative barriers, technical barriers and economic and financial barriers. The authors of the reports citing these barriers came from all sectors of the remediation field, indicating widespread belief that numerous obstacles exist in the marketplace affecting the adoption of innovative treatment technologies.

Many of these factors are particularly important for biological technologies and affect the prospects for use of GMOs. Advanced bio-technology R&D can be expensive and time-consuming, with long lead times needed to develop new bacterial strains or plant lines. It is very difficult for remediation companies to justify the costs and timelines of such research programs, because the low profit margins will make it tough to recoup R&D costs. In addition, biological methods suffer additional constraints not shared by physical or chemical techniques: the inherent limitations of biological systems and enzyme-based catalysis places limits on the efficiency of biological remediation methods. A microorganism or plant may well be able to remove or convert 98-99% of a given contaminant, but will often be unable to achieve the much higher standards set by regulation (e.g., "6 logs" or 99.9999% reduction). In many cases, particularly with phytoremediation, biological processes can be slower than competing technologies, particularly energy-intensive physicochemical methods. Although biological processes have advantages that in many cases outweigh the disadvantages (e.g., lower cost, complete destruction of wastes, esthetically pleasing as a "green" technology), these disadvantages play into the conservative nature of site owners and regulators, leading to increased barriers to the use of biological methods at any specific site.

There have been other reasons why GMOs have not yet been used in commercial bio- or phytoremediation. One widely-believed reason has to do with government regulation and public perceptions of the environ-mental uses of GMOs: many in the environmental business community have come to believe that, because of the public controversies over such uses in the 1980s and the resulting government regulations, it is either impossible or prohibitively expensive to test or use GMOs in the open environment. As is discussed below, this is not true (Glass 1994, 1997), and GMOs are beginning to be used in phytoremediation and certain preliminary tests have taken place with GMOs for bioremediation. Yet the perception persists and has been a powerful disincentive against the use of GMOs for environmental remediation. For example, Dümmer and Bjornstad (2004), while documenting numerous regulatory and institutional barriers generally affecting the use of bioremediation at U.S. DOE remediation sites, nevertheless consider biotechnology regulations to be an obstacle to the use of GMOs at DOE sites, that will "undoubtedly call for expanding risk-related information bases and assessment protocols".

However, it is true that in many cases, the technological need to improve organisms, especially microorganisms, intended for remediation has been lacking. As discussed below, most uses of bioremediation today involve methods to stimulate the growth or activity of indigenous microorganisms at contaminated sites, and so inoculant organisms are not needed at all. Even for those applications where it might be plausible to use an introduced culture, investigators have been able to find naturally-occurring organisms, or to create strains using classical techniques of mutagenesis, having the desired activity. The R&D necessary to create or isolate such microbial strains would be expected to be less expensive than a genetic engineering approach, and using such strains avoids any issues relating to the use of GMOs, including the added costs of GMO-specific regulations, and so this strategy has clearly been favored by those in the industry developing new remedial strains.

Nevertheless, there are still many unmet needs in commercial remediation, including many scenarios where available remedial technologies for a given contaminant are either too expensive or too inefficient to be broadly adopted for commercial use. These offer opportunities for the introduction of innovative technologies, including biological methods using GMOs. The power of the new biotechnologies makes it quite plausible that biological solutions can be found for many of these needs, through the creation of new plants or microorganisms having novel biochemical traits or enzymatic activities that might be useful for remediation. The following section will explore those areas where GMOs are likeliest to be used in commercial remediation to address these unmet needs.

Use of Genetic Engineering to Address Unmet Needs in Site Remediation

There continue to be opportunities where novel technologies can be introduced in the remediation market, in spite of flat market growth and the abundance of other available technologies. In particular, although regulatory and economic factors continue to exert their influence in slowing the pace of clean-ups in the U.S., it is likely that the riskiest sites will, over time, continue to be remediated. Specifically, there are a number of contaminant classes which pose unusually high health or environmental risks, or which the general public believes to be dangerous and therefore demands be remediated. Remediation of many of these types of compounds has historically been hindered by a lack of affordable and/or effective remediation options, and therefore many of these represent opportunities for the development of effective, low-cost remedial strategies, and for development of biological methods in particular. Examples include:

- Pervasive, toxic chlorinated solvents like TCE.
- Recalcitrant, long-persistent compounds like PCBs, dioxins, and other high molecular weight chlorinated compounds.
- Xenobiotics and other hazardous materials which have only recently been recognized as environmental contaminants, such as MTBE and perchlorate.
- Heavy metals, particularly ones recognized as health threats like mercury, lead, chromium or arsenic, or for which adequate or affordable remediation methods do not exist.
- Radioisotopes and mixed radioactive/hazardous contaminants.

It is reasonable to believe that demand for remediation of these contaminants will continue to be high in the foreseeable future, and that effective remedial technologies will be accepted in the market and can be implemented at premium (rather than commodity) prices, thus potentially justifying the high costs of biotechnology R&D. To the extent such contaminants are amenable to biological remediation or containment approaches, these pollutants might be good targets for development of new remedial methods through the use of advanced biotechnology. Strategies to address these needs with advanced biotechnology would, in general, involve enhancing existing degradative pathways to be faster or more efficient (i.e., to do the job at a time and cost that are commercially feasible), or to create biological treatment options that do not exist in nature. Such strategies are discussed below, first for microorganisms for use in bioremediation and then for plants for use in phytoremediation.

Prospects for Commercial Bioremediation Using Genetically Engineered Microorganisms

Existing Bioremediation Technologies

Bioremediation is generally considered to include a number of specific applications, as summarized below and as described in detail elsewhere in this volume. Most *in situ* bioremediation methods practiced today rely on the stimulation of indigenous microbial populations at the site of contamination, by addition of appropriate nutrients, principally carbon, oxygen, phosphorus and nitrogen, and by maintaining optimum conditions of pH, moisture and other factors, to trigger increased growth and activity of indigenous biodegradative microorganisms. Applications of this strategy are sometimes referred to by the umbrella term "biostimulation", with the most commonly practiced variants being:

For in situ treatment of groundwater contamination:

- *Bioventing*: the injection of oxygen into the unsaturated zone above a water table, in order to stimulate biodegradation by indigenous organisms in the groundwater while also volatilizing ("stripping") certain of the contaminants.

- *Biosparging*: the injection of oxygen into the saturated zone (i.e., below the water table), so that oxygen bubbles can rise into the unsaturated zone, where natural biodegradation can be stimulated and volatile contaminants stripped.

- *Bioslurping*: the combination of soil vapor extraction/bioventing with removal of liquid hydrocarbons from the surface of the aquifer (NAPLs -- nonaqueous phase liquids).

For in situ or ex situ treatment of soil contamination:

- *Land-farming*: the application of soil bioremediation in which adequate oxygenation is ensured by frequent turning or disking of the soil.

- *Ex situ or solid-phase bioremediation*: in which soil is excavated and placed in a pile where biodegradation is stimulated by addition of nutrients, water, and sometimes added bacterial cultures, surfactants, etc.

In addition to "biostimulation" approaches, soil or groundwater contamination can also be addressed by natural attenuation: the method of allowing contaminant levels to decline over time due to the natural biodegradative capabilities of indigenous microflora. It is important to note that natural attenuation and the various biostimulation approaches share the common feature that nonindigenous microbial populations are generally not utilized, and that no bacterial cultures are added to the site in any manner.

Remediation technologies in which selected microbial cultures or consortia are introduced to contaminated sites are sometimes referred to as "bioaugmentation". Bioaugmentation may utilize selected, laboratory-bred microbial strains or microbial consortia that are believed to have enhanced biodegradative capabilities, often against specific compounds or contaminant categories. Bioaugmentation approaches can be carried out either *in situ* or *ex situ*, however bioaugmentation is not widely practiced in commercial remediation. Although there are several reasons for this bias, one major issue is the concern that introduced cultures will not compete well with indigenous species in the environment, and may not survive long enough to carry out their intended purpose.

Another bioremediation (or "biotreatment") application is the use of bioreactors or biofilters in which indigenous or added microorganisms are immobilized on a fixed support, to allow continuous degradation of contaminants. These reactors can be used either with aqueous wastes or slurries or with contaminated vapor phase wastestreams, and in fact microbial biofilters are becoming better accepted within the odor control market and other markets for treatment of contaminated off-gases. Although most often utilizing indigenous microflora, bioreactors can be used with select, pure microbial cultures, particularly if the reactor is intended for use with a specific contaminant or well-characterized wastestream. One possible use for bioreactors would be the use of microorganisms for biosorption of metals from aqueous wastestreams (discussed below).

Most of the bioremediation technologies described above not only utilize naturally-occurring organisms, but more specifically they rely on species and populations indigenous to the site of contamination. More importantly for the prospects of using GMOs in remediation, these applications generally do not involve the use or introduction of well-defined, selected single-species cultures. It would seem to be an essential prerequisite for the potential use of GMOs in bioremediation that there be accepted, plausible uses for introduction of single-species plants or microbial inocula; otherwise the engineered organisms created in the laboratory would likely not be accepted in the commercial marketplace.

Most microbial inoculants or additives sold for use in bioaugmentation approaches have historically been blends or consortia of microorganisms, purportedly tailored for the types of compounds found in the target waste stream. Initial products were used for municipal waste water treatment or for biotreatment of restaurant grease traps and sewer lines. Several companies have sold microbial blends purported to be active against hazardous compounds, including use against industrial effluents and for *in situ* waste remediation, as well as products rich in lipases, proteases and cellulases for use in activated sludge treatment lagoons or on-line biological reactors for waste water treatment. The most common products for *in situ* remediation are formulations for degradation of hydrocarbons and petroleum distillates. The earlier of these strains have been used to clean oily bilges in tankers and other ships since the 1960s, and have also attracted attention for their possible usefulness against oil spills on land and sea, although the efficacy of such cultures for this purpose was never proven.

More recently, a number of single-species products have been identified or investigated, and some have been used in commercial remediation. For example, there are several microbial isolates capable of degrading chlorinated aliphatics. These microbes generally utilize unrelated pathways that fortuitously can metabolize the contaminants of interest. Trichloroethylene (TCE; the most common pollutant of groundwater) is the most important chlorinated compound that can be biodegraded by such serendipitous pathways. One of the earliest TCE degrading strains to be identified is a pseudomonad (now known as *Burkholderia cepacia*) named G4 (Shields *et al.* 1989), that was investigated for commercial use in the early 1990s and continues to be useful in research to this day. Two different strains of *Dehalococcoides* are now sold commercially for use in bioaugmentation approaches for the dechlorination of TCE or PCE: strain BAV-1, identified at Georgia Tech (He *et al.* 2003), and now being commercialized by Regenesis Corporation; and KB-1, developed and being sold by DuPont.

Other more recent examples are two microbial cultures that are being used for treatment of methyl tertiary-butyl ether (MTBE). Strain PM1, a member of the β1 subgroup of Proteobacteria, was isolated by Kate Scow and colleagues at UC Davis from a mixed microbial culture originally enriched from a compost biofilter (Hanson *et al.* 1999). This strain is now being commercialized by Regenesis Corporation for use both for *in situ* bioaugmentation strategies and also in bioreactors. Salanitro and colleagues isolated a mixed bacterial culture, called BC-1, from chemical plant bioreactor sludge. The culture can be maintained in culture for long periods of time, and can grow on aqueous waste streams with MTBE concentrations of 120-200 ppm. (Salanitro *et al.* 1994). This strain has been marketed by Shell Global Solutions under the trade name BioRemedy®, and it can be used in the direct inoculation of contaminated groundwater, for intercepting a spreading pollution plume, or for treatment of ground-water in an aboveground reactor.

The fact that most *in situ* applications of bioremediation involve indigenous microorganisms rather than introduced cultures places a barrier in the path of potential uses of genetically modified microorganisms in bioremediation, that is likely to be a major factor affecting market adoption of GMOs. Many observers feel that a more plausible use for GMOs in remediation will be in bioreactors, designed for use with defined wastestreams. Not only does this avoid the widespread release of the GMO into the environment and avoids the problem of competition with indigenous microflora, but it allows the microorganism to be maintained at controlled temperatures and other growth conditions, and to be used with relatively well-defined wastestreams containing one or a small number of specific contaminants.

Bioremediation Research Needs

Early in the adoption of bioremediation within the commercial marketplace, even as it became clear that indigenous microflora could be a powerful tool in the clean-up of easily biodegradable contaminants, the limitations of such methods were also recognized, and many were calling attention to how much additional research was needed to make bioremediation more viable commercially. Several reports were published in the early to mid 1990s analyzing bioremediation research needs, and several of the recommendations of these reports can also be seen as potential strategies for the improvement of bioremedial microorganisms through genetic engineering or other methods. An excellent review of some of these efforts can be found in an online publication by the U.S. Department of Energy's Natural and Accelerated Bioremediation Research (NABIR) program (U.S. DOE undated) . A recurring theme among many of these assessments was the need for integrated multidisciplinary approaches (e.g., microbiology, engineering, etc.) to understand how bioremediation works in the field and how these processes can be optimized for commercial use. In addition, these reports often called for expanded field research and better abilities to model and monitor field remediation. Among recommendations relating to the fundamental biology of bioremediation mechanisms are the following (citations and more information on these reports can be found in U.S. DOE undated):

- Factors limiting degradation rates in bioremediation applications need to be adequately identified and addressed (from a 1991 Rutgers University workshop).
- Identification of microbial capability of biotransformation (from a 1992 EPA report).

- Understand microbial processes in nature and how they are interrelated within a microniche; promote more efficient contact between the contaminant and the microorganism (from a 1993 National Research Council report).
- Examine bioremedial catalytic systems of microorganisms not previously well studied; focus on diverse metabolic pathways of anaerobic microorganisms; explore use of combined aerobic/anaerobic systems; assess the bioavailability of contaminants and catalysis in nonaqueous phase contamination (from a 1994 U.S./European workshop).
- Develop an understanding of microbial communities; develop an understanding of biochemical mechanisms involved in aerobic and anaerobic degradation of pollutants; extend the understanding of microbial genetics as a basis for enhancing the capabilities of microorganisms to degrade pollutants (from a 1995 National Science and Technology Council subcommittee report).

Although some of these objectives have been met in the years since these reports were issued, many remain as useful goals for the improvement of microbial bioremediation.

Potential Approaches to Use Genetic Engineering to Improve Microorganisms for Bioremediation

Potential strategies for improving bioremediation that arise from such recommendations are summarized in Table 1, and these general approaches are also reviewed elsewhere (Keasling and Bang 1998, Lau and de Lorenzo 1999, Timmis and Pieper 1999, Menn et al. 2000, Pieper and Reineke 2000, de Lorenzo 2001, DEFRA 2002, Morrissey et al. 2002). Many of these strategies can be addressed through the use of recombinant DNA genetic engineering. For example, expression of key catabolic enzymes can be enhanced through use of constitutive or stronger promoters; new biodegradative pathways can be created using transformation of one or more genes encoding degradative enzymes into microorganisms already possessing a complementary pathway; genes encoding transport proteins or metal-sequestering molecules can be introduced into microorganisms to enhance contaminant uptake or sequestration.

Other strategies can be accomplished using classical techniques: for example, novel pathways can be created by conjugal matings of different bacterial strains, resulting in the transfer of entire plasmid-encoded pathways into novel organisms (Timmis and Pieper 1999, Pieper and Reineke 2000). On the other hand, newer biotechnologies may also lead to promising new strategies. Several approaches to improving the efficiency of

Table 1. Potential strategies for use of genetic engineering to improve microbial bioremediation.

- Enhancing expression or activity of existing catabolic enzymes.
  - o   Modified or new promoters
  - o   Enhanced protein translation
  - o   Improved protein stability or activity

- Creation of new biodegradative pathways.
  - o   Pathway construction (introduction of heterologous enzymes).
  - o   Modifications to enzyme specificity, affinity, to extend the scope of existing pathways.

- Enhancing contaminant bioavailabilty.
  - o   Surfactants to enhance bioavailability in soil.
  - o   Transport proteins to enhance contaminant uptake.

- Enhancing microbial survival or competitiveness.
  - o   Resistance to toxic contaminants.
  - o   Resistance to radioactivity.
  - o   Enhanced oxygen, nutrient uptake.

- Improvements in bioprocess control (e.g., for contained bioreactors).

- Creation of organisms for use as biosensors (e.g., for detection, monitoring).

Sources : Menn *et al.* (2000), Pieper and Reineke (2000), DEFRA (2002).

biodegradative enzymes are offered by technologies such as protein engineering (see Ornstein 1991 for an early example), site-directed mutagenesis, DNA shuffling (e.g., Dai and Copley 2004), and three-dimensional modeling of protein structure (reviewed in Timmis and Pieper 1999 and Pieper and Reineke 2000). And finally, an increasing number of microbial genomes are being sequenced, including genomes from thermophiles and other extremophiles as well as from unculturable microorganisms, and this could lead to the identification of new biodegradative enzymes (and their genes) having previously-unsuspected but useful properties.

In recent years, there has been an explosion of research aimed at addressing many of the identified "research needs" discussed above, including discovery of previously-unknown species and strains having useful degradative properties, research on catabolic pathways and their individual enzyme components, microbial competitiveness, contaminant bioavailability, and others. This research is far too voluminous to be reviewed here, but there are a number of recent references that provide useful summaries (e.g., Timmis and Pieper 1999, Menn *et al.* 2000, Pieper and Reineke 2000, DEFRA 2002). The following is a brief summary of some of the research strategies that are being pursued for several of the contaminant categories that we feel are the likeliest to be effectively pursued on a commercial level using GMOs, including contaminant classes shown in Table 2.

Table 2. Contaminants for which microbial genetic engineering strategies are being investigated.

- Chlorinated compounds.
  - o   TCE.
  - o   Chlorobenzoates.
  - o   Chlorinated herbicides and other pesticides.

- Polychlorinated biphenyls (PCBs) and chlorobiphenyls.

- Hydrocarbons, BTEX.

- Nitroaromatics.

- Sulfur compounds.

- Heavy metals.
  - o   Sequestration.
  - o   Transformation to less toxic form.
  - o   Precipitation from solution.

Sources: Menn et al. (2000), DEFRA (2002).

Trichloroethylene

Naturally occurring microorganisms exist which can break down TCE through the use of pathways evolved for catabolism of other compounds. Specifically, several species can use toluene degradation pathways for the breakdown of TCE; however these organisms often require the presence of an inducer molecule in order to activate the pathway. Because this is clearly not an optimal situation for commercial remediation, TCE was a natural early target for the use of genetic engineering. One early effort was undertaken by Winter *et al.* (1989), who expressed the toluene mono-oxygenase gene from *Pseudomonas mendocina* in *Escherichia coli* under the control of a constitutive promoter and also a temperature-inducible promoter, and created recombinant strains that were capable of degrading

TCE and toluene without any chemical inducer. Rights to this system were acquired in the early 1990s by Envirogen, which spent some time developing these strains for possible use in vapor phase bioreactors for TCE treatment (Glass 1994), but GMOs were never commercially used in this system. Other approaches to creating recombinant microorganisms for TCE degradation have involved the cloning and expression of toluene dioxygenase (*tod*) genes (Zylstra *et al.* 1989, Furukawa *et al.* 1994), as well as the phenol catabolic genes (*phe*A, B, C, D and R) from *P. putida* BH (Fujita *et al.* 1995).

PCBs

Polychlorinated biphenyls (PCBs) have also been an early target for genetic engineering work, because there did not appear to be microorganisms naturally possessing a complete pathway for the enzymatic mineralization of these complex molecules, which appeared to be quite recalcitrant to natural biodegradation. PCBs are now known to be degradable by a combination of anaerobic and aerobic reactions, where the aerobic pathway involves the insertion of an oxygen molecule into one aromatic ring to form a chlorinated cis-dihydrodiol, and the anaerobic steps include the reductive dehalogenation of the more highly-chlorinated congeners (Wackett 1994, Mondello *et al.* 1997, Pieper and Reineke 2000, DEFRA 2002). The genes controlling the aerobic pathway are found in the *bph* operon (Mondello 1989, Erickson and Mondello 1992, Dowling and O'Gara 1994), and these genes encode a multicomponent dioxygenase that degrades the biphenyl residue, ultimately to benzoic acid and a pentanoic acid (see references in DEFRA 2002). These genes have been introduced and expressed in recombinant bacteria that have been shown to be capable of degrading chlorobiphenyls (Menn *et al.* 2000). The dehalogenase genes have largely been studied by the Tiedje laboratory, which has expressed the genes encoding enzymes for ortho- and para-dechlorination of chlorobenzenes in a bacterial strain having the capability to degrade biphenyls, resulting in a recombinant strain that could completely dechlorinate 2, and 4-chlorobiphenyl (Hrywna *et al.* 1999). The Tiedje lab has also identified bacterial strains capable of reductively dehalogenating trichloroacetic acid (De Wever *et al.* 2000) and 1, 1, 1-trichloroethane (Sun *et al.* 2002).

Chlorobenzoates and other aromatic compounds

A great deal of research has gone into pathways for breakdown of aromatic compounds, in particular the TOL pathway found on a plasmid of *Pseudomonas putida* (Ramos *et al.* 1987). Ramos et al. modified the TOL pathway to enable the degradation of 4-ethylbenzoate, by addition of mutant bacterial genes, one of which encoded a modified form of a key pathway enzyme. In one of the first efforts to construct an artificial pathway, (Rojo *et al.* 1987) combined enzymes from five different catabolic pathways found in three different soil microorganisms to create a pathway for the degradation of methylphenols and methylbenzoates. A version of this organism in which the heterologous genes were stably integrated into the chromosome was shown to be able to reduce the toxicity of phenol-containing wastestreams (Erb *et al.* 1997).

Heavy Metals and Inorganics

There has been continuing interest in using microorganisms for the remediation of metals, in spite of the fact that, as elemental contaminants, metals cannot be chemically degraded as organic molecules can. Using microbes for clean-up of metals would involve either (a) sequestration of metal ions within microbial biomass (sometimes called biosorption); (b) precipitation of the metal ions on the surface of the cell; or (c) electrochemical transformation of metals into less toxic forms (DEFRA 2002). In many cases, particularly for strategies (b) and (c), microorganisms, including GMOs, would best be used in flow-through bioreactors in which the metal ions can be removed from an aqueous waste stream and captured on or in microbial biomass.

DEFRA (2002) provides a good review of efforts to improve these metal-remediating processes using genetic engineering. This report describes efforts to express in bacteria a variety of metal-binding proteins and peptides, many of which (e.g., metallothionein) are also being investigated in phytoremediation strategies (see below). DEFRA (2002) also describes the existing use of sulfate-reducing bacteria to precipitate various metals from aqueous solutions (a method being investigated for treatment of acid mine waste and other metals-contaminated waters), and discusses efforts to improve this activity, e.g., through overexpression of the genes encoding the key enzyme thiosulfate reductase.

To use one metal pollutant as an example, there has been a fair amount of work constructing genetically engineered microorganisms for bio-sorption of mercury. Most of this research has revolved around a well-studied cluster of bacterial genes that encode mercury resistance, which are also being investigated for phytoremediation purposes (see below). These genes are found in an operon called *mer*TPABD, under the control of a regulatory protein encoded by *mer*R (Summers 1986, Meagher 2000). *Mer*A encodes mercuric ion reductase, an enzyme that catalyzes the electro-chemical reduction of ionic mercury [Hg(II)] to metallic or elemental mercury [Hg(0)]; and *merB* encodes a bacterial organomercury lyase which mediates the reduction of methylmercury and other forms of organic mercury to ionic mercury. *MerT* encodes a membrane transport protein and *merP* encodes a periplasmic Hg binding protein, and together the genes in this operon, when expressed in a bacterial host, allow the host to tolerate high Hg concentrations in the growth media, by taking up Hg(II) or methylmercury and converting it into the less toxic elemental form (Summers 1986).

Horn *et al.* (1994) created strains of *P. putida* that had an enhanced ability to detoxify mercury, through constitutive overexpression of the *mer*TPAB genes. In this report, overexpression of the *mer* genes was accomplished by linking the gene cluster to transposon Tn501, transferring this construct into the host organisms, and selection of transformants where the gene cluster was inserted downstream of proximal host promoters. Another group (Chen and Wilson 1997a, b, Chen *et al.* 1998) reported the construction of *E. coli* strains that accumulated high concentrations of Hg(II) through over-expression of the transport proteins encoded by *merT* and *merP* as well as a glutathione-S-transferase/metallothionein fusion protein. These recombinant strains were used in hollow fiber bioreactors to remove Hg from aqueous wastestreams.

Mixed Hazardous/Radioactive Wastes

Many organic and inorganic hazardous materials are found at contaminated sites that also include radionuclides or other radioactive wastes. Therefore, there has been some interest in developing microorganisms that can remediate the hazardous contaminants and possibly the radionuclides while also being able to withstand the high radiation levels found at some of these sites. This has directed attention to the unusual microorganism *Deinococcus radiodurans* and related *Deinococcus* species that are naturally able to withstand enormous doses of radiation - up to 5 Mrad of gamma irradiation. One approach to clean-up of mixed wastes would be to engineer a *Deinococcus* strain to have the ability to degrade organic contaminants and/or to sequester or precipitate heavy metals. This has been done for two types of hazardous contaminant. Lange *et al.* (1998) has engineered *D. radiodurans* to express the TOD gene cluster, thus expressing toluene dioxygenase, enabling this strain to metabolize toluene, chlorobenzene and other aromatic compounds. The same group has also created *D. radiodurans* expressing the *E. coli merA* gene, creating a strain that was capable of growing in the presence of radiation as well as high levels of Hg(II), and reducing Hg(II) to elemental mercury (Brim *et al.* 2000). The entire genome of *D. radiodurans* has now been sequenced (White *et al.* 1999), leading many to hope that this will lead to additional potential remedial strategies.

Regulation of Genetically Engineered Microorganisms for Bioremediation

As discussed above, a major (but not the only) factor that has hindered the use of GMOs in commercial remediation has been the specter of government regulation. The technologies collectively known as genetic engineering have attracted public attention and government scrutiny since their development in the 1970s, and in particular the use of engineered plants and microbes in the open environment, and subsequent use of transgenic plants in foods, has at times been quite controversial in the U.S. and elsewhere in the world. Regulatory schemes adopted in the 1980s primarily to regulate agricultural uses of the new genetic technologies have instituted new layers of government oversight specific for the uses of GMOs in the environment. It is a widespread perception in the environmental industry that these regulations make it impossible or impractical to use GMOs in the open environment (see, for example, the closing comments of Glick (2004) relating to the "current political impediments ... to using either GM plants or GM bacteria in the environment"); but in reality tens of thousands of field tests of transgenic plants and hundreds of field trials of modified microorganisms have taken place under these regulations all over the world, with numerous GMOs, both microbes and plants, approved for commercial sale in agriculture.

Although many in the regulated community feel that regulation of engineered microorganisms is excessive and not necessarily science-based, it is true that there are potential environmental risks that should be assessed for any proposed introduction of a new microorganism into a novel environment. Such questions might include an evaluation of the potential survivability and competitiveness of the microorganism in the environment, its possible effects on target plants and non target species, and on dispersal of the microbe or transfer of the introduced genetic material (i.e., horizontal gene flow) to other organisms (e.g., as discussed in Alexander 1985 and National Research Council 1989 and in many other more recent references such as DEFRA 2002). Detailed discussion of the issues that should be considered in a biotechnology risk assessment are beyond the scope of this article, except to say that the regulatory schemes adopted in most countries to cover uses of GMOs in the environment include scientific assessments addressing questions such as these (see Glass 2002 for more details). It should also be noted that it has often been proposed that GMOs designed for environmental use include features that would enfeeble the organism, making it less likely to survive in the environment, or to include "suicide" features such that microbial populations would die out after their desired task has been carried out; however, in our view such an approach is not required by the current regulations or by any realistic risk scenario.

The discussion below of regulatory requirements for use of engineered microorganisms and transgenic plants in the environment will largely cover the situation in the United States. However, many other nations and jurisdictions around the world have adopted or created regulatory programs for the same purpose, which often are based on the same or similar scientific issues, but which address proposed uses in different ways (see Conner *et al.* 2003 and Nap *et al.* 2003 for recent discussions of risk assessment issues and a summary of GMO regulations in a number of countries). For example, the European Union recently adopted revised regulations for environmental uses of GMOs, replacing a directive first promulgated in 1990 (see Morrissey *et al.* 2002 for a summary of these regulations). The use of GMOs in the environment, particularly for agricultural purposes, has become widespread and commonplace throughout the world, and most countries having significant agricultural activities are grappling with the same regulatory and scientific issues as those discussed here in the context of the U.S. regulatory scheme.

Overview of U.S. Regulation of Genetically Engineered Microorganisms

The products of biotechnology are regulated in the U.S. under the so-called Coordinated Framework. It was decided in 1986 that the products of biotechnology would be regulated under existing laws and in most cases under existing regulations, based on the intended end-use of each product, rather than under any newly-enacted, broad-based biotechnology legislation. The term "Coordinated Framework" refers to the matrix of existing laws and regulations that have served to regulate the biotechnology industry since its publication in the Federal Register in June 1986 (see Glass 1991 and Glass 2002 for a more detailed history).

Most of the products of biotechnology have been drugs or other health care products, and these have been regulated by the U.S. Food and Drug Administration. However, those commercial products that consist of living microorganisms (and in some cases killed or inactivated microorganisms) are regulated under a number of product-specific laws (see Glass 2002 for a more comprehensive review). For example, microorganisms, including GMOs, designed to act as pesticides would be regulated by the U.S. Environmental Protection Agency (EPA) under the Federal Insecticide,

Fungicide and Rodenticide Act (FIFRA). Most of the genetically engineered microorganisms that have been used in agriculture have fallen into this category, and as of the early years of this decade, several dozen pesticidal GMOs had been approved by the EPA (Glass 2002).

Under the Coordinated Framework, genetically modified micro-organisms used in bioremediation would be subject to regulation by the EPA under a different federal law. This is the Toxic Substances Control Act (TSCA), and it is a law that EPA has used since the mid-1980s to regulate microorganisms intended for environmental use for purposes other than as a pesticide. EPA has also used this law to regulate certain engineered microorganisms used in commercial manufacturing, as well as certain agricultural bacteria engineered for enhanced nitrogen fixation (Glass 1991, 1994, 2002). Although there have not yet been any commercial uses of GMOs in bioremediation, there have been several field tests regulated by EPA under this program (see below).

EPA Biotechnology Regulation Under the Toxic Substances Control Act

EPA is using TSCA to regulate the microbial production of certain chemicals or enzymes not regulated elsewhere in the government, as well as those planned introductions of microorganisms into the environment that are not regulated under other federal statutes. TSCA (15 U.S. Code 2601) is a law requiring manufacturers to notify EPA at least 90 days before commencing manufacture of any "new" chemical, i.e., one that is not already in commerce, for purposes not subject to regulation as a pesticide or under the food and drug laws. In the Coordinated Framework, EPA decided to use TSCA in this same "gap-filling" way, to capture those microorganisms that were not regulated by other federal agencies. The primary areas which therefore became subject to the TSCA biotechnology regulations were (a) microorganisms used for production of non-food-additive industrial enzymes, other specialty chemicals, and in other bioprocesses; (b) microorganisms used as, or considered to be, pesticide intermediates; (c) microorganisms used for nonpesticidal agricultural purposes; and (d) microorganisms used for other purposes in the environment, such as bioremediation (Glass 1994, 2002).

Because of political difficulties and in-fighting (Glass 2002), EPA was not able to promulgate final biotechnology regulations under TSCA until April 11, 1997 (62 Federal Register 17910-17958). These rules amended the existing TSCA regulations to specify the procedures for EPA oversight over commercial use and research activities involving microorganisms subject to TSCA. The net result was to institute reporting requirements specific for microorganisms (but which paralleled the commercial notifications used for traditional chemicals), while also creating new requirements to provide suitable oversight over outdoor uses of genetically modified micro-organisms.

Procedures under the TSCA biotechnology regulations are similar to existing practice for new chemical compounds. Note that TSCA is a "screening" statute that allows EPA to be notified of all new chemicals so that it can identify those which might pose an environmental or public health risk and therefore require further regulatory review. Manufacturers of chemicals new to commerce must file Pre-Manufacture Notices (PMNs) with EPA at least 90 days prior to the first intended commercial sale or use or importation and must submit all relevant health and safety data in their possession. The large majority of chemical PMNs are approved within the 90 day period after only brief agency review.

The biotechnology rule requires premanufacture reporting for new organisms, but it was a long-running challenge in the development of the regulations to adequately define "new organism" (see Glass 1991, 1994, 2002 for historical background). The final rule defines a "new organism" as an "intergeneric organism", as instituted in the Coordinated Framework and used continuously since then in EPA's interim policy. Intergeneric organisms are those that include coding DNA sequences native to more than one taxonomic genus, and EPA chose this definition under the assumption that genetic combinations within a genus are likely to occur in nature but that combinations across genus lines are less likely to occur naturally, so that intergeneric organisms are likely to be "new" (Glass 2002). Organisms that are not new, including naturally occurring and classically mutated or selected microbes, are exempt from reporting requirements under TSCA.

New microorganisms used for commercial purposes subject to TSCA's jurisdiction require premanufacture reporting 90 days in advance of the commercial activity, using a new procedure called a Microbial Commercial Activity Notification (MCAN) that is analogous to the previous bio-technology PMN procedures under the interim policies, and to long-existing PMN practice for chemical entities. However, several exemptions from MCAN reporting are possible for specific organisms that qualify and a procedure was also put into place for EPA to create new exemption categories based on appropriate scientific evidence.

Generally speaking, research activities involving new microbes are exempt from reporting if conducted only in "contained structures". The rule specifically contemplates that this exemption would apply broadly to many types of structures, including greenhouses, fermenters and bioreactors. Outdoor experimentation with GMOs remains potentially subject to some sort of reporting, with only limited exemptions at this time that mostly do not pertain to bioremediation. Those field tests not qualifying for an exemption can be conducted under a reduced reporting requirement known as TSCA Environmental Release Application (TERA). The TERA process replaced the previous (voluntary) policy under which all outdoor uses of intergeneric microorganisms were reviewed under PMN reporting, regardless of the scale or potential risks of the field experiment. The regulations specify that TERAs would be reviewed by EPA within 60 days, although the agency could extend the review period by an additional 60 days. In approving TERAs, EPA has the authority to impose conditions or restrictions on the proposed outdoor use of GMOs.

The biotechnology rule specified the types of information and data that applicants should submit to accompany MCANs and TERAs. The basic information for MCANs constitutes a description of the host micro-organism, the introduced genes and the nature of the genetic engineering, and information related to health and safety impacts of the organism. For those applications pertaining to environmental releases, including TERAs, information about the possible environmental impacts of the microbe must be submitted (see Glass 2002 for more details).

Interestingly, because TSCA is a statute covering "commercial" introductions of new chemicals (i.e., into commerce), EPA in the final rule decided that noncommercial research would be exempt from TSCA, meaning that many academic research activities, unless clearly supported by or done for the benefit of a for-profit entity, would be exempt from TSCA reporting.

EPA has been receiving PMNs and other notifications of biotechnology products under TSCA since 1987. Most of the notifications received were for contained applications: uses of intergeneric microorganisms for manu-facturing products for commercial purposes not regulated by other federal agencies, primarily including industrial enzymes and pesticide intermediates (Glass 2002). Since the adoption of the final rules in 1997, several MCANs have been received for such products. There have also been numerous PMNs (and more recently, TERAs) received for environmental introductions of altered microorganisms. Most of these have been for genetically altered nitrogen-fixing bacteria (*Rhizobium* or *Bradyrhizobium*) and in fact strains of engineered *R. meliloti* for improved nitrogen fixation are the only recombinant microorganisms used in the open environment approved for commercial sale under TSCA. In addition to these agricultural tests, there have been a small number of notifications relating to bioremediation, for R&D projects that are discussed below. There have been no PMNs or MCANs submitted to the EPA for uses of microorganisms in bioremediation.

EPA's biotechnology regulations under TSCA are unique to the United States, but a somewhat similar system has been adopted in Canada. In November 1997, Environment Canada issued regulations under the Canadian Environmental Protection Act that allow that agency to conduct risk assessments of certain biotechnology products that are new to commerce in Canada and which are not regulated by other federal agencies. Among products that would fall under this law's scope would be microbial cultures used for bioremediation. Differing from the U.S. EPA, Environment Canada would consider a microorganism to be subject to "New Substance Notification" under these regulations if it was intended for introduction into commerce but was not explicitly listed as having been used in commerce between January 1, 1984 and December 31, 1986. In this way, the Canadian CEPA regulations are broader than those of the U.S. EPA, in subjecting a larger class of microorganisms to regulation, including naturally occurring or classically mutated strains (see Glass 2002 for more details).

Field Uses of Genetically Engineered Microorganisms for Bioremediation

There are no documented uses of live genetically modified microorganisms (i.e., microorganisms altered using recombinant DNA) in any commercial project or process for hazardous waste bioremediation. This is certainly true in the United States, and it appears to be the case in the rest of the world as well. There is some anecdotal evidence that specific companies had investigated the use of GMOs in either field remediation or in contained bioreactors (e.g., Envirogen's investigation of recombinant bacteria for TCE degradation in vapor-phase bioreactors; Winter *et al.* 1989, Glass 1994). In addition, a killed strain of *E. coli*, engineered to overexpress the enzyme atrazine chlorohydrolase, has been used in the field to remediate atrazine at the site of an accidental spill (Strong *et al.* 2002; see also "Atrazine Soil Remediation Field Test", at http://biosci.umn.edu/cbri/lisa/web/index.html). However from the available public record it seems that no living GMOs have ever been used in an actual bioremediation project.

However, there have been two live strains of recombinant microorganisms that have been used in the field for bioremediation research purposes, after having been reviewed and approved by the U.S. EPA under the TSCA biotechnology regulations. The field trials using these organisms were designed as research experiments, more to validate molecular detection methodology than for any intended remedial purpose. As shown in Table 3, these are as follows.

Table 3. Genetically modified microorganisms approved by the U.S. EPA for field testing for bioremediation purposes.

EPA Case Number (TERA unless noted)	Date	Institution	Microorganism	Phenotype	Location(s)
PMN P95-1601	6/28/95	University of Tennessee	*Pseudomonas fluorescens* strain HK44	Naphthalene degradation gene and bioluminescent reporter gene	Tennessee
R98-0004	07/21/98	NEWTEC and ORNL	*Pseudomonas putida* strain RB1500	Luminesces in presence of TNT	South Carolina
R98-0005	07/21/98	NEWTEC and ORNL	*Pseudomonas putida* strain RB1501	Fluoeresces in presence of TNT	South Carolina
R01-0002	03/28/01	ORNL	*Pseudomonas putida*	Detection of TNT	California
R01-0003	04/25/01	ORNL	*Pseudomonas putida*	Detection of TNT	Ohio
R01-0004	04/25/01	ORNL	*Pseudomonas putida*	Detection of TNT	Ohio

Source: U.S. Environmental Protection Agency, http://www.epa.gov/opptintr/biotech/submain.htm

Gary Sayler of the University of Tennessee and collaborators created a modified strain of *Pseudomonas fluorescens* HK44 that contained a plasmid encoding genes for naphthalene catabolism as well as an transposon-introduced *lux* gene under the control of a napththalene catabolic promoter (Ripp *et al.* 2000, Sayler and Ripp 2000). With both the catabolic genes and the bioluminescent *lux* gene under the control of the same promoter, this strain could be induced to degrade naphthalene and to bioluminesce by exposure to naphthalene or certain salicylate metabolites. This modified strain was tested in subsurface lysimeters in an experiment at Oak Ridge National Laboratory (ORNL) that lasted from October 1996 to December 1999 (Sayler and Ripp 2000). The microbial inoculant showed enhanced naphthalene gene expression and adequate survival in the lysimeters, however due to heterogeneity in the contaminant concentrations in the lysimeters, it was not possible to make any precise conclusions about the efficacy of using such strains in an actual bioremediation project.

The second set of genetically engineered microbial strains used in EPA-approved field testing were created for the purpose of monitoring and detecting contaminants in the field. These are strains of *Pseudomonas putida* created by Robert Burlage and colleagues of Oak Ridge National Laboratory. The parent strains are capable of catabolyzing nitroaromatics like TNT, and Burlage et al. engineered these strains so that a TNT-responsive promoter also controlled expression either of a *lux* gene or a gene encoding green fluorescent protein. As a result of this engineering, when the microbes are exposed to TNT in the soil, they are expected not only to begin degrading the contaminant, but also to either fluoresce or bioluminesce. The goal is to use such microorganisms to detect land mines, unexploded ordinance, or other leaking sources of TNT contamination. These strains were first field tested in October 1998 at the National Explosives Waste Technology and Evaluation Center in South Carolina. The recombinant organisms were sprayed onto a site containing simulated mine targets, and then later that day, after dark, the field was surveyed using ultraviolet light to detect areas of microbial activity. According to accounts of the test published on the ORNL website (see "Microbial Minesweepers" at http://www.ornl.gov/info/ornlreview/meas_tech/threat.htm   and "Green Genes: Genetic Technologies for the Environment" at http://www.ornl.gov/info/ornlreview/v32_2_99/green.htm), the bacteria were able to detect the location of all five simulated mine targets in a 300 square meter field. EPA approval was also obtained for subsequent tests at Edwards Air Force Base in California and the Ravenna Army Ammunition Plant in Ohio.

Plans were made for one field test in Europe of a GMO for bioremediation. The research consortium funded by the European Union under the project acronym RHIZODEGRADATION planned to conduct a research field test to document the safety of bioremediation using engineered versions of *Pseudomonas fluorescens* F113. This strain of *P. fluorescens* is a well-known root-colonizing microorganism that has been used in the field. The investigators created a mutant form of F113 with the "*lac*"ZY reporter genes inserted into the chromosome, and then derived a rifampicin-resistant strain by spontaneous mutation. This strain was to be used as a control against another strain, also with a spontaneous rifampicin-resistance mutation, but into which the *bph* genes from *B. cepacia* LB400 have been inserted, giving the microbes the abiltiy to use biphenyl as a carbon source. A field test of these two strains was planned to take place at a petroleum hydrocarbon-contaminated site in Arhus, Denmark, however, the test did not receive the needed regulatory approvals and so was never carried out (U. Karlson, personal communication).

Although there has not yet been a commercial use of a GMO in microbial bioremediation, there is no reason to believe this will not someday occur. The amount of research taking place using recombinant methods to improve biodegradative microorganisms is staggering, and, at least in the U.S., it is clearly possible to conduct outdoor field trials of GMOs with the proper preparation. What has been missing is the commercial and technological need to use a GMO as opposed to an approach involving naturally-occurring microorganisms. Although economic and other factors may yet hold back such proposed uses, others of the commonly perceived barriers may not be significant factors should the right application come along.

Prospects for Commercial Phytoremediation Using Transgenic Plants

Existing Phytoremediation Technologies

Phytoremediation is the use of plants (including trees, grasses and aquatic plants) to remove, degrade or sequester hazardous contaminants from the environment. Although some phytoremediation applications are believed to work through stimulation of rhizosphere bacteria by the growing plant root, the focus of phytoremediation is to use plants as the driving force behind the remediation. As currently practiced, phytoremediation has used a variety of naturally-occurring plant and tree species, including several tree species selected for their abilities to remove prodigious amounts of water from the subsurface. But often, the plant species to be used at a given site are carefully selected for that site based on the soil, climate and other characteristics of the site. The following is a summary of the major potential uses for phytoremediation (see also Glass 1999, U.S. EPA 2000b, and ITRC 2001, for more complete descriptions).

For remediation of soil:

- *Phytoextraction:* the absorption of contaminants from soil into roots, often utilizing plants known as "hyperaccumulators" that have evolved the ability to take up high concentrations of specific metals. Inside the plant, the contaminants are generally transported into shoots and leaves, from which they must be harvested for disposal or recycling.
- *Phytostabilization:* the stabilization of contaminants in soil, through absorption and accumulation into the roots, the adsorption onto the roots, or precipitation or immobilization within the root zone, by the action of the plants or their metabolites.
- *Phytostimulation (also called Rhizostimulation):* the stimulation of contaminant biodegradation in the rhizosphere, through the action of

rhizosphere microorganisms or by enzyme exudates from the plants.

- *Phytovolatilization:* the uptake and release into the atmosphere of volatile compounds by transpiration through the leaves.
- *Phytotransformation:* the uptake of contaminants into plant tissue, where they are degraded by the plant's catabolic pathways.

For remediation or treatment of water:

- *Rhizofiltration:* the absorption of contaminants from aqueous solutions into roots, a strategy primarily being investigated for metal-contaminated wastestreams.
- *Hydraulic Barriers:* the removal of large volumes of water from aquifers by trees, using selected species whose roots can extend deep into an aquifer to draw contaminated water from the saturated zone.
- *Vegetative Caps:* the use of plants to retard leaching of hazardous compounds from landfills, by intercepting rainfall and promoting evapotranspiration of excess rain.
- *Spray Irrigation:* the spraying of wastewater onto tree plantations to remove nutrients or contaminants.

All commercial applications of phytoremediation to date have involved naturally-occurring plant species. Often the chosen plants are indigenous to the region or climate where the remediation is taking place, but this is not always the case. In addition, remediation is sometimes accomplished through the use of a single plant species, but often a site is planted with a variety of different species, either to address different contaminants or simply to better simulate a "natural" ecosystem. Among the more important categories of plants used in phytoremediation are the following:

Natural Metal Hyperaccumulators. Plants naturally capable of accumulating large amounts of metals ("hyperaccumulators") were first described by Italian scientists in 1948. This work was later repeated and expanded upon by Baker and Brooks (1989), who defined hyper-accumulators as those plants that contain more than 1,000 mg/kg (i.e., 0.1% of dry weight) of Co, Cu, Cr, Pb or Ni, or more than 10,000 mg/kg (1.0% of dry weight) of Mn or Zn in their dry matter. Hyperaccumulators have often been isolated from nature in areas of high contamination or high metal concentration (see Reeves and Baker 2000 and Salt and Kramer 2000 for recent reviews). Examples of species that are being used commercially are Indian mustard (*Brassica juncea*), being used for remediation of lead and other metals (Raskin *et al.* 1997, Blaylock and Huang 2000) and Chinese brake fern (*Pteris vittata* L.), which has been discovered to be an efficient hyperaccumulator of arsenic (Ma *et al.* 2001).

Stimulators of Rhizosphere Biodegradation. Many types of plants are effective at stimulating rhizosphere degradation. The most commonly used have been alfalfa and different types of grasses, which have fibrous root systems which form a continuous, dense rhizosphere. Other plants that have been used include crested wheatgrass, rye grass and fescue (see the reviews by Anderson *et al.* 1993 and Hutchinson *et al.* 2003).

Trees. Because of their ability to pump large amounts of water from aquifers, trees of the Salicaceae family are used for phytoremediation of aqueous media. Although hybrid poplar is by far the most common tree species to be used in phytoremediation activities, at least in the United States, other species selected include willow, black willow, juniper and cottonwood. These species are phreatophytic plants, which are capable of extending their roots into aquifers in order to remove water from the saturated zone. Examples of compounds which have been remediated by poplars include inorganics like nitrates and phosphates, and many organic compounds including TCE, PCE, carbon tetrachloride, pentachlorophenol, and methyl tert-butyl ether (MTBE) (Newman 1998, Newman *et al.* 1999, Shang *et al.* 2003).

Plants and Trees with Biodegradative Capabilities. A number of trees and plants have enzymatic activities suitable for degrading environmental contaminants (McCutcheon and Schnoor 2003, Wolfe and Hoehamer 2003). Among these enzyme systems are nitroreductase, useful for degrading TNT and other nitroaromatics, dehalogenases, for degradation of chlorinated solvents and pesticides, and laccases, for metabolism of anilines (e.g., triaminotoluene) (Schnoor *et al.* 1995, Boyajian and Carreira 1997). Among the plants possessing such enzyme systems are hybrid poplars (*Populus* sp.), which have been shown to be able to degrade TCE (Newman *et al.* 1997) and atrazine (Burken and Schnoor 1997), parrot feather (*Myriophyllium spicatum*) and Eurasian water milfoil, capable of degrading TNT, and others.

The nature of phytoremediation technologies make them potentially more amenable to use with GMOs than is the case for microorganisms. In virtually all cases where phytoremediation is practiced in the field, it is done with introduced plant species, and although this may include species indigenous to the site or the region where the project is taking place, and it may involve mixed combinations of plant species, the plants or trees are almost always brought to the site for installation (i.e., planting) at the location of the contamination. Transgenic plants can be quite plausibly used in such a scenario, taking into account the likely need to engineer different varieties of a given species, for use in different climactic zones.

Phytoremediation Research Needs

The possible need to create transgenic plants for phytoremediation must be viewed in the context of the capabilities and limitations of naturally-occurring plant and tree species that have been used in phytoremediation. Although native plant species having the capability to remediate almost every major class of contaminant have been identified, in many cases these species grow too slowly or produce too little biomass to provide commercially-useful remediation times. Among other obstacles to the greater adoption and larger-scale use of phytoremediation, van der Lelie *et al.* (2001) cited the long timeframes often needed for remediation, the need for plants and trees to tolerate the high toxin levels found at contaminated sites, and the fact that phytoremediation only addresses the bioavailable fraction of the contamination. These shortcomings are targets to be addressed by further research and creation of improved plant varieties.

With the possible exception of some systems that are already widely studied and understood (e.g., the use of deep rooted poplars for groundwater control), all of phytoremediation's major applications still require further basic and applied research in order to optimize in-field performance. A workshop held by the U.S. Department of Energy in 1994 articulated the following areas where research is needed (U.S. DOE 1994):

- Mechanisms of uptake, transport and accumulation: Better understand and utilize physiological, biochemical, and genetic processes in plants that underlie the passive adsorption, active uptake, translocation, accumulation, tolerance and inactivation of pollutants.
- Genetic evaluation of hyperaccumulators: Collect and screen plants growing in soils containing elevated levels of metals or other pollutants for traits useful in phytoremediation.
- Rhizosphere interactions: Better understand the interactive roles among plant roots, microbes, and other biota that make up the rhizosphere, and utilize their integrative capacity in contaminant accumulation, containment, degradation and mineralization.

A more recent, influential report on phytoremediation (ITRC 2001) summarized the following categories of needs to be addressed by research into new phytotechnologies:

- Expanding phytoremediation mechanisms through plant bio-chemistry.
- Expanding phytoremediation mechanisms through genetic engi-neering.
- Applying phytoremediation to new contaminants.
- Applying phytoremediation to new media (i.e., sediments, greenhouse gases).
- Combining phytoremediation with other treatment technologies.

All of these recommendations are primarily directed towards basic research, aimed at understanding the mechanisms that underlie the biological processes central to phytoremediation. However, gaining this knowledge will provide the means to manipulate or control these processes to improve commercial performance, whether simply through selection and use of optimal plants for given waste scenarios, or through more advanced techniques. These and other general strategies for improving phytoremediation's efficacy are summarized in Table 4.

Table 4. Strategies to improve phytoremediation.

Agronomic Enhancements

- Improving metal solubility in soils through the use of chelators.
- Combining phytoremediation with other *in situ* technologies (e.g., electro-osmosis)
- Enhancing phytoremediation processes by using exogenous modulators or inducers, or soil amendments that enhance plant growth.
- Enhancing plant growth and biomass accumulation by improved crop management practices.

Genetic Enhancements

- Creating improved plants through classical plant breeding
- Creating improved plants through genetic engineering.

Source: Glass (1999), adapted from the framework of Cunningham and Ow (1996).

A number of agronomic enhancements are possible, ranging from traditional crop management techniques (use of pesticides, soil amendments, fertilizers, etc.) to approaches more specific to phytoremediation, such as soil chelators. Metal chelators such as EDTA and hydroxyethylethylene diaminetriacetic acid (HEDTA) can cause a thousand-fold enhancement in soil solubility of metals such as Pb and can result in significant increases in plant uptake of metals (Cunningham and Ow 1996).

Efforts to improve the plants used for phytoremediation have involved either classical genetics or genetic engineering. Traditional plant breeding is a well-understood process for improving plant germplasm. However, it is best practiced with those commodity crops (particularly food or oilseed crops) that have long been cultivated on a large scale and whose genetics are well understood. Many plant species used in phytoremediation do not have this long history of use, nor is there an accumulated base of knowledge of genetics that would allow breeding to proceed smoothly. Traditional crop breeding can also be time-consuming, with several generations needed to introduce stably inherited traits into an existing genetic background.

Industrial and food-producing crop plants created by recombinant DNA methods are now being used on a large scale in commercial agriculture in the U.S., Europe and elsewhere in the world. Although engineered microorganisms have not yet been used in commercial bioremediation, it is nevertheless reasonable to expect that genetic engineering will have a significant impact on phytoremediation. This is because there is a clear need to improve the performance of naturally-occurring plant species to obtain commercially-significant performance; genetic engineering of plants is quicker, easier, and more routine than genetic engineering of soil microorganisms; phytoremediation processes are likely to be simpler and easier to understand and manipulate than microbial biodegradative pathways where consortia of organisms are sometimes needed; and regulatory and public acceptance barriers are substantially less severe for the use of transgenic plants than they are for engineered microbes.

Potential Approaches to Use Genetic Engineering to Improve Plants for Phytoremediation

Progress in creating transgenic plants for phytoremediation has been recently reviewed by several authors, including several reviews focusing on phytoremediation of metals or other inorganics (Meagher 2000, Kramer and Chardonnens 2001, Terry 2001, DEFRA 2002, Pilon-Smits and Pilon 2002). Research on the use of transgenics for remediation of organic contaminants is at an earlier stage and has not been reviewed in any one location, except for the excellent discussion in DEFRA (2002). Rather than reviewing the growing body of academic research in this field, we will summarize those research projects that appear to be closest to commercial use or which actually have been tested in the field. Possible strategies for the use of genetic engineering to improve phytoremediation are shown in Table 5, and the following discussion follows the format of that table.

Metals, Metalloids and Inorganics

Enhancing bioavailabilty of metals : For phytoremediation of certain metals, one important rate-limiting step is often the ability to mobilize metal ions from the soil particles to which they are tightly bound, so that they can be made available to plant roots. This has especially proven to be a problem for lead remediation: although natural lead hyperaccumulators are known, their effectiveness is often limited by the poor availability of lead from the soil (Blaylock and Huang 2000).

Table 5. Strategies to improve phytoremediation using genetic engineering.

Metals

- Enhancing bioavailability and mobilization of metals in the soil (e.g., expression of chelators).
- Enhancing metal uptake into the root (e.g., expression of transport proteins).
- Enhancing translocation of metals to aboveground biomass.
- Enhancing the ability of the plant to sequester metals (e.g. expression of metal-sequestering proteins and peptides).
- In certain cases, enhancing chemical or electrochemical transformation of metals into less toxic forms.

Organics

- Introduce genes encoding key biodegradative enzymes (plant and microbial origin).
  - Laccases
  - Dehalogenases
  - Nitroreductases
- Introduce genes for the stimulation of rhizosphere microflora.

General

- Introduce genes to enhance:
  - growth rates/biomass production rates
  - enhancement of root depth, penetration
- Introduce genes encoding insect resistance, disease resistance, etc. to reduce costs of agricultural chemical input, enhance biomass yield.

Sources: Raskin (1996), Cunningham and Ow (1996), Glass (1997), Glass (1999), Kramer and Chardonnens (2001), Pilon-Smits and Pilon (2002).

Several groups have experimented with addition of organic acids such as citric acid, and a recombinant approach has also been tried. De la Fuente *et al.* (1997) created transgenic tobacco and papaya constitutively expressing the citrate synthase (CS) gene from *Pseudomonas aeruginosa*, and showed that the resulting plants had increased aluminum tolerance, perhaps due to extracellular complexation of aluminum by citrate that had been excreted from plant roots into soil. More recently, López-Bucio *et al.* (2000) showed that these plants took up more phosphorus than wild type, and Guerinot (2001) reported that the plants became resistant to iron deficiency. This is a potentially promising approach to reducing the costs of lead phytoremediation, and Edenspace Corporation, in collaboration with Neal Stewart of the University of Tennessee, is planning 2004 field tests of transgenic tobacco expressing CS at a Pb-contaminated site (M. Elless, personal communication, also discussed below).

Pilon-Smits and Pilon (2002) and Kramer and Chardonnens (2001) review other strategies being undertaken to enhance metal mobilization in the soil, for example, involving the use of ferric reductases, expressed in plants, to reduce insoluble ferric ion to the more soluble ferrous form, or through the expression of enzymes in the biosynthetic pathways for phytosiderophores.

Enhancing Metal Uptake into Roots : The next critical step is the uptake of metal ions into the roots of plants. This requires transport of the metals across the root cell membrane into the root symplasm, and often this is mediated by transport proteins of various kinds, generally located in cell membranes, which have an affinity for metal ions or which create favorable energetic conditions to allow metals to enter the cell. According to Pilon-Smits and Pilon (2002) and authors referenced therein, there are over 150 different cation transporters that have been found in the model plant species *Arabidopsis thaliana* alone, and so there are likely to be many possible metal transport proteins that one could envision engineering into plants to enhance phytoremediation. Several of these have been well-studied in recent years, although to our knowledge none have been used in the field or are contemplated for commercial use in the near future.

The best-studied of these transporter proteins are the ZIP family, including IRT1 and other related IRT proteins. The ZIP family has been identified in *Arabidopsis*, and these proteins apparently regulate the uptake of a number of cations including Cd^{2+}, Fe^{2+}, Mn^{2+} and Zn^{2+} (Eide *et al.* 1996, Eng *et al.* 1998). Other transporter genes and gene products are the MRP1 gene encoding an Mg-ATPase transporter, also from *Arabidopsis* (Lu *et al.* 1997); NtCBP4 from tobacco, a putative cyclic-nucleotide and calmodulin-regulated cation channel that caused increased sensitivity to lead and increased nickel tolerance when overexpressed in tobacco (Arazi *et al.* 1999); the wheat LCT1 gene encoding a low-affinity cation transporter and the Nramp family of transporters from *Arabidopsis* (both reviewed in Kramer and Chardonnens 2001 and Pilon-Smits and Pilon 2002); and MTP1, encoding metal tolerance protein 1, isolated from the nickel/zinc hyperaccumulator *Thlaspi goesingense* (Persans *et al.* 2001, Kim *et al.* 2004) that appears to be a member of the cation diffusion facilitator (CDF) family. MTP1 likely has activity in transporting metal ions into plant cell vacuoles; another necessary step in creating a hyperaccumulator. Another vacuolar metal ion transporter, the yeast protein YCF1 (yeast cadmium factor 1) has been discovered and studied by Song *et al.* (2003), who expressed this protein in *Arabidopsis* and showed enhanced tolerance and accumulation of lead and cadmium.

Enhancing the ability of the plant to sequester metals : Another important general strategy is to express within plant cells proteins, peptides or other molecules that have high affinity for metals. The two categories of such molecules that have been investigated to date are metallothioneins and phytochelatins.

The metallothioneins (MTs) are a class of low-molecular weight (approx. 7 kilodalton) proteins with a high cysteine content and a generally high affinity for metal cations such as cadmium, copper and zinc (Cobbett and Goldsbrough 2000). MTs are known to exist in all organisms, and transgenic plants have been created in which MTs of animal origin have been constitutively expressed in plants. These experiments were not designed to test a phytoremediation approach, but instead to prevent metal accumulation in plant shoots by having it sequestered in the roots. One such plant, a transgenic tobacco, was field tested under two of the earliest permits to be issued by the U.S. Department of Agriculture for transgenic plants, granted to the Wagner group at the University of Kentucky (see below). When grown in the field, however, significant differences were not seen in either cadmium uptake or plant growth, when transgenics were compared to wild type (Yeargan *et al.* 1992). Kramer and Chardonnens (2001) summarize many experiments in which MTs were overexpressed to increase cadmium tolerance in plants by saying "The overexpression of MTs can increase plant tolerance to specific metals, for example cadmium or copper. However, this remains to be confirmed under field conditions. Only in a few instances did MT overexpression result in slight increases in shoot metal accumulation". Kramer and Chardonnens (2001) conclude that these results imply a limited role for MTs in phytoremediation. A more recent study, however (Thomas *et al.* 2003), reported that tobacco plants expressing the yeast metallothionein gene CUP1 were capable of accumulating high levels of copper but not cadmium, providing hope that this may someday be a viable phytoremediation strategy for that metal.

More recent attention has been devoted to the phytochelatins (PCs), which are small cysteine-rich metal binding peptides containing anywhere from 5 to 23 amino acids (Cobbett and Goldsbrough 2000, Pilon-Smits and Pilon 2002). PCs are believed to exist in all plants and are induced under metal stress conditions, probably to impart metal tolerance. PCs are synthesized non-ribosomally, by a three-step enzymatic pathway. In the first step, glutamate and cysteine are joined by the enzyme gamma glutamyl cysteine synthetase (gamma-ECS), to create gamma-glutamylcysteine. In the second step, a glycine residue is added by the enzyme glutathione synthetase (GS), to create glutathione. Finally, the enzyme phytochelatin synthetase (PCS), adds a variable number of additional gamma-glutamylcysteine units to create phytochelatins. The genes encoding these enzymes have been cloned from several organisms: the gamma-ECS enzyme is encoded by the *gsh*1 gene of *E. coli* and the CAD2 gene of *Arabidopsis*; GS is encoded by *E. coli gsh*2; and PCS is encoded by *Arabidopsis* CAD1 and by wheat TaPCS1. (Meagher 2000, Kramer and Chardonnens 2001, Terry 2001, Pilon-Smits and Pilon 2002).

As described in Terry (2001), Pilon-Smits and Pilon (2002), and Kramer and Chardonnens (2001), transgenic plants expressing these enzymes have been created and have shown promising results in either metal tolerance or metal uptake. The Terry group overexpressed GS enzyme from the *gsh*2 gene and ECS from the *gsh*1 gene in *Brassica juncea*, and in both cases found enhanced tolerance to cadmium and 2-3-fold greater cadmium uptake (Zhu *et al.* 1999a, b). Other groups that have created transgenic plants overexpressing PCs are Xiang *et al.* (2001), who created *Arabidopsis* overexpressing gamma-ECS and saw increased glutathione levels; Harada *et al.* (2001), who overexpressed cysteine synthase in tobacco and saw enhanced PC levels, enhanced Cd tolerance, but lower Cd concentrations in plant biomass; and Freeman *et al.* (2004), who over-expressed *Thlaspi goesingense* serine acetyltransferase in *Arabidopsis*, causing accumulation of glutathione and increased nickel tolerance. Clemens *et al.* (1999) expressed TaPCS1 from *Arabidopsis* and *Schizosaccharomyces pombe* in *Saccharomyces cerevisiae*, and showed that the gene product conferred enhanced cadmium tolerance in the host yeast. This same group (Gong *et al.* 2003) showed that organ-specific expression of wheat TaPCS1 in *Arabidopsis* could affect cadmium sensitivity and root-to-shoot transport.

More recently, Richard Meagher's group at the University of Georgia (Dhankher *et al.* 2002) created *Arabidopsis* plants that expressed gamma-ECS constitutively while also expressing an arsenate reductase in leaf tissues, and these plants showed enhanced tolerance to arsenic and the ability to accumulate high concentrations of this metal in plant tissue. In this study, plants expressing ECS alone from a strong constitutive promoter were moderately tolerant to arsenic compared to wild type. Li *et al.* (submitted), from the same group, created ECS-expressing *Arabidopsis* and showed these transgenic plants to have increased arsenic and mercury resistance, but with cadmium sensitivity. The Meagher group, working with Scott Merkle and colleagues, has also constitutively expressed ECS in cottonwood (*Populus deltoides*; A. Heaton and R. Meagher, personal communication). These groups are investigating the utility of ECS-expressing plants in phytoremediation strategies for mercury and arsenic.

The Terry group has field tested *Brassica juncea* overexpessing phytochelatins (see below), and Applied PhytoGenetics, in collaboration with the Meagher group, has applied for a USDA permit for field testing ECS-expressing cottonwood in 2004 or 2005.

Introduce genes to chemically or electrochemically transform metals or metalloids : As discussed above, the elemental nature of metals limits the possible biological remediation strategies to sequestration in plant biomass and transformation into less reactive or less toxic species. There are two major systems for the latter that have been explored to date: the use of bacterial genes governing the reduction of methylmercury or ionic mercury into elemental mercury (Meagher 2000); or the use of genes encoding enzymes that can methylate selenium into dimethylselenate (Hansen *et al.* 1998). In both these cases, the resulting form of the metal/metalloid is volatile, so that one can create plants capable of metal remediation by phytovolatilization.

Work on mercury phytoremediation has largely been done by the laboratory of Richard Meagher at the University of Georgia. This work involves the bacterial system discussed above: the gene encoding mercuric ion reductase (*merA*) and the gene encoding the bacterial organomercury lyase (*merB*) (Meagher 2000).

In the Meagher group's initial experiments, *Arabidopsis thaliana* was engineered to constitutively express a *merA* gene that had been modified for optimal expression in plants, and seeds and plants derived from the T_2 and subsequent generations showed stable resistance to high levels of mercuric ion in growth media (Rugh *et al.* 1996). Similar resistance data was also seen when *merA* transgenics of other species were constructed, including tobacco (Heaton *et al.* 1998; Heaton *et al.* submitted), yellow poplar (*Liriodendron tulipifera*) (Rugh *et al.* 1998), cottonwood (*Populus deltoides*) (Che *et al.* 2003), and rice (*Oriza sativa*) (Heaton *et al.* 2003). In many of these studies, evidence was seen that suggested that ionic mercury was taken up from the growth media, converted to Hg(0), and was transpired into the atmosphere from plant biomass. In fact, *merA* plants grown in ionic mercury showed significantly less mercury accumulation in plant tissue as compared to wild type plants, showing that *merA* plants could efficiently process ionic mercury into elemental mercury in plants (Meagher, personal communication).

Meagher and his colleagues have also demonstrated that transgenic *merB*-expressing *Arabidopsis* plants efficiently take up methylmercury and transform it to ionic mercury (Bizily *et al.* 1999, 2000, 2003, Bizily 2001). The Meagher lab has also constructed cottonwood and tobacco plants expressing *merB* and have shown these plants to also be resistant to organic mercury.

Ruiz *et al.* (2003) pursued a different approach and expressed a native merAB operon in chloroplasts of tobacco, and showed the resulting transgenic plants to be highly resistant to an organomercurial compound.

Chloroplast expression potentially offers several advantages over traditional nuclear expression, because it avoids the need for the codon optimization pursued by Meagher and colleagues, and because it lessens or eliminates the possibility that the transgene could spread beyond the engineered plant through pollen flow.

The Meagher laboratory conducted a limited-scale field trial of *merA*-expressing tobacco at an industrial site in New Jersey in 2001. In collaboration with Applied PhytoGenetics, Inc., two field tests of *merA*-expressing cottonwood were begun in 2003 (discussed below). Because of concerns over mercury emissions into the atmosphere, the use of the *merA* gene, which results in volatilization of low levels of Hg(0), may not be a favored remedial strategy. Meagher and his collaborators are hoping to create mercury hyperaccumulators by combinations of the *merA*/*merB* genes with ECS and other genes that could lead to mercury accumulation in plant tissue (Meagher, personal communication).

Research on phytoremediation strategies for selenium has been carried out by the laboratory of Norman Terry at the University of California, Berkeley, and several collaborators including Gary Banuelos of the USDA (de Souza *et al.* 2000), and this work led to a field test of transgenic plants in 2003 (discussed below). There are two possible phytoremediation strategies for selenium. There are plants that are naturally capable of hyper-accumulating Se, and although these species grow too slowly for commercial use, engineered hyperaccumulators might be more useful. In addition, pathways exist in which Se can be converted into dimethyl-selenate (de Souza *et al.* 1998), a compound which is volatilized into the atmosphere. In contrast to concerns over mercury volatilization, Se volatilization may be an effective strategy because selenium is a required nutrient and volatilized Se would be expected to be redposited on selenium-deficient soils. In addition, dimethylselenate is 600 times less toxic than selenate or selenite (Terry 2003). In one possible strategy to engineer an efficient selenium volatilizing plant, Van Huysen *et al.* (2003) overexpressed cystathionine-gamma-synthase, an enzyme believed to catalyze the first step in the pathway converting Se-cysteine to volatile dimethylselenide, in *Brassica juncea*, and showed enhanced selenium volatilization in the resulting transgenic plants.

Selenate is generally believed to be taken up by plants using pathways intended for uptake and assimilation of sulfate. The first step in this pathway is the transport of sulfate (or selenate) into plant tissue, mediated by the enzyme ATP sulfurylase. The Terry group expressed ATP sulfurylase in *Brassica juncea*, and created plants that showed somewhat increased tolerance to selenate while also accumulating 2- to 3-fold more selenate than wild type (Pilon-Smits *et al.* 1999). This could be the first step in creating a selenium hyperaccumulator. The same group (Wangeline *et al.* 2004) showed that overexpression of ATP sulfurylase in *B. juncea* also conferred tolerance to other metals including arsenic, cadmium, copper and zinc.

One problem that must be overcome in creating selenium hyper-accumulators is to avoid the nonspecific incorporation of seleno-amino acids into plant proteins, which is believed to be a mechanism for selenium toxicity. One way to achieve this is to divert selenium into molecules that are not incorporated into protein, such as selenomethylcysteine. One of the hyperaccumulating species referred to above, *Astragalus bisulcatus*, expresses an enzyme, selenocysteine methyltransferase, that is a key component of the methylation pathway of selenate/sulfate, with a preference for selenate. Two groups have expressed this enzyme in *Brassica juncea*, and have shown that the resulting plants can tolerate and accumulate selenium (Ellis *et al.* 2004, LeDuc *et al.* 2004). Another strategy to prevent selenium incorporation into protein is to over-express the gene encoding selenocysteine lyase, an enzyme that catalyzes the decomposition of selenocysteine into alanine and elemental selenium. Pilon *et al.* (2003) expressed mouse selenocysteine lyase in *Arabidopsis* and showed enhanced selenium tolerance and uptake.

Organics

Strategies for enhancing phytoremediation of organics are potentially more straightforward. In fact, because the goals of organic phytoremediation are to degrade and mineralize contaminants, strategies in this sector parallel some of the objectives discussed above for enhancing microbial bioremediation. Genes encoding biodegradative enzymes can be introduced and/or overexpressed in transgenic plants, leading to enhanced biodegradative abilities. In general, one can try to enhance or augment an existing pathway, or to create new biodegradative pathways or capabilities that do not exist in nature. Furthermore, one can use genetic engineering to impart degradative capability into fast-growing plants, or into species that are otherwise favored for use in the field. The following are some examples of projects in progress.

Degradation of Trichloroethylene and Other Volatile Organics : As discussed above, bioremediation approaches, including ones involving genetically modified organisms, have been investigated for trichloroethylene but concerns over the possible need for stimulatory cometabolites and the frequent occurrence of vinyl chloride as an intermediate in some microbial degradation pathways have hindered use of biological technologies for this purpose (Doty *et al.* 2000). TCE has been a target of phytoremediation from the earliest days of the technology's development (Chappell 1997, Shang *et al.* 2003), with hybrid poplars often being used to intercept groundwater streams contaminated with TCE. A team of investigators from the University of Washington have demons-trated that poplars are able to take up and metabolize TCE from groundwater in the field (Newman *et al.* 1999). This same group (Doty *et al.* 2000) has more recently created transgenic tobacco plants expressing cytochrome P450 2E1, a mammalian cytochrome that is capable of catalyzing the oxidation of a broad range of compounds including TCE, ethylene dibromide (EDB) and vinyl chloride (Guengerich *et al.* 1991). Doty et al. placed the gene encoding P450 2E1 under the control of a constitutive plant promoter that is active in all plant tissues, particularly including roots, and transformed tobacco plants with this construct. Transgenic tobacco plants grown hydroponically in the greenhouse had up to 640-fold higher ability to metabolize TCE and also were capable of debrominating EDB. Transgenic plants engineered in this way have not yet been used in the field (L. Newman, personal communication).

An interesting approach to phytoremediation of volatile organic compounds has recently been demonstrated. Barac *et al.* (2004) introduced the pTOM toluene-degradation plasmid found in *B. cepacia* G4 into the L.S.2.4 strain of *B. cepacia*, which is a microbial endophyte of yellow lupine, and showed that the transformed strain could grow within this plant species and exhibit strong degradation of toluene. The authors suggest that the use of modified endophytic bacteria could be a potentially powerful strategy towards creating plant/microbe systems with biodegradative capabilities, while avoiding the regulatory problems of introducing altered microorganisms to the open environment (Barac *et al.* 2004, Glick 2004).

Trinitrotoluene and Other Explosives : Trinitrotoluene (TNT), used for decades as an explosive, is a pervasive contaminant at many military sites around the world. Because of the need to treat TNT-contaminated sites with care, non-invasive *in situ* technologies like phytoremediation are being investigated, and a number of naturally-occurring plants have been shown to have the ability to degrade TNT and other nitroaromatic compounds through the activity of enzymes such as nitrate reductases (Subramanian and Shanks 2003, Wolfe and Hoehamer 2003); however, the possible creation of toxic byproducts by such plant systems has limited their potential usefulness in commercial remediation (French *et al.* 1999).

Neil Bruce and his colleagues have now constructed two different lines of transgenic plants that demonstrate the potential feasibility of the use of genetically modified plants for TNT remediation. French *et al.* (1999)

created transgenic tobacco plants constitutively expressing the *omr* gene from *Enterobacter cloacae* PB2, which encodes pentaerythritol tetrantirate (PETN) reductase, an enzyme that catalyzes the dentiration of explosive compounds like PETN and glycerol trinitrate (GTN). These transgenic tobacco were shown to be able to successfully germinate in concentrations of TNT and GTN that are toxic to wild type plants, and to show more rapid and complete denitration of GTN than wild type. More recently, the Bruce lab created transgenic tobacco plants expressing a nitrate reductase from the *nfsI* gene of a different *E. cloacae* strain (NICMB10101) and showed that these plants are greatly increased in their ability to tolerate, take up and detoxify TNT (Hannink *et al.* 2001). Degradation of TNT in these plants follows a pathway different than that of the PETN pathway: TNT is reduced to hydroxyaminodinitrotoluene which is subsequently reduced to aminodinitrotoluene derivatives.

Because explosive compounds are often found as contaminants in aquatic environments or in sediments, Donald Cheney and his colleagues are creating seaweed (*Porphyra*) transformed with the *E. cloacae* nfsI *gene*, to enable degradation of TNT in aquatic environments (Cheney *et al.* 2003). *Porphyra* plants transformed with this gene can survive extended periods of time in concentrations of TNT in seawater that kill wild type plants within days, and the engineered plants appear to be metabolizing the TNT (D. Cheney, personal communication). The Bruce lab has also cloned an gene cluster from *Rhodococcus rhodochrous* whose gene products can degrade the explosive compound hexahydro-1,3,5-trinitro-1,3,5-triazine, known as RDX (Seth-Smith *et al.* 2002).

Regulation of Transgenic Plants for Phytoremediation

Genetically engineered plants are regulated in the United States by the U.S. Department of Agriculture (USDA) under regulations first promulgated in 1987 (52 <u>Federal Register</u> 22892-22915). Similar regulations exist in many other countries around the world (Nap *et al.* 2003). Although these regulations arose from the debates over "deliberate releases" of genetically engineered organisms in the mid 1980s, field tests of plants have never been unusually controversial (see Glass 1991 and Glass 1997 for a historical review). Today these rules present only a minimal barrier against research field tests, and also allow commercial use of transgenic plants under a reasonable regulatory regime.

Under these regulations, USDA's Animal and Plant Health Inspection Service (APHIS) uses the Federal Plant Protection Act to regulate outdoor uses of transgenic plants. Originally, permits were required for most field tests of genetically engineered plants. Applications for these permits were required to include a description of the modifications made to the plant, data characterizing the stability of these changes, and a description of the proposed field test and the procedures to be used to confine the plants in the test plot, and submitters also had to assess potential environmental effects.

These regulations were substantially relaxed in 1993 (58 Federal Register 17044-17059) to create two procedures to exempt specific plants. Under the first, transgenic plants of six specific crops (corn, soybean, tomato, tobacco, cotton and potato) were able to be field tested merely upon notifying the agency 30 days in advance, provided the plants did not contain any potentially harmful genetic sequences and the applicant provided certain information and submitted annual reports of test results. The second procedure allowed applicants to petition that specific transgenic plant varieties be "delisted" following several years of safe field tests, to proceed to commercial use and sale without the need for yearly permits (Glass 1997). This delisting procedure would be the way specific transgenic plant varieties would be approved for widespread commercial use in phytoremediation.

The situation in the United States was further simplified by a 1997 amendment to the regulations (62 Federal Register 23945-23958) that now allows almost all transgenic plants to be field tested without a permit, merely upon 30 days advance notice to APHIS. The only exceptions under the regulations are transgenic plants derived from noxious weeds, which would need a permit for field testing. However, more recently, in response to proposed new industrial uses for transgenic plants (e.g., for the production of pharmaceutical products), USDA has begun requiring permits (rather than notifications) for those proposed field uses of transgenic plants for which it lacks significant experience. Phytoremediation is among these uses (see http://www.aphis.usda.gov/brs/letters/011404%20.pdf for details), and beginning in 2003, all field tests of transgenic plants for phytoremediation have been conducted under permits rather than under the notification process.

Field tests of transgenic plants have generated far less public controversy than have field uses of engineered microbes (note that we distinguish concerns over field testing from the current concerns in some European countries over food use of transgenic plants, an issue which, while serious, should not affect use of transgenics in phytoremediation). The APHIS regulations have allowed a large number of field tests to be carried out with moderate levels of government oversight: through June 2004, APHIS had received over 10,000 permits or notifications for field tests (9,984 of which were approved, and many of which covered multiple test sites), of well over 100 different plant species, in every state of the U.S., the

Virgin Islands and Puerto Rico. Through June 2004, 60 different transgenic varieties had been delisted for commercial use in the U.S.. (All U.S. statistics can be found at the website "Information Systems for Biotechnology",. http://www.isb.vt.edu/cfdocs/fieldtests1.cfm). Field tests of transgenic plants have also taken place in at least 34 countries other than the United States (see directory of Internet field test databases at http://www.isb.vt.edu/cfdocs/globalfieldtests.cfm).

As is the case with the field uses of engineered microorganisms, there are legitimate questions that must be assessed concerning the possible environmental impacts of proposed uses of transgenic plants in phytoremediation. These issues, as they generally apply to agricultural uses of transgenic plants, have been thoroughly presented and analyzed since the 1980s (e.g., National Research Council 1989, 2000); and Glass (1997) presents a detailed discussion of how these questions might affect uses of transgenics in phytoremediation.

Briefly, the two most important environmental issues relate to possible enhancement of the weediness of the transgenic plant and its potential to outcross (and spread the introduced gene) to related species. Single gene changes can enhance weediness, although more often multiple changes are needed (Keeler 1989, National Research Council 1989). Crops that have been subject to extensive agricultural breeding are less likely to revert to a weedy phenotype by simple genetic changes (National Research Council 1989), but those plant species used in phytoremediation may not be as well-characterized or as long-cultivated as agricultural crop species, and some may be related to weeds. It might be necessary to consider whether genes encoding an enhanced hyperaccumulation phenotype would confer on the recipient plant any growth advantage or enhance weediness, particularly if the transgene were introduced via cross-pollination into a weedy relative.

Almost all plants have wild relatives (National Research Council 1989), so every plant species of commercial utility would have some potential to interbreed with wild, perhaps weedy, species. In many transgenic field trials, the possibility of cross-pollination has been mitigated by preventing pollination, for example, by bagging or removing the pollen-producing organs or harvesting biomass before flowering, and this should be possible for many phytoremediation projects. Some phytoremediation projects will utilize trees that would not be expected to set pollen during the course of the test. For phytoremediation, one must also be concerned over transfer of a hyperaccumulation phenotype into crop plants, possibly causing contaminants to enter the food chain.

For all proposed field tests, regulatory agencies would want to be certain that the products of the introduced genes are not toxic or pathogenic. One concern unique to phytoremediation might be the potential risks to birds and insects who might feed on plant biomass containing high concentrations of hazardous substances, particularly metals. Questions relating to the proper disposal of plants after use would also arise, and commercial approvals may require restrictions on the use of the harvested plant biomass for human or animal food.

Field Uses of Transgenic Plants for Phytoremediation

The situation with transgenic plants for commercial phytoremediation is somewhat more advanced than is the case with modified microorganisms. After several early academic field tests of model plant species engineered for enhanced heavy metal accumulation, the first field tests of transgenic plants of commercially-relevant species began in 2003. As noted above, the Wagner group at the University of Kentucky field tested tobacco plants expressing the mouse metallothionein gene in the late 1980s, and the Meagher lab conducted a small field trial of tobacco plants expressing the *merA* gene at a contaminated site in New Jersey in 2001 (Meagher and Heaton, personal communication). However, all these early tests involved a model species, and there were no field uses of plants belonging to any species better suited for commercial remediation, until 2003, when three such field tests were begun under permits granted from the USDA (See Table 6).

The first field test of commercially-relevant transgenic plants for phytoremediation was planted in the spring of 2003, and carried out as a collaboration between Norman Terry of the University of California at Berkeley and Gary Bañuelos of the USDA Agricultural Research Service. Three transgenic Indian mustard [*Brassica juncea* (L.) Czern.] lines were tested at a California field site for their ability to remove selenium from Se- and boron-contaminated saline soil. The three transgenic lines overexpressed genes encoding the enzymes ATP sulfurylase, gamma-glutamyl-cysteine synthetase, and glutathione synthetase, respectively (all discussed above). In what is likely the first report showing that plants genetically engineered for phytoremediation can perform successfully under field conditions, the transgenic lines exhibited superior abilities for Se accumulation and for tolerance to highly contaminated saline soil (Bañuelos *et al.* 2005).

In July 2003, Applied PhytoGenetics, Inc. (APGEN) began its first pilot field project of its technology for phytoremediation of mercury. This field test features transgenic cottonwood trees expressing the *merA* gene, encoding mercuric ion reductase (discussed above), and is taking place at an urban mercury-contaminated site in Danbury, Connecticut. APGEN is undertaking this project as a collaboration with the City of Danbury, researchers at Western Connecticut State University and the Meagher

Table 6. Transgenic plants reviewed or approved by the U.S. Department of Agriculture for field testing for phytoremediation purposes.

Year of APHIS submission	Institution	Organism	Gene(s)	Location(s)
1989	U. of Kentucky	Tobacco	Mouse Metallothionein	KY
1990	U. of Kentucky	Tobacco	Mouse Metallothionein	KY
2000	U. of Georgia	Tobacco	E. coli Mercuric ion reductase	NJ
2001	U. of Georgia	Poplar	E. coli Mercuricion reductase	NJ (test not conducted)
2003	Agricultural Research Service	Brassica	Genes expressing enzymes for selenium phytoremediation	CA
2003	Applied PhytoGenetics, Inc.	Cottonwood (Populus deltoides)	E. coli Mercuric ion reductase and Organomercury lyase	AL, CT, IN (test conducted Al, CT only)
2003	Applied PhytoGenetics, Inc.	Cottonwood (Populus deltoides)	E. coli Mercuric ion reductase and Organomercury lyase	NY, TN (test not conducted)
2003	Applied PhytoGenetics, Inc.	Rice	E. coli Mercuric ion reductase and Organomercury lyase	IN (test not conducted)

Source: "Information Systems for Biotechnology", http://www.isb.vt.edu/cfdocs/fieldtests1.cfm).

laboratory at the University of Georgia. The site of the field test is one of several properties in and around Danbury that has mercury contamination arising from their prior use in the manufacture of hats. In October 2003, APGEN began a similar pilot project at a private mercury-contaminated industrial site in Alabama. These are believed to be the first transgenic phytoremediation projects in the United States carried out by a commercial (for-profit) entity. These are intended to be multi-year tests, and APGEN expects to obtain the first data on mercury removal from the soil at the end of the 2004 growing season.

Additional field test permit requests are pending at USDA as of June 2004. APGEN has applied for permission to field test gamma-ECS-expressing cottonwood at several sites, and Edenspace Corporation has requested approval for a test of citrate synthase-expressing tobacco plants at a lead-contaminated site (both mentioned above).

Because the regulatory barriers to getting transgenic plants into the field are low and relatively easy to overcome (even for academic groups), we expect transgenic plants to be used commercially in remediation sooner than will engineered microorganisms. However, transgenic plants will face other obstacles, primarily because phytoremediation is still establishing itself as a viable technology in the market, and this may make it harder to convince site owners and regulators to take a chance on the use of an engineered plant. Anecdotally to date, there does not appear to have been any significant resistance to the use of GMOs in phytoremediation on the part of stakeholders, giving some comfort that transgenics will be adopted when their efficacy is proven for specific applications.

Conclusions

There are many compelling reasons to use genetic engineering to improve the plants or microorganisms that might be used in commercial remediation, and there has been an enormous amount of research in the past ten to fifteen years directed at the basic research or the applied innovations needed to accomplish this. These facts alone might lead one to the conclusion that commercial use of GMOs in remediation is inevitable and imminent; but consideration of other factors, including economic, regulatory, public relations, and even technical concerns, should give reason for caution in such predictions.

We believe that the more recalcitrant and/or most toxic contaminants will continue to be targets beckoning the development of innovative technologies like bio- or phytoremediation, and that efforts to develop and utilize GMOs against such contaminants will continue. Of the two sectors, we feel that microbial GMO products are less likely to come to the market soon. This is largely for technical reasons: it will usually be possible to use classical strategies for strain improvement, or even to discover previously-unknown microbial strains, to develop a biological approach to any given contaminant, and such approaches are likely to be quicker and less expensive than using genetic engineering. Combined with the uncertainties (and possible added costs) of the regulatory situation for microbial GMOs, it seems likely that workers in the field will continue to favor the use of naturally-occurring or classically-mutated microbial strains. We are more optimistic with regard to transgenic plants for phytoremediation: here, a genetic engineering strategy to improve a plant variety is likely to be quicker, more powerful, more straightforward, and perhaps cheaper than trying to do the same using classical breeding; and the regulatory and public perception problems are far less daunting for plants than they are for microbes. This is borne out by the fact that, at this writing, three field tests of commercially-relevant transgenic plants for phytoremediation have taken place in the U.S., as opposed to none for commercially-relevant engineered microorganisms, and at least two U.S. companies are intending to use transgenic plants in commercial remediation projects in the near future.

Economic and marketplace barriers will remain as obstacles to overcome. In particular, it is hard to achieve meaningful returns on investment in the environmental field for innovative technologies that are costly to develop, and in addition, it is very hard to obtain venture capital or other "seed" funding for innovative technologies in the envirotech sector. One possible way to surmount this problem would be for companies to in-license and commercialize technologies invented at universities or other non-profit laboratories, where the earliest stages of research would have been funded by government grants and other sources, thus leveraging the investment made by such research sponsors, so that the company need only recoup its own development (and licensing) costs, rather than recoup the costs of the entire R&D process. A good portion of the research described in this chapter was conducted at academic institutions and is available for commercial licensing, and so may ultimately be used in the marketplace under favorable economic circumstances.

Finally, it comes down to the technical and market need. Should there be any contaminant or specific contamination scenario for which traditional techniques do not work or do not meet the market's needs, and for which biological methods cannot be developed using native or classically-mutated organisms, then a GMO approach may well reach the commercial market. From that point, the free market will decide the future applicability of GMOs to commercial site remediation.

REFERENCES

Alexander, M. 1985. Genetic engineering: Ecological consequences. *Issues Sci. Technol.* 1: 57-68.

Anderson, T.A., E.A. Guthrie, and B.T. Walton. 1993. Bioremediation in the Rhizosphere. *Env. Sci. Technol.* 27: 2530-2636.

Arazi, T., R. Sunkar, B. Kaplan, and H. Fromm. 1999. A tobacco plasma membrane calmodulin-binding transporter confers Ni2+ tolerance and Pb2+ hypersensitivity in transgenic plants. *Plant J.* 20: 171-182.

Baker, A.J.M., and R.R. Brooks. 1989. Terrestrial higher plants which hyperaccumulate metallic elements -- a review of their distribution, ecology and phytochemistry. *Biorecovery* 1: 81-126.

Bañuelos, G., N. Terry, D. L. LeDuc, E. A. H. Pilon-Smits, and B. Mackey. 2005. Field trial of transgenic Indian mustard plants shows enhanced phytoremediation of selenium-contaminated sediment. *Environ. Sci. Technol.*, ASAP Article 10.1021/es09035f S0013-936X (04)09035-2, Web Release Date: February 1, 2005.

Barac, T., S. Taghavi, B. Borremans, Provoost, L. Oeyen, J.V. Colpaert, J. Vangronsveld, and D. van der Lelie. 2004. Engineered endophytic bacteria improve phytoremediation of water-soluble, volatile, organic pollutants. *Nat. Biotechnol.* 22: 583-588.

Bizily, S. 2001. Genetic engineering of plants with the bacterial genes *merA* and *merB* for the phytoremediation of methylmercury contaminated sediments. University of Georgia: Genetics Department 145, Athens, GA.

Bizily, S., T. Kim, M.K. Kandasamy, and R.B. Meagher. 2003. Subcellular targeting of methylmercury lyase enhances its specific activity for organic mercury detoxification in plants. *Plant Physiol.* 131: 463-471.

Bizily, S., C.L. Rugh, and R.B. Meagher. 2000. Phytodetoxification of hazardous organomercurials by genetically engineered plants. *Nat. Biotechnol.* 18: 213-217.

Bizily, S., C.L. Rugh, A.O. Summers, and R.B. Meagher. 1999. Phytoremediation of methylmercury pollution: *merB* expression in *Arabidopsis thaliana* confers resistance to organomercurials. *Proc. Natl. Acad. Sci.* USA 96: 6808-6813.

Blaylock, M.J., and J.W. Huang, 2000. Phytoextraction of metals. Pages 53-70 in *Phytoremediation of Toxic Metals - Using Plants to Clean Up the Environment*, I. Raskin and B.D. Ensley, eds., Wiley, New York.

Boyajian, G.E. and L.H. Carreira. 1997. Phytoremediation: a clean transition from laboratory to marketplace? *Nat. Biotechnol.* 15: 127-128.

Brim, H., S.C. McFarlan, J.K. Fredrickson, K.W. Minton, M. Zhai, L.P. Wackett, and M.J. Daly. 2000. Engineering *Deinococcus radiodurans* for metal remediation in radioactive mixed waste environments. *Nat. Biotechnol.* 18: 85-90.

Burken, J.G., and J.L. Schnoor. 1997. Uptake and metabolism of atrazine by poplar trees. *Environ. Sci. Technol.* 31: 1399-1406.

Chappell, J. 1997. Phytoremediation of TCE using *Populus*. http://clu-in.com/phytotce.htm.

Che, D.S., R.B. Meagher, A.C.P. Heaton, A. Lima, and S. A. Merkle. 2003. Expression of mercuric ion reductase in eastern cottonwood confers mercuric ion reduction and resistance. *Plant Biotech.* 1: 311-319.

Chen, S.L., E.K. Kim, M.L. Shuler, and D.B. Wilson. 1998. Hg2+ removal by genetically engineered *Escherichia coli* in a hollow fiber bioreactor. *Biotechnol. Prog.* 14: 667-671.

Chen, S.L., and D.B. Wilson. 1997a. Construction and characterization of *Escherichia coli* genetically engineered for bioremediation of Hg2+ contaminated environments. *Appl. Environ. Microbiol.* 63: 2442-2445.

Chen, S.L., and D.B. Wilson. 1997b. Genetic engineering of bacteria and their potential for Hg2+ bioremediation. *Biodegradation* 8: 97-103.

Cheney, D., P. Bernasconi, B. Curtis, G. Rorrer, and N. Bruce 2003. *Phytoremediation of Mercury and TNT in our Oceans.* The Annual International Conference on Contaminated Soils, Sediments and Water, Amherst, MA, http://www.umasssoils.com/posters2003/phytoposter.htm.

Clemens, S., E.J. Kim, D. Neumann, and J.I. Schroeder. 1999. Tolerance to toxic metals by a gene family of phytochelatin synthases from plants and yeast. *EMBO J.* 18: 3325-3333.

Cobbett, C.S., and P.B. Goldsbrough. 2000. Mechanisms of Metal Resistance: Phytochelatins and Metallothioneins. Pages 247-270 in *Phytoremediation of Toxic Metals - Using Plants to Clean Up the Environment,* I. Raskin and B.D. Ensley, eds., Wiley, New York.

Conner, A.J., T.R. Glare, and J.P. Nap. 2003. The release of genetically modified crops into the environment. Part II. Overview of ecological risk assessment. *Plant J.* 33: 19-46.

Cunningham, S.D., and D.W. Ow. 1996. Promises and prospects of phytoremediation. *Plant Physiol.* 110: 715-719.

Dai, M., and S.D. Copley. 2004. Genome shuffling improves degradation of the anthropogenic pesticide pentachlorophenol by *Sphingobium chlorophenolicum* ATCC 39723. *Appl. Environ. Microbiol.* 70: 2391-2397.

De la Fuente, J.M., V. Ramírez-Rodríguez, J.L. Cabreraponce, and L. Herrera-Estrella. 1997. Aluminum tolerance in transgenic plants by alteration of citrate synthesis. *Science* 276: 1566-1568.

de Lorenzo, V. 2001. Cleaning up behind us. The potential of genetically modified bacteria to break down toxic pollutants in the environment. *EMBO Rep.* 2: 357-359.

de Souza, M., E. Pilon-Smits, and N. Terry. 2000. The physiology and biochemistry of selenium volatilization by plants. Pages 171-190 in *Phytoremediation of Toxic Metals - Using Plants to Clean Up the Environment.* I. Raskin and B.D. Ensley, eds., Wiley, New York.

de Souza, M.P., E.A. Pilon-Smits, C.M. Lytle, S. Hwang, J. Tai, T.S. Honma, L. Yeh, and N. Terry. 1998. Rate-limiting steps in selenium assimilation and volatilization by Indian mustard. *Plant Physiol.* 117: 1487-1494.

De Wever, H., J.R. Cole, M.R. Fettig, D.A. Hogan, and J.M. Tiedje. 2000. Reductive dehalogenation of trichloroacetic acid by *Trichlorobacter thiogenes* gen. nov., sp. nov. *Appl. Environ. Microbiol.* 66: 2297-2301.

DEFRA (U.K. Department for Environment, Food and Rural Affairs). 2002. Genetically modified organisms for the bioremediation of organic and inorganic pollutants http://www.defra.gov.uk/environment/gm/research/epg-1-5-142.htm.

Dhankher, O.P., Y. Li, B.P. Rosen, J. Shi, D. Salt, J.F. Senecoff, N.A. Sashti, and R.B. Meagher. 2002. Engineering tolerance and hyperaccumulation of arsenic in plants by combining arsenate reductase and gamma-glutamylcysteine synthetase expression. *Nat. Biotechnol.* 20: 1140-1145.

Doty, S.L., T.Q. Shang, A.M. Wilson, J. Tangen, A.D. Westergreen, L.A. Newman, S. E. Strand, and M. P. Gordon. 2000. Enhanced metabolism of halogenated hydrocarbons in transgenic plants containing mammalian cytochrome P450 2E1. *Proc. Natl. Acad. Sci. USA* 97: 6287-6291.

Dowling, D.N., and F.O'Gara. 1994. Genetic manipulation of ecologically adapted *Pseudomonas* strains for PCB degradation. *Curr. Topics Mol. Genet. Life Sci. Adv.* 2: 1-8.

Dümmer, C., and D.J. Bjornstad, 2004. Regulatory policy towards organisms produced through biotechnology: evolution of the framework and relevance for DOE's bioremediation program. Joint Institute for Energy & Environment, Knoxville, TN, January, 2004.

Eide, D., M. Broderius, J. Fett, and M.L. Guerinot. 1996. A novel iron-regulated metal transporter from plants identified by functional expression in yeast. *Proc. Natl. Acad. Sci. USA* 93: 5624-5628.

Ellis, D.R., T.G. Sors, D.G. Brunk, C. Albrecht, C. Orser, B. Lahner, K.V. Wood, H. H. Harris, I.J. Pickering, and D.E. Salt. 2004. Production of Se-methylselenocysteine in transgenic plants expressing selenocysteine methyltransferase. *BMC Plant Biol.* 4: 1.

Eng, B.H., M.L. Guerinot, D. Eide, and M.H. Saier, Jr. 1998. Sequence analyses and phylogenetic characterization of the ZIP family of metal ion transport proteins. *J. Membr. Biol.* 166: 1-7.

Environmental Business Journal. 2003. Industry Overview 2003, Volume XVI, Number 5/6.

Erb, R.W., C.A. Eichner, I. Wagner-Dobler, and K. N. Timmis. 1997. Bioprotection of microbial communities from toxic phenol mixtures by a genetically designed pseudomonad. *Nat. Biotechnol.* 15: 378-382.

Erickson, B.D., and F.J. Mondello. 1992. Nucleotide sequencing and transcriptional mapping of the genes encoding biphenyl dioxygenase, a multicomponent polychlorinated-biphenyl-degrading enzyme in *Pseudomonas* strain LB400. *J. Bacteriol.* 174: 2903-2912.

Freeman, J.L., M.W. Persans, K. Nieman, C. Albrecht, W. Peer, I.J. Pickering, and D. Salt. 2004. Increased glutathione biosynthesis plays a role in nickel tolerance in *Thlaspi* nickel hyperaccumulators. *The Plant Cell*, in press.

French, C.E., S.J. Rosser, G.J. Davies, S. Nicklin, and N.C. Bruce. 1999. Biodegradation of explosives by transgenic plants expressing pentaerythritol tetranitrate reductase. *Nat. Biotechnol.* 17: 491-494.

Fujita, M., M. Ike, J.I. Hioki, K. Kataoka, and M. Takeo. 1995. Trichloroethylene degradation by genetically-engineered bacteria carrying cloned phenol catabolic genes. *J. Ferment. Bioengineering* 79: 100-106.

Furukawa, K., J. Hirose, S. Hayashida, and K. Nakamura. 1994. Efficient degradation of trichloroethylene by a hybrid aromatic ring dioxygenase. *J. Bacteriol.* 176: 2121-2123.

Glass, D.J. 1991. Impact of government regulation on commercial biotechnology. Pages 169-198 in *The Business of Biotechnology: From the Bench to the Street,* R.D. Ono, ed., Butterworth-Heinemann, Stoneham, MA.

Glass, D.J. 1994. Obtaining Regulatory Approval and Public Acceptance for Bioremediation Projects with Engineered Organisms in the United States. Pages 256-267 in *Applied Biotechnology for Site Remediation.* R.E. Hinchee, D.B. Anderson, F.B. Metting, and G.D. Sayles, eds. Lewis Publishers, Ann Arbor, MI.

Glass, D.J. 1997. Prospects for use and regulation of transgenic plants. Pages 51-56 in *Phytoremediation: In Situ and On-Site Bioremediation,* Vol. 4, B.C. Alleman and A. Leeson, eds., Battelle Press, Columbus, Ohio

Glass, D.J. 1999. U.S. and International Markets for Phytoremediation, 1999-2000, D. Glass Associates, Inc., Needham, MA.

Glass, D.J. 2000. International Remediation Markets: Perspectives and Trends. *The Second International Conference of Remediation of Chlorinated and Recalcitrant Compounds.* G.B. Wickramanayake, ed. Columbus, OH, Battelle Press. Volume 1, "Risk, Regulatory and Monitoring Considerations": 33-40.

Glass, D.J. 2002. Regulation of the commercial uses of microorganisms. Pages 2693-2707 in *Encyclopedia of Environmental Microbiology,* G. Bitton, ed., John Wiley and Sons, New York.

Glick, B.R. 2004. Teamwork in phytoremediation. *Nat. Biotechnol.* 22: 526-527.

Gong, J.M., D.A. Lee, and J.I. Schroeder. 2003. Long-distance root-to-shoot transport of phytochelatins and cadmium in *Arabidopsis. Proc. Natl. Acad. Sci.* USA 100: 10118-10123.

Guengerich, F.P., D.H. Kim, and M. Iwasaki. 1991. Role of human cytochrome P-450 IIE1 in the oxidation of many low molecular weight cancer suspects. *Chem. Res. Toxicol.* 4: 168-179.

Guerinot, M.L. 2001. Improving rice yields--ironing out the details. *Nat. Biotechnol.* 19: 417-418.

Hannink, N., S.J. Rosser, C.E. French, A. Basran, J.A. Murray, S. Nicklin, and N.C. Bruce. 2001. Phytodetoxification of TNT by transgenic plants expressing a bacterial nitroreductase. *Nat. Biotechnol.* 19: 1168-1172.

Hansen, D., P.J. Duda, A. Zayed, and N. Terry. 1998. Selenium removal by constructed wetlands: role of biological volatilization. *Environ. Sci. Technol.* 32: 591-597.

Hanson, J.R., C.E. Ackerman, and K.M. Scow. 1999. Biodegradation of methyl tert-butyl ether by a bacterial pure culture. *Appl. Environ. Microbiol.* 65: 4788-4792.

Harada, E., Y.E. Choi, A. Tsuchisaka, H. Obata, and H. Sano. 2001. Transgenic tobacco plants expressing a rice cysteine synthase gene are tolerant to toxic levels of cadmium. *J. Plant Physiol.* 158: 655-661.

He, J., K.M. Ritalahti, K.L. Yang, S.S. Koenigsberg, and F.E. Loffler. 2003. Detoxification of vinyl chloride to ethene coupled to growth of an anaerobic bacterium. *Nature* 424: 62-65.

Heaton, A., C. Rugh, and R. Meagher. submitted. Responses of *merA*-expressing *Nicotiana tabacum* to mercury exposure: A model for engineered phytoremediation. *Water Air Soil Pollut.*

Heaton, A.C.P., C.L. Rugh, T. Kim, N.J. Wang, and R.B. Meagher. 2003. Toward detoxifying mercury-polluted aquatic sediments using rice genetically-engineered for mercury resistance. *Environ. Toxicol. Chem.* 22: 2940-2947.

Heaton, A.C.P., C.L. Rugh, N.-J. Wang, and R.B. Meagher. 1998. Phytoremediation of mercury and methylmercury polluted soils using genetically engineered plants. *J. Soil Contam.* 7: 497-509.

Horn, J.M., M. Brunke, W.-D. Deckwer, and K.N. Timmis. 1994. *Pseudomonas putida* strains which constitutively overexpress mercury resistance for detoxification of organomercurial pollutants. *Appl. Environ. Microbiol.* 60: 357-362.

Hrywna, Y., T.V. Tsoi, O.V. Maltseva, J.F. Quensen, 3rd, and J.M. Tiedje. 1999. Construction and characterization of two recombinant bacteria that grow on ortho- and para-substituted chlorobiphenyls. *Appl. Environ. Microbiol.* 65: 2163-2169.

Hutchinson, S.L., A.P. Schwab, and M.K. Banks 2003. Biodegradation of petroleum hydrocarbons in the rhizosphere. Pages 355-386 in *Phytoremediation: Transformation and Control of Contaminants,* S.C. McCutcheon and J.L. Schnoor, eds., John Wiley and Sons, New York.

ITRC (Interstate Technology and Regulatory Cooperation) 2001. Technical and Regulatory Guidance Document: Phytotechnologies, Washington, DC.

Keasling, J.D., and S.W. Bang. 1998. Recombinant DNA techniques for bioremediation and environmentally-friendly synthesis. *Curr. Opin. Biotechnol.* 9: 135-140.

Keeler, K.H. 1989. Can genetically engineered crops become weeds? *Bio/Technol.* 7: 1134-1139.

Kim, D., J.L. Gustin, B. Lahner, M.W. Persans, D. Baek, D.-J. Yun, and D.E. Salt. 2004. The plant CDF family member TgMTP1 from the Ni/Zn hyperaccumulator *Thlaspi goesingense* acts to enhance efflux of Zn at the plasma membrane when expressed in *Saccharomyces cervisiae. Plant J.,* in press.

Kramer, U. and A.N. Chardonnens. 2001. The use of transgenic plants in the bioremediation of soils contaminated with trace elements. *Appl. Microbiol. Biotechnol.* 55: 661-672.

Lange, C.C., L.P. Wackett, K.W. Minton, and M.J. Daly. 1998. Engineering a recombinant Deinococcus radiodurans for organopollutant degradation in radioactive mixed waste environments. *Nat. Biotechnol.* 16: 929-933.

Lau, P., and V. de Lorenzo. 1999. Genetic engineering: the frontier of bioremediation. *Environ. Sci. Technol.* 4: 124A-128A.

LeDuc, D.L., A.S. Tarun, M. Montes-Bayon, J. Meija, M.F. Malit, C.P. Wu, M. AbdelSamie, C.Y. Chiang, A. Tagmount, M. DeSouza, B. Neuhierl, A. Bock, J. Caruso, and N. Terry. 2004. Overexpression of selenocysteine methyltransferase in *Arabidopsis* and Indian mustard increases selenium tolerance and accumulation. *Plant Physiol.* 135: 377-383.

Li, Y., O.P. Dhankher, L. Carriera, D. Lee, J.I. Shroeder, R.S. Balish, and R.B. Meagher. Arsenic and mercury resistance and cadmium sensitivity in Arabidopsis plants expressing bacterial glutamylcysteine synthetase. *J. Biol. Chem., submitted.*

López-Bucio, J., O. Martinez de la Vega, A. Guevara-García, and L. Herrera-Estrella. 2000. Enhanced phosphorus uptake in transgenic tobacco plants that overproduce citrate. *Nature Biotechnol.* 18: 450-453.

Lu, Y.P., Z.S. Li, and P.A. Rea. 1997. AtMRP1 gene of *Arabidopsis* encodes a glutathione S-conjugate pump: isolation and functional definition of a plant ATPbinding cassette transporter gene. *Proc. Natl. Acad. Sci.* USA 94: 8243-8248.

Ma, L.Q., K.M. Komar, C. Tu, W. Zhang, Y. Cai, and E.D. Kennelley. 2001. A fern that hyperaccumulates arsenic. *Nature* 409: 579.

McCutcheon, S.C., and J. Schnoor 2003. Overview of phytotransformation and control of wastes. Pages 3-58 in *Phytoremediation: Transformation and Control of Contaminants.* S.C. McCutcheon and J.L. Schnoor, eds., John Wiley and Sons, New York.

Meagher, R.B. 2000. Phytoremediation of toxic elemental and organic pollutants. *Curr. Opin. Plant. Biol.* 3: 153-162.

Menn, F.M., J.P. Easter, and G.S. Sayler 2000. Genetically engineered micro-organisms and bioremediation. Pages 443-460 in *Biotechnology.* Volume 11b, Environmental Processes II, J. Klein, John Wiley, New York.

Mondello, F.J. 1989. Cloning and expression in *Escherichia coli* of *Pseudomonas* strain LB400 genes encoding polychlorinated biphenyl degradation. *J. Bacteriol.* 171: 1725-1732.

Mondello, F.J., M.P. Turcich, J.H. Lobos, and B.D. Erickson. 1997. Identification and modification of biphenyl dioxygenase sequences that determine the specificity of polychlorinated biphenyl degradation. *Appl. Environ. Microbiol.* 63: 3096-3103.

Morrissey, J.P., U.F. Walsh, A.O'Donnell, Y. Moenne-Loccoz, and F.O'Gara. 2002. Exploitation of genetically modified inoculants for industrial ecology applications. *Antonie Van Leeuwenhoek* 81: 599-606.

Nap, J.P., P.L. Metz, M. Escaler, and A.J. Conner. 2003. The release of genetically modified crops into the environment. Part I. Overview of current status and regulations. *Plant J.* 33: 1-18.

National Research Council 1989. *Field Testing Genetically Engineered Organisms: Framework for Decisions.* National Academy Press., Washington, DC.

National Research Council 2000. *Genetically Modified Pest-Protected Plants: Science and Regulation.* National Academy Press, Washington, DC.

Newman, L.A., *et al.* 1998. Phytoremediation of Organic Contaminants: a Review of Phytoremediation Research at the University of Washington. *J. Soil Contam.* 7: 531-542.

Newman, L.A., M.P. Gordon, P. Heilman, E. Lory, K. Miller, and S.E. Strand. 1999. MTBE at California Naval Site. *Soil and Groundwater Cleanup* February/March 1999: 42-45.

Newman, L.A., S.E. Strand, N. Choe, J. Duffy, G. Kuan, M. Rusxaj, R.B. Shurtleff, J. Wilmoth, P. Heilman, and M. Gordon. 1997. Uptake and biotransformation of trichloroethylene by hybrid poplars. *Environ. Sci. Technol.* 31: 1062-1067.

Ornstein, R.L. 1991. *Rational Redesign of Biodegradative Enzymes for Enhanced Bioremediation: Overview and Status Report for Cytochrome P450.* National Research and Development Conference on the Control of Hazardous Materials '91, Hazardous Materials Control Research Institute.

Persans, M.W., K. Nieman, and D.E. Salt. 2001. Functional activity and role of cation-efflux family members in Ni hyperaccumulation in *Thlaspi goesingense. Proc. Natl. Acad. Sci. USA* 98: 9995-10000.

Pieper, D.H. and W. Reineke. 2000. Engineering bacteria for bioremediation. *Curr. Opin. Biotechnol.* 11: 262-270.

Pilon, M., J.D. Owen, G.F. Garifullina, T. Kurihara, H. Mihara, N. Esaki, and E.A. Pilon-Smits. 2003. Enhanced selenium tolerance and accumulation in transgenic *Arabidopsis* expressing a mouse selenocysteine lyase. *Plant Physiol.* 131: 1250-1257.

Pilon-Smits, E., and M. Pilon. 2002. Phytoremediation of metals using transgenic plants. *Crit. Rev. Plant Sci.* 21: 439-456.

Pilon-Smits, E.A., S. Hwang, C. Mel Lytle, Y. Zhu, J.C. Tai, R.C. Bravo, Y. Chen, T. Leustek, and N. Terry. 1999. Overexpression of ATP sulfurylase in indian mustard leads to increased selenate uptake, reduction, and tolerance. *Plant Physiol.* 119: 123-132.

Ramos, J.L., A. Wasserfallen, K. Rose, and K.N. Timmis. 1987. Redesigning metabolic routes: manipulation of TOL plasmid pathway for catabolism of alkylbenzoates. *Science* 235: 593-596.

Raskin, I. 1996. Plant genetic engineering may help with environmental cleanup. *Proc. Natl. Acad. Sci. USA* 93: 3164-3166.

Raskin, I.I., R.D. Smith, and D.E. Salt. 1997. Phytoremediation of metals: using plants to remove pollutants from the environment. *Curr. Opin. Biotechnol.* 8: 221-226.

Reeves, R.D., and A. Baker 2000. Metal-accumulating plants. Pages 193-230 in *Phytoremediation of Toxic Metals—Using Plants to Clean Up the Environment,* and I. Raskin and B.D. Ensley, eds., Wiley, New York.

Ripp, S., D.E. Nivens, Y. Ahn, C. Werner, J. Jarre, J.P. Easter, C.D. Cox, R.S. Burlage, and G.S. Sayler. 2000. Controlled field release of a bioluminescent genetically engineered microorganism for bioremediation process monitoring and control. *Environ. Sci. Technol.* 34: 846-853.

Rojo, F., D.H. Pieper, K.H. Engesser, H.J. Knackmuss, and K.N. Timmis. 1987. Assemblage of ortho cleavage route for simultaneous degradation of chloro- and methylaromatics. *Science* 238: 1395-1398.

Rugh, C.L., J.F. Senecoff, R.B. Meagher, and S.A. Merkle. 1998. Development of transgenic yellow poplar for mercury phytoremediation. *Nat. Biotechnol.* 16: 925-928.

Rugh, C.L., D. Wilde, N.M. Stack, D.M. Thompson, A.O. Summers, and R.B. Meagher. 1996. Mercuric ion reduction and resistance in transgenic *Arabidopsis thaliana* plants expressing a modified bacterial *merA gene. Proc. Natl. Acad. Sci. USA* 93: 3182-3187.

Ruiz, O.N., H.S. Hussein, N. Terry, and H. Daniell. 2003. Phytoremediation of organomercurial compounds via chloroplast genetic engineering. *Plant Physiol.* 132: 1344-1352.

Salanitro, J.P., L.A. Diaz, M.P. Williams, and H.L. Wisniewski. 1994. Isolation of a bacterial culture that degrades methyl t-butyl ether. *Appl. Environ. Microbiol.* 60: 2593-2596.

Salt, D., and U. Kramer, 2000. Mechanisms of metal hyperaccumulation in plants. Pages 231-246 in *Phytoremediation of Toxic Metals - Using Plants to Clean Up the Environment,* I. Raskin, and B.D. Ensley, eds., Wiley, New York,

Sayler, G.S. and S. Ripp. 2000. Field applications of genetically engineered microorganisms for bioremediation processes. *Curr. Opin. Biotechnol.* 11: 286-289.

Schnoor, J.L., L.A. Licht, S.C. McCutcheon, N.L. Wolfe, and L.H. Carreira. 1995. Phytoremediation of organic and nutrient contaminants. *Env. Sci. Technol.* 29: 318A-323A.

Seth-Smith, H.M., S.J. Rosser, A. Basran, E.R. Travis, E.R. Dabbs, S. Nicklin, and N.C. Bruce. 2002. Cloning, sequencing, and characterization of the hexahydro-1,3,5-Trinitro-1,3,5-triazine degradation gene cluster from *Rhodococcus rhodochrous. Appl. Environ. Microbiol.* 68: 4764-4771.

Shang, T.Q., L.A. Newman, and M.P. Gordon 2003. Fate of tricholorethylene in terrestrial plants. Pages 529-560 in *Phytoremediation: Transformation and Control of Contaminants,* S.C. McCutcheon and J.L. Schnoor, eds., John Wiley and Sons, New York.

Shields, M.S., S.O. Montgomery, P.J. Chapman, S.M. Cuskey, and P.H. Pritcchard. 1989. Novel pathway of toluene catabolism in the trichloro-ethylene-degrading bacterium G4. *Appl. Environ. Microbiol.* 55: 1624-1629.

Song, W.Y., E.J. Sohn, E. Martinoia, Y.J. Lee, Y.Y. Yang, M. Jasinski, C. Forestier, I. Hwang, and Y. Lee. 2003. Engineering tolerance and accumulation of lead and cadmium in transgenic plants. *Nat. Biotechnol.* 21: 914-919.

Strong, L.C., C. Rosendahl, G. Johnson, M.J. Sadowsky, and L.P. Wackett. 2002. *Arthrobacter aurescens* TC1 metabolizes diverse s-triazine ring compounds. *Appl. Environ. Microbiol.* 68: 5973-5980.

Subramanian, M., and J.V. Shanks. 2003. Role of plants in the transformation of explosives. Pages 389-408 in *Phytoremediation: Transformation and Control of Contaminants*, S.C. McCutcheon and J.L. Schnoor, eds., John Wiley and Sons, New York.

Summers, A.O. 1986. Organization, expression, and evolution of genes for mercury resistance. *Annu. Rev. Microbiol.* 40: 607-634.

Sun, B., B.M. Griffin, H.L. Ayala-del-Rio, S.A. Hashsham, and J.M. Tiedje. 2002. Microbial dehalorespiration with 1,1,1-trichloroethane. *Science* 298: 1023-1025.

Terry, N. 2001. Enhancing the phytoremediation of toxic trace elements through genetic engienering, http://www.calacademy.org/education/bioforum/bioforum2001-2/geneticengineering/terrysummary.htm.

Terry, N. 2003. Phytoextraction and phytovolatilization of selenium from se-contaminated environments, http://www.clu-in.org/studio/2003phyto/agenda.cfm.

Thomas, J.C., E.C. Davies, F.K. Malick, C. Endreszl, C.R. Williams, M. Abbas, S. Petrella, K. Swisher, M. Perron, R. Edwards, P. Ostenkowski, N. Urbanczyk, W.N. Wiesend, and K.S. Murray, 2003. Yeast metallothionein in transgenic tobacco promotes copper uptake from contaminated soils. *Biotechnol. Prog.* 19: 273-280.

Timmis, K.N., and D.H. Pieper. 1999. Bacteria designed for bioremediation. *Trends Biotechnol.* 17: 200-204.

U.S. DOE (Department of Energy) 1994. Summary report of a workshop on phytoremediation research needs, DOE/EM-0224.

U.S. DOE (Department of Energy) undated. Bioremediation Research Needs http://www.er.doe.gov/production/ober/nabir/needs.html.

U.S. EPA (Environmental Protection Agency) 1997. Cleaning up the Nation's Waste sites: Markets and Technology Trends, 1996 Edition, EPA 542/R/96/005.

U.S. EPA (Environmental Protection Agency) 1999. Treatment technologies for site cleanup: Annual Status Report (Ninth Edition), EPA-542-R99-001.

U.S. EPA (Environmental Protection Agency) 2000a. An analysis of barriers to innovative treatment technologies: summary of existing studies and current initiatives, EPA 542-B-00-003.

U.S. EPA (Environmental Protection Agency) 2000b. Introduction to Phyto-remediation, EPA/600/R-99/107.

van der Lelie, D., J.P. Schwitzguebel, D.J. Glass, J. Vangronsveld, and A. Baker. 2001. Assessing phytoremediation's progress in the United States and Europe. *Environ. Sci. Technol.* 35: 446A-452A.

Van Huysen, T., S. Abdel-Ghany, K.L. Hale, D. LeDuc, N. Terry, and E.A. Pilon-Smits. 2003. Overexpression of cystathionine-gamma-synthase enhances selenium volatilization in *Brassica juncea*. *Planta* 218: 71-78.

Wackett, L.P. 1994. Dehalogenation in environmental biotechnology. *Curr. Opin. Biotechnol.* 5: 260-265.

Wangeline, A.L., J.L. Burkhead, K.L. Hale, S.D. Lindblom, N. Terry, M. Pilon, and E.A. Pilon-Smits. 2004. Overexpression of ATP sulfurylase in Indian mustard: effects on tolerance and accumulation of twelve metals. *J. Environ. Qual.* 33: 54-60.

White, O., J.A. Eisen, J.F. Heidelberg, E.K. Hickey, J.D. Peterson, R.J. Dodson, D.H. Haft, M.L. Gwinn, W.C. Nelson, D L. Richardson, K.S. Moffat, H. Qin, L. Jiang, W. Pamphile, M. Crosby, M. Shen, J.J. Vamathevan, P. Lam, L. McDonald, T. Utterback, C. Zalewski, K.S. Makarova, L. Aravind, M.J. Daly, C.M. Fraser *et al.* 1999. Genome sequence of the radioresistant bacterium *Deinococcus radiodurans* R1. *Science* 286: 1571-1577.

Winter, R.B., K.-M. Yen, and B.Ensley. 1989. Efficient degradation of trichloroethylene by a recombinant *Escherichia coli*. *Bio/Technol.* 7: 282-285.

Wolfe, N.L., and C.F. Hoehamer 2003. Enzymes used by plants and microorganisms to detoxify organic compounds. Pages 159-188 in *Phytoremediation: Transformation and Control of Contaminants.* S.C. McCutcheon and J.L. Schnoor, eds., John Wiley and Sons, New York.

Xiang, C., B.L. Werner, E.M. Christensen, and D.J. Oliver. 2001. The biological functions of glutathione revisited in *Arabidopsis* transgenic plants with altered glutathione levels. *Plant Physiol.* 126: 564-574.

Yeargan, R., I.B. Maiti, M.T. Nielsen, A.G. Hunt, and G.J. Wagner. 1992. Tissue partitioning of cadmium in transgenic tobacco seedlings and field grown plants expressing the mouse metallothionein I gene. *Transgenic Res.* 1: 261-267.

Zhu, Y., E.A.H. Pilon-Smits, L. Jouanin, and N. Terry. 1999a. Overexpression of glutathione synthetase in *Brassica juncea* enhances cadmium tolerance and accumulation. *Plant Physiol.* 119: 73-79.

Zhu, Y. L., E.A. Pilon-Smits, A.S. Tarun, S.U. Weber, L. Jouanin, and N. Terry. 1999b. Cadmium tolerance and accumulation in Indian mustard is enhanced by overexpressing gamma-glutamylcysteine synthetase. *Plant Physiol.* 121: 1169-1178.

Zylstra, G.J., L.P. Wackett, and D.T. Gibson. 1989. Trichloroethylene degradation by *Escherichia coli* containing the cloned *Pseudomonas putida* F1 toluene dioxygenase genes. *Appl. Environ. Microbiol.* 55: 3162-3166.

Bioremediation of Heavy Metals Using Microorganisms

Pierre Le Cloirec and Yves Andrès

Ecole des Mines de Nantes, GEPEA UMR CNRS 6144,
BP 20722, 4 rue Alfred Kastler, 44307 Nantes cedex 03, France

Introduction

Due to natural sources or human activities, heavy metal ions are found in surface water, wastewater, waste and soils. Attention is being given to the potential health hazard presented by heavy metals in the environment. Various industries use heavy metals due to their technological importance and applications: metal processing, electroplating, electronics and a wide range of chemical processing industries. Table 1 presents some sources in water, waste and soil and their effects on human health.

However, in order to control heavy metal levels before they are released into the environment, the treatment of the contaminated wastewaters is of great importance since heavy metal ions accumulate in living species with a permanent toxic and carcinogenic effect (Sitting 1981, Liu *et al.* 1997, Manahan 1997). The most common treatment processes used include chemical precipitation, oxidation/reduction, ion exchange, membrane technologies, especially reverse osmosis, and solvent extraction. Each process presents advantages, disadvantages and ranges of applications depending on the metal ion, initial concentration, flow rate or raw water quality. In the past few years, a great deal of research has been undertaken to develop alternative and economical processes. Agricultural by-products, such as biosorbents for heavy metals, also offer a potential alternative to existing techniques and are a subject of extensive study. Biosorbents, including not only microorganisms (bacteria, yeast and fungi) but also soybean hulls, peanut hulls, almond hulls, cottonseed hulls and corncobs, have been shown to remove heavy metal ions (Brown *et al.* 2000, Marshall *et al.* 2000, Wartelle and Marshall 2000, Gardea-Torresdey *et al.* 2001, Reddad *et al.* 2002a, b).

Biosorption or bioaccumulation onto microorganisms or biofilm has

Table 1. Occurrence and significance of some heavy metal ions in the environment (adapted from Manahan 1997).

Heavy metal	Sources	Effect
Beryllium	Coal, nuclear power and space industries	Acute and chronic toxicity, possibly carcinogenic
Cadmium	Industrial discharge, mining waste, metal plating, water pipes	Replaces zinc biochemically, causes high blood pressure and kidney damage, destroys testicular tissue and red blood cells, toxic to aquatic biota
Chromium	Metal plating, cooling-tower water additive (chromate) normally found as Cr(VI) in water (soluble species)	Essential trace element (glucose tolerance factor, possibly carcinogenic as Cr(VI))
Copper	Metal plating, industrial and domestic waste, mining mineral leaching	Essential trace element, not very toxic for animals, toxic for plants and algae at moderate levels
Iron	Corroded metal, industrial waste, natural minerals	Essential nutrient (component of hemoglobin), not very toxic, damages materials
Lead	Industrial sources, mining, plumbing, fuels (coals), batteries	Toxicity (anemia, kidney disease, nervous system), wildlife destruction
Manganese	Mining, industrial waste, acid mine drainage, microbial action on manganese mineral at low pE	Relatively non-toxic to animals, toxic to plants at higher levels, stains materials
Mercury	Industrial waste, mining coal	Acute and chronic toxicity
Molybdenum	Industrial waste, natural sources, cooling-tower water additive	Toxic to animals, essential for plants
Silver	Geological sources, mining, electroplating, film-waste processing wastes	Causes blue-gray discoloration of skin, mucous membranes, eyes
Zinc	Industrial waste, metal	Essential element in many metalloenzymes, aids wound healing, toxic to plants at higher levels; major component of sewage sludge, limiting land disposal of sludge

emerged as a potential and cost-effective option for heavy metal removal from aqueous solution, polluted soil or solid waste after aqueous extraction (Eccles 1999). From the literature, Veglio and Beolchini (1997) have presented a large number of the metal ion adsorption capacities of several microorganisms. The use of algae was reviewed some time ago by Volesky (1990). Some pilot plant studies have been carried out to investigate the potential of microorganisms to remove metal ions from liquid and, in the past 20 years, a few systems have been commercialized. However, more effort has to be made in the application of bacteria and/or biofilms, both low cost adsorbents, in metal removal processes.

The objective of this chapter is to present the remediation of metal ions by microorganisms. First, some mechanisms of interactions between ions and microorganisms are discussed. Then, the use of these kinds of adsorbent to remove heavy metals in water and wastewater in mixed batch contactors or in fixed packed beds in continuous flow operations is described. Soil and solid waste remediation is also considered. For each paragraph, multi-scale approaches, integrating the mechanisms, the design of the adsorbers and operating conditions are given and illustrated by some examples (Le Cloirec 2002).

Mechanisms of microbial interaction processes

Microorganisms (bacteria, yeast and fungi) may have a direct action on metal mobility through biosorption, bioaccumulation or resistance/ detoxification processes (Fig. 1). In addition, they may influence the environment by producing compounds from metabolic reactions such as acids or chelating agents such as siderophores. In this part, some examples of microbial interaction mechanisms are presented including biosorption, metabolism by-product complexation and indirect metal use for microbial life, bioaccumulation and resistance/detoxification systems. Indirect influences of microorganisms on the speciation of heavy metals and/or radionuclides are also presented.

Biosorption

Biosorption is a physico-chemical mechanism including sorption, surface complexation, ion exchange and entrapment, which is relevant for living and dead biomass as well as derived products. Biosorption can be considered as the first step in the microorganism-metal interaction. It encompasses the uptake of metals by the whole biomass (living or dead) through physico-chemical mechanisms such as sorption, surface complexation, ion exchange or surface precipitation. These mechanisms take place on the cell wall (Shumate and Strandberg 1985) which is a rigid

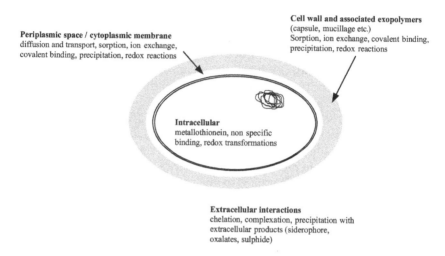

Periplasmic space / cytoplasmic membrane
diffusion and transport, sorption, ion exchange,
covalent binding, precipitation, redox reactions

Cell wall and associated exopolymers
(capsule, mucillage etc.)
Sorption, ion exchange, covalent binding,
precipitation, redox reactions

Intracellular
metallothionein, non specific
binding, redox transformations

Extracellular interactions
chelation, complexation, precipitation with
extracellular products (siderophore,
oxalates, sulphide)

Figure 1. Microorganisms / metal relationships (adapted from Gadd and White 1993).

layer around the cell (Fig. 1) and they have fast kinetics. One dominant factor affecting the capacity of the microbial cell wall to "trap" the metal ion is the composition of this outer layer. For a better understanding of the biosorption mechanisms, the cell wall structure of microorganisms will be briefly and simply presented.

Cell wall structure

In the prokaryotic world (bacteria), the wall is composed by a peptidoglycan structure bound to a techoic acid (Gram-positive bacteria) or to a lipopolysaccharide (Gram-negative bacteria). These two groups are differentiated using a coloration reaction. The cell wall of Gram-positive bacteria is 50 to 150 nm thick and mainly consists of 40 to 90 % peptidoglycan. It is a rigid, porous, amorphous material, made of linear chains of the disaccharide N-acetylglucosamine-b-1,4-N-acetylmuramic acid. The cell wall of Gram-negative bacteria appears to be somewhat thinner, usually 30 to 80 nm thick, and only 10 % of the material is made up of peptidoglycan (Remacle 1990). The cell wall composition of archaebacteria differs from the eubacteria by the lack of muramic acid and peptidoglycan.

The cells of many bacteria groups are often covered by an additional surface layer non-covalently associated with the cell wall. This structure, called the S-layer, is usually composed of regular arrays of homogeneous polypeptides and sometimes of carbohydrates.

In the eukaryotic world (fungi and yeast), the cell wall is made up of various polysaccharides arranged in a multilaminate microfibrillar structure. Ultrastructural studies reveal two phases: (i) an outer layer constituted of glucans, mannans or galactans and (ii) an inner microfibrillar layer. The crystalline properties of this latter are given by the parallel arrangement of chitin or sometimes of cellulose chains or, in some yeasts, of non-cellulosic glucan chains. There is a continuous transition between these two layers (Remacle 1990).

Cell wall characteristics and biosorption

A large variety of chemical microenvironments is present on the bacterial surface (Table 2). These include phosphate, carboxyl, hydroxyl and amino functional groups, among others. Various methods have been investigated to identify the bacterial surface functional groups involved in metal uptake. A first approach consisted of performing metal binding studies on extracted cell wall polymers, such as peptidoglycan and teichoic acid, to determine the types of cell wall components responsible for metal binding (Beveridge and Fyfe 1985). In addition, selective chemical modifications of the various functional groups were carried out to assess their contribution to the metal uptake (Beveridge and Murray 1980, Doyle *et al.* 1980). The major inconvenience in the use of this kind of technique is the rather heavy experimental protocol, which does not allow the study of intact cells for adsorption investigations. The potentiometric titration technique provides a simple and efficient method to measure and determine the different functional groups available to bind metallic ions. Consequently, the use of this method is interesting for the surface characterization of algae, fungi (Deneux-Mustin *et al.* 1994, Schiewer and Wong 2000), and bacteria (Van

Table 2. Functional groups of microbial complexing compounds (Birch and Bachofen 1990).

	Basic		Acidic
- NH_2	amino	- CO_2H	carboxylic
= NH	imino	- SO_3H	sulphonic
- N =	tertiary acyclic or heterocyclic nitrogen	- $PO(OH)_2$	phosphonic
= CO	carbonyl	- OH	enolic, phenolic
- O -	ether	= N - OH	oxime
- OH	alcohol	- SH	thioenolic and thiophenolic
- S -	thioether		
- PR_2	substituted phosphine		

der Wal *et al.* 1997, Texier *et al.* 1999). For example, Fein *et al.* (1997) have characterized the acid/base properties of the cell wall of *Bacillus subtilis* and have shown three distinct types of surface organic acid functional groups with pKa of 4.82, 6.9 and 9.4. These various values are generally attributed to carboxyl, phosphate and hydroxyl moieties respectively. Furthermore, various spectroscopic methods, including IR spectroscopy, XANES spectroscopy (X-ray absorption near-edge structure), EXAFS spectroscopy (extended X-ray absorption fine structure) and NMR spectroscopy amongst many others, can provide information about the chemical environment of the sorbed metallic ions on biological material.

Until recently, the emphasis has been placed on the use of such spectroscopic methods to characterize the surfaces of algae (Kiefer *et al.* 1997), bacteria (Schweiger 1991), fungi (Sarret *et al.* 1998) and plant cells (Tiemann *et al.* 1999, Salt *et al.* 1999). Drake *et al.* (1997), Texier *et al.* (2000) and Markai *et al.* (2003) have investigated the binding of europium to a biomaterial derived respectively from the plant *Datura innoxia*, from *Pseudomonas aeruginosa* and from *Bacillus subtilis*. They characterized the functional groups answerable for metal ion uptake with the help of laser-induced spectrofluorometry. A simultaneous determination of emission wavelength and fluorescence lifetime provided two-dimensional information about fluorescing ions. These spectroscopic approaches have confirmed many times that the fixation occurs with the free functional groups present in the cell wall layer of the microorganisms. For Gram-negative bacteria, the functional groups are, for example, present in the lipopolysaccharide of the outer layer and in the peptidoglycan and for Gram-positive bacteria in the techoic acid. Mullen *et al.* (1989) indicated, after electronic microscopy studies, that lanthanum was accumulated at the surface of *P. aeruginosa* inducing crystalline precipitation.

Biosorption capacities

Biosorption capacities of microorganisms for metal ions generally depend on the metal concentration, the pH of the solution, the contact time, the ionic strength and the presence of competitive ions in the solution. Significant differences were observed in the uptake capacities of gadolinium ions by the various microorganisms used and no general relationship was applicable to all microbial species. These differences could be related to the nature, the structure and the composition of the cell wall layers and the specific surface developed by the sorbents in suspension. Morley and Gadd (1995) concluded for fungal biomass that the different cell wall polymers have various functional groups and differing charge distributions and therefore different metal-binding capacities and affinities. Schiewer and

Wong (2000) described a decrease in the biosorption capacities in relation to the algae species. Furthermore, the physiological stage of the bacteria seems to be important in the case of *Mycobacterium smegmatis* (Andrès *et al.* 2000) This observation could be explained by the fact that cell starvation leads to a modification in the composition of the cell wall layer. Penumarti and Khuller (1983) measured effectively an increase in the total amount of mannosides with the age of culture from *Mycobacterium smegmatis*. These observations could be correlated with the variation in the composition of the macromolecular compounds or in their quantity at the microbial surface and with the growth conditions. Daughney *et al.* (2001) have shown that the number of functional groups present at the cell surface, their pKa values and, related to these, the electronegativity of the cells wall could be changed according to the physiological state of the bacteria. Various authors (Volesky 1994, Andrès *et al.* 2000, Goyal *et al.* 2003) have shown that biosorption on bacterial, fungal and yeast biomass is a function of the growth medium composition and the culture age of the cells. McEldowney and Fletcher (1986) concluded that the macromolecular compounds of bacterial surfaces varied in quantity and in composition with the growth conditions and the growth rate.

Complexing substances

Bacteria and fungi can produce many complexing or chelating substances. The mobilization capacities of a bacterial and fungal iron-chelating agent, for plutonium and uranium, have been studied by Brainard *et al.* (1992). They used two siderophores: the first one produced by *Escherichia coli* (enterochelin) with catechol functions and the second one by *Streptomyces pilosus* (desferrioxamin B), with hydroxamate groups. They showed that these molecules could solubilize plutonium and uranium oxides. A review was published by Fogarty and Tobin (1996) about the complexation between fungal melanins and metal ions (Ni, Cu, Zn, Cd, Pb). These compounds are dark brown or black pigments located in the cell walls. Fungi are also able to produce small organic acids (gluconic, oxalic) which can react with metallic oxides and lead to their solubilization.

Indirect influences

Two main indirect mechanisms of interaction are related to the change in pH or redox conditions of the medium. In the presence of air, sulfur-oxidizing bacteria (SOB) such as *Thiobacillus* sp. use sulfide minerals (FeS_2, Cu_2S, PbS) for their growth.

   Under reducing conditions, sulfate-reducing bacteria (SRB) such as *Desulfovibrio* sp. are able to reduce sulfate to sulfide, which reacts with metal ions to precipitate highly insoluble sulfides (Ehrlich 1996) as shown by their solubility potential: NiS, $3 \cdot 10^{-21}$; Cu_2S, $2.5 \cdot 10^{-50}$; CoS, $7 \cdot 10^{-23}$; PbS, 10^{-29}; HgS, $3 \cdot 10^{-53}$). In addition, dissolved sulfide ions can directly reduce metals including U(VI), Cr(VI) and Tc(VII). Reduction of sulfate requires organic carbon, like natural organic matter or more simple compounds such as lactate, ethanol and acetate or H_2 as an electron donor.

Indirect metal use for microbial life

Some microorganisms are able to grow under anaerobic conditions by coupling the oxidation of simple organic substances with the reduction of metallic compounds. For example, *Shewanella putrefaciens* could reduce uranyl ions U(VI) to U(IV) in the presence of hydrogen (Lovley *et al.* 1991). Many metal ions (U, Cr, Fe, Mn, Mo, Hg, Co, V) and metalloids (As, Se) can be reduced by a variety of metal- (for example *Geobacter metallireducens*) and sulfate-reducing bacteria (for example *Desulfovibrio*) (Lovley 1993, 1995). Pure or mixed cultures of these bacteria can couple the oxidation of organic compounds (lactate, formate, ethanol) or H_2 and lead to the reduction of the metal. The reduced U precipitates as the highly insoluble mineral uranite (UO_2). Abdelouas *et al.* (1999) showed that subsurface sulfate-reducing bacteria from a mill-tailing site were able to reduce U(VI), which precipitated at the periphery of the cell. Enzymatic reduction of U was shown by Lovley *et al.* (1993). The authors showed that the cytochrome c_3 enzyme, which is located in the soluble fraction of the periplasmic region of *Desulfovibrio vulgaris*, reduced U(VI) in the presence of excess hydrogenase and H_2. In natural reducing environments, metal- and sulfate-reducing bacteria are expected to play a significant role in uranium immobilization.

Geochemical and microbiological evidence suggests that the reduction of Fe(III) may have been an early form of respiration on earth. Moreover, recent studies have shown that some xenobiotic compounds could be degraded under anaerobic conditions by Fe(III)- and Mn(IV)-reducing microorganisms. The metal is the electron acceptor and the organic substances, like toluene, phenol or benzoate, are used as electron donors (Lloyd 2003). A wide range of facultative anaerobes, including *Escherichia coli* and *Pseudomonas*, reduce Cr(VI) to Cr(III) for their growth.

In many cases, the metal reduction enzyme is located in the periplasmic space, in the outer membrane or at the cell wall surface.

Bioaccumulation

Bioaccumulation is a possible interaction between microorganisms and metal ions in relation to metabolic pathways; in this case, living cells are required. Metal ions are involved in all aspects of microbial growth. Many metals are essential, whereas others have no known essential biological function. Accumulation of radionuclides through the pathways of their stable isotopes or of chemically homologous elements can be considered as bioaccumulation. It is well known that cesium ions are accumulated by the potassium channel (Avery 1995).

Resistance and detoxification mechanisms

In a polluted environment, microorganisms develop a great diversity of resistance and detoxification systems. The most important mechanism is the transformation of toxic species into inactive forms by reduction, methylation or precipitation. For example, the predominant redox states of selenium in the natural environment are Se(VI) (selenate, SeO_4^{2-}) and Se(IV) (selenite, SeO_3^{2-}), which are reduced to elemental selenium Se(0) by telluric bacteria (*Clostridium, Citobacter, Flavobacterium, Pseudomonas*) or by bacteria in anoxic aquatic sediment (Lovley 1993). The oxianion species are potentially electron acceptors for the microbial metabolism. Another transformation route is the biomethylation of inorganic selenium compounds in dimethylselenide [$(CH_3)_2Se$]. The methylated species are generally volatile compounds in environmental conditions (Gadd 1993) and have a great influence on heavy metal migration.

Heavy metal removal in water and wastewater

Free or immobilized microorganisms are used to remove heavy metal ions. Among the different types of process configurations, batch reactors or fixed bed reactors have been widely investigated (Atkinson *et al.* 1998). In this section, mechanisms and processes to control metal ions in aqueous emissions will be developed.

Metal ion removal in stirred reactors

Some technologies

Stirred reactors are simple systems to transfer metal ions present in wastewater onto bacteria, biosorbent or biofilm coated particles (Levenspiel 1979). Figure 2 presents some technologies useful for this kind of treatment. The wastewater is put in contact with biosorbent in a stirred reactor until an equilibrium between the concentration in the liquid phase and the concentration onto the solid adsorbent is reached. After the mass transfer, the liquid and the solid are separated using classical processes like a settling tank, a clarifier or membrane microfiltration. Veglio *et al.* (2003) propose a standardization of heavy metal biosorption using a stirred batch

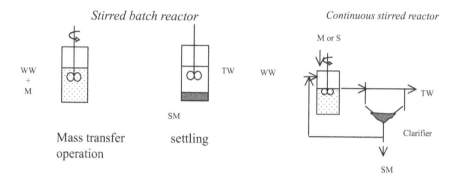

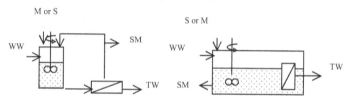

Figure 2. Some continuously stirred processes for metal ion removal in wastewater by biofilm particles.
WW: Wastewater   TW: Treated Water
M: Microorganisms   S: Substrate
SM: Saturated Microorganisms

reactor methodology. Pagnanelli *et al.* (2003a) consider mechanisms of heavy metal biosorption using batch and membrane reactor systems.

Operating conditions and metal removal

In this part, the various conditions affecting the adsorption of a solute onto a surface are briefly presented and discussed :
  – the specific surface area of microorganisms and the porous volume of biofilm are important characteristics and the adsorption capacities of a metal ion are directly proportional to them,
  – pore diameters of the biofilm or the bacterial aggregate control the accessibility of metal ions as a function of their size,
  – the metal ion radius or the solvated metal ion size are important factors affecting the diffusion and adsorption capacities,
  – in a multi-component solution, the species compete for available active sites and induce a reduction in the amount adsorbed for a given solute,

- pH is extremely important for metal ion species present in the aqueous solution and for the overall microorganism or biofilm surface charge,
- rinsing temperature has an influence on the kinetics due to an increase in diffusion coefficients,
- ionic force affects the adsorption. Investigators have shown that the other cations and anions in the solution compete with active sites in the bacteria walls (Kratochvil and Volesky 2000).

Kinetics - Equilibria - Adsorption capacities

Consider a volume of solution loaded with a metal ion, which is in contact with a mass of bacteria or a biofilm coated on a particle. The system is continuously stirred for a time. Assuming there is no chemical or biological (constant mass of bacteria) reaction but only a mass transfer from the liquid phase to the solid surface, the mass balance can be written:

$$m(q_t - q_0) = V(C_0 - C_t) \tag{1}$$

where

m : mass of adsorbent (g)
q_t : concentration of the solute on the solid at time t (mg g^{-1})
q_0 : concentration of the solute on the solid at t = 0 (mg g^{-1})
     For a virgin adsorbent $q_0 = 0$
V : volume of the solution (L)
C_0 : initial concentration in the solution (mg L^{-1})
C_t : concentration at time t in the solution (mg L^{-1})

The metal ion concentration is analyzed as a function of time. A kinetic curve is obtained for the cation being removed from the solution. From the previous data and the mass balance equation, the adsorption capacity is found as a function of time. The Adams Bohart Thomas theory assumes that the adsorption is an equilibrated reaction between a solute (A) and a

surface (σ) following the equation: $A + \sigma \underset{k1}{\overset{k2}{\longrightarrow}} A\sigma$

and proposes a relation to model the evolution of the amount adsorbed:

$$\frac{dq}{dt} = k_1 C(q_m - q) - k_2 q \tag{2}$$

where

k_1 : adsorption kinetic coefficient (L mg^{-1}h^{-1})
k_2 : desorption kinetic coefficient (h^{-1})
q_m : maximal adsorption capacity (mg g^{-1})

From this overall equation, the initial velocity is extracted.
When $t \to 0$ $\quad C \to C_0$ and $\quad q \to 0$

then, the previous kinetic relation (2) becomes:

$$\left(\frac{dq}{dt}\right)_{t\to0} = k_1 C_0 q_m \tag{3}$$

i.e. a straight line equation. Brasquet and Le Cloirec (1997) proposed a normalized initial velocity coefficient $\gamma = \left(\dfrac{dq}{dt}\right)_{t\to0} \dfrac{V}{mC_0}$

An example is given in Figure 3. In this case, the adsorption is very quick and the equilibrium is reached after 1 or 2 hours contact time.

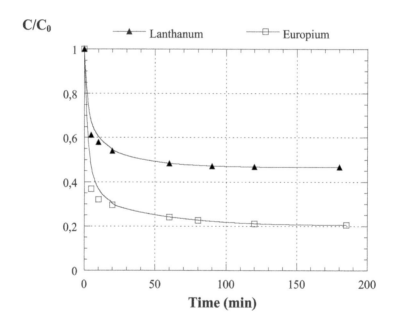

Figure 3. Adsorption kinetic curves of lanthanum onto dry *Mycobacterium smegmatis* biomass (C_0 = 0.05 mM; initial mass: 1.25 g dried at 37°C, V = 100 mL; stirring velocity = 500 rpm; T = 20 ± 5°C).

Langmuir equation : Another specific zone of the kinetic curve is when $\tau \to \infty$

$$\text{then } \frac{dq}{dt} = 0 \quad C \to C_e \quad \text{and} \quad \theta \to \theta_\varepsilon$$

Equation (2) becomes:

$$k_1 C_e (q_m - q_e) = k_2 q_e \qquad (4)$$

or

$$q_e = \frac{b q_m C_e}{1 + b C_e} \qquad (5)$$

with $b = \dfrac{k_1}{k_2}$ the equilibrium constant and $\theta = \dfrac{q_e}{q_m}$ the fraction of the surface covered. This relation is applied to adsorption on a completely homogeneous surface with negligible interactions between adsorbed molecules. Pagnanelli et al. (2003b) proposed an empirical model based on the Langmuir equation and applied it to the adsorption equilibrium of lead, copper, zinc and cadmium onto Sphaerotilus natans.

From an experimental data set (Ce, qe), the constant b and q_m are determined by plotting $1/q_e$ vs. $1/C_e$. The straight-line slope is $1/bq_m$ and the intercept is $1/q_m$. Examples for different bacteria and several heavy metals are given in Tables 3 and 4.

Freundlich equation

Tien (1994) mentions various expressions that can be used to describe adsorption isotherms. An empirical relation, the so-called Freundlich isotherm equation, has been proposed in order to fit the data on adsorption:

$$q_e = K_F C_e^{1/n} \qquad (6)$$

where K_F and $1/n$ constants depend on the solute-adsorbent couple and temperature. When $1/n < 1$, the adsorption is favorable. On the contrary, $1/n > 1$ shows an unfavorable adsorption. This relation could correspond to an exponential distribution of adsorption heat. However, the form of the equation shows that there is no limit for q_e as C_e increases, which is physically impossible. This means that the Freundlich equation is useful for low C_e values. The logarithms of each side of equation (6) give: $L_n(q_e) = L_n(K_F) + \dfrac{1}{n} l_n(C_e)$. With the straight line $Ln(q_e)$ vs. $Ln(C_e)$, one obtains the slope $1/n$ and the intercept $Ln(K_F)$. Table 3 gives a set of Freundlich equation parameters. When the amount adsorbed (q) is far smaller than the maximum adsorption capacity (q_m), the Freundlich equation is reduced to the Henry type equation:

$$q_e = K_F C_e \qquad (7)$$

Some examples

In order to illustrate heavy metal ion removal onto bacteria and biofilm, an example of an adsorption curve is presented in Figure 4.

The applications of the equilibrium model are proposed in Tables 3 and 4. Good adsorption capacities for several microorganisms and different metal ions can be noted. However, the results obtained are a function of operating conditions such as pH, the evolution phase of the bacteria or the initial concentration.

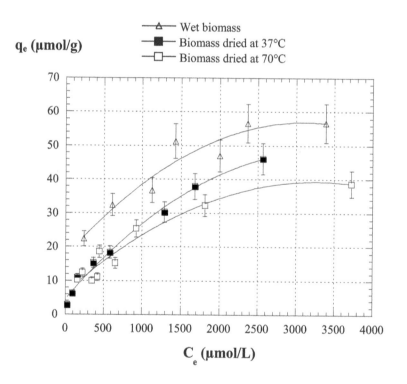

Figure 4. Adsorption isotherm curves of lanthanum onto *Mycobacterium smegmatis* ($C_0 = 0.05 - 4$ mM; initial mass: 0.25 g at 37°C, V = 20 mL; stirring velocity = 300 rpm; T = 20 ± 5°C).

Table 3. Adsorption capacity of several heavy metal ions onto some bacteria, microorganisms or a mixture of microorganisms.

Element	Bacteria	Biosorption ($\mu mol.g^{-1}$)	References
Ag$^+$	Streptomyces noursei	358	Mattuschka and Straube 1993
Au$^+$	Aspergillus niger	862	Kapoor et al. 1995
	Sargassum natans	2132	Kuyucak and Volesky 1988
Cd^{2+}	Activated sludge	325	Solaris et al. 1996
	Gram-positive bacteria	164	Gourdon et al. 1990
	Gram-negative bacteria	120	
	Alcagines sp.	89	Veglio and Beolchini 1997
	Arthrobacter gloformis	2	Scott and Palmer 1988
	Ascophyllum nodosum	1112-1735	Holan et al. 1993
	Penicillium digitatum	31	Galum et al. 1987
	Pseudomonas aeruginosa PU 21	516	Chang et al. 1997
	Saccharomyces cerevisiae	632	Volesky et al. 1993
	Sargassum natans	1023	Volesky 1992
	Streptomyces noursei	28	Mattuschka and Straube 1993
	Rhizopus arrhizus	267	Kapoor and Viraraghavan 1995
Cr(III)	Streptomyces noursei	204	Mattuschka and Straube 1993
	Halimeda opuntia	769	Volesky 1992
Cr(VI)	Activated sludge	461	Aksu et al. 1991
	Zoogloea ramigera	57	Nourbakhsh et al. 1994
	Rhizopus arrhizus	86	
	Saccharomyces cerevisiae	57	
	Chlorella vulgaris	67	
	Chlodophara crispata	57	
Co^{2+}	Arthrobacter simplex	186	Sakagushi and Nakajima 1991
	Pseudomonas saccharophilia	186	
	Aspergillus niger	41	
	Rhizopus arrhizus	49	
	Saccharomyces cerevisiae	98	Mattuschka and Straube 1993
	Streptomyces noursei	20	Kuyucak and Volesky 1988
	Aspergillus niger	1610	
	Ascophyllum nodosum	2644	
Cu^{2+}	Arthrobacter sp.	2329	Veglio and Beolchini 1997
	Chlorella vulgaris	667	Aksu et al. 1992
	Penicillium digitatum	47	Galum et al. 1987

(Contd.)

Table 3. (*cont.*)

Element	Bacteria	Biosorption ($\mu mol.g^{-1}$)	References
	Pseudomonas aeruginosa PU 21	362	Chang et al. 1997
	Pseudomonas syringae	399	Cabral 1992
	Rhizopus arrhizus	252	Kapoor et al. 1995 Tonex vpq
	Streptomyces noursei	141	Mattuschka and Straube 1993
	Zoogloea ramigera	536	Sag and Kutsal 1995
	Aurebasidium pullulans	94	Gadd and De Rome 1988
	Clasdospoium resinae	252	
	Saccharomyces cerevisiae	6 - 12	Huang et al. 1990
	Activated sludge	789	Aksu et al. 1991
Eu^{3+}	Mycobacterium smegmatis (CIP 73.26)	101	Texier et al. 1997
	Pseudomonas aeruginosa (CIP A 22)	290	Texier et al. 1997
Hg^{2+}	Pseudomonas aeruginosa PU 21	969	Chang and Hong 1994
	Rhizopus arrhizus	289	Kapoor et al. 1995
Gd^{3+}	Mycobacterium smegmatis (CIP 73.26)	110 - 190	Andrès et al. 1993
	Pseudomonas aeruginosa (CIP A 22)	322	
	Saccharomyces cerevisiae	5	Andrès et al. 2000
	Ralstonia metallidurans CH34	40 - 147	
	Bacillus subtilis	350	
La^{3+}	Mycobacterium smegmatis (CIP 73.26)	57	Texier et al. 1997
	Pseudomonas aeruginosa (CIP A 22)	397	Andrès et al. 2000
Ni^{2+}	Activated sludge	630	Aksu et al. 1991
		369	Solaris et al. 1996
	Pseudomonas syringae	102	Cabral 1992
	Streptomyces noursei	14	Kuyucak and Volesky 1988
	Arthrobacter sp.	221	Veglio and Beolchini 1997
	Rhizopus arrhizus	318	Fourest and Roux 1992
	Ascophyllum nodosum	1192	Holan and Volesky 1994
	Fucus vesiculosus	289	
Pb^{2+}	Arthrobacter sp.	628	Veglio and Beolchini 1997
	Ascophyllum nodosum	1351	Holan and Volesky 1994
	Fucus vesiculosus	1621	Holan and Volesky 1994

(Contd.)

Table 3. (*cont.*)

Element	Bacteria	Biosorption ($\mu mol.g^{-1}$)	References
	Pseudomonas aeruginosa PU 21	531	Chang et al. 1997
	Penicillium chrysogenum	559	Kapoor et al. 1995
	Penicillium digitatum	26	Galum et al. 1987
	Rhizopus arrhizus	502	Kapoor et al. 1995
	Saccharomyces cerevisiae	13	Huang et al. 1990
	Sargassum natans	1496	Volesky 1992
	Streptomyces noursei	482	Friis and Myers-Keith 1986
	Streptomyces noursei	176	Mattuschka and Straube 1993
	Zoogloea ramigera	392	Sag and Kutsal 1995
Th^{4+}	Mycobacterium smegmatis (CIP 73.26)	187	Andrès et al. 1993
	Saccharomyces cerevisiae	500	Gadd 1990
	Rhizopus arrhizus	733	Tzesos and Volesky
	Pseudomonas fluorescens	64	1982a, b
	Streptomyces niveus	146	
	Aspergillus niger	93	
	Penicillium chrysogenum	635	
UO_2^{2+}	Mycobacterium smegmatis (CIP 73.26)	170	Andrès et al. 1993
	Pseudomonas aeruginosa	630	Strandberg et al. 1981
	Saccharomyces cerevisiae	630	Strandberg et al. 1981
	Penicillium chrysogenum	336	Jilek et al. 1975
	Rhizopus arrhizus	756	Tzesos and Volesky 1982a, b
	Chlorella regularis	16.5	Sakagushi and Nakajima 1991
	Arthrobacter simplex	243	
	Aspergillus niger	122	
Yb^{3+}	Mycobacterium smegmatis (CIP 73.26)	103	Andrès et al. 1993
	Pseudomonas aeruginosa (CIP A 22)	326	Texier et al. 1999
Zn^{2+}	Activated sludge	392	Solaris et al. 1996
	Pseudomonas syringae	122	Cabral 1992
	Saccharomyces cerevisiae	260	Volesky 1994
	Rhizopus nigricans	220	
	Rhizopus arrhizus	306	Kapoor et al. 1995
	Aspergillus niger	210	Volesky 1994
	Streptomyces noursei	24	Mattuschka and Straube 1993

Table 4. Applications of equilibrium models to sorption isotherm curves for some microorganisms and heavy metal ions.

Microorganism	Metal ions and operating conditions	Model parameters	References
Pseudomonas aeruginosa	pH = 7.4 Cd(II) Zn(II)	Langmuir: q_m = 12.4 mg/g q_m = 13.7 mg/g	Savvaidis et al. 1992
P. cepacia	Cd (II) Zn (II) Ni (II)	Langmuir : q_m = 14.6 mg/g q_m = 13.1 mg/g q_m = 7.63 mg/g	
Pseudomonas aeruginosa PU 21 (Rip 64)*	Hg 10-500 mg/L 40 h pH = 6.8	Langmuir: q_m = 194.4 mg/g b = 0.055 L/mg Freundlich: $K : mg^{1-1/n}L^{1/n}g^{-1}$ K = 32.4 1/n = 0.32	Chang and Hong 1994
Pseudomonas aeruginosa ATCC 14886	pH = 4.0; 2 h Cd Cu	Freundlich: $K : \mu g^{1-1/n}L^{1/n}g^{-1}$ K= 43.7 1/n = 0.77 K = 159 1/n = 0.67	Mullen et al. 1989
Pseudomonas aeruginosa PU 21	Pb pH 5.5 Cu pH 5.0 Cd pH 6.0 Pb pH 5.5 Cu pH 5.0 Cd pH 6.0	Langmuir: (q_m in mg/g b in L/mg) q_m = 110; b = 0.3 q_m = 23; b = 0.22 q_m = 58; b = 0.8 q_m = 79; b = 0.02 q_m = 23; b = 0.06 q_m = 42; b = 20	Ledin et al. 1997
Pseudomonas putida CCUG 28920	PH = 6.4 0.01 M KCl $10^{-4} - 10^{-8}$ Cs Sr Eu Zn Cd Hg	Freundlich: $K : \mu g^{1-1/n}L^{1/n}g^{-1}$ K = 50.5; L/n = 1.01 K = 23.0; L/n = 0.76 K = 480.5; L/n = 0.83 K = 23.2; L/n = 0.74 K = 60.4; L/n = 0.76 K = 112.8; L/n = 0.73	Ledin et al. 1997

Metal ion removal in fixed beds

Some processes have been developed in fluidized beds (Coulson *et al.* 1991) or in a membrane biofilm reactor in a helical fixed bed (Wobus *et al.* 2003) but, generally, biofilm particles or biosorbents are packed in a fixed bed. Immobilization of microorganisms is carried out with a material such as calcium alginate gel, polyacrylamide gel, polyacrylonitrile polymer or a polysulfone matrix (Zouboulis *et al.* 2003, Beolchini *et al.* 2003, Arica *et al.* 2003). The water loaded with metal ions goes through the packing material in a continuous flow operation. In order to get a general approach of the process, flow (pressure drop) and efficiency (performance) have to be determined and modeled (Le Cloirec 2002, Baléo *et al.* 2003).

Pressure drop

The head loss in a filter packed with particles of biofilm can be given by different relations. Recently, Trussell and Chang (1999) reviewed the relations useful for calculating the clean bed head loss in water filters. Some semi-empirical models are presented in this section.

Darcy's law

In 1830 in Dijon (France), Darcy determined a relation between the pressure drop and operating conditions by examining the rate of water flow through beds of sand. This equation, confirmed by a number of researchers, can be written:

$$\frac{\Delta P}{H} = \frac{\mu}{B} U_0 \tag{8}$$

ΔP : pressure drop (Pa)
H : bed thickness (m)
μ : dynamic viscosity of fluid (10^{-3} Pl for water at 20 °C)
B : permeability coefficient (m^2)
U_0 : empty bed velocity (m s^{-1})

The permeability coefficient (B) values are a function of the material used in the adsorbers but typical data range between 10^{-8} to 10^{-10} m^2 (Coulson *et al.* 1991). The Darcy equation applies only to laminar flow (Re < 1, equation 9).

Carman-Kozeny-Ergun equations

In order to obtain general expressions for pressure drop, operating conditions and characteristics of the packing material, a new concept of flow through beds has been proposed by Carman and co-workers. The flow is defined by the modified Reynolds number:

$$Re = \frac{d_p U_0 \rho}{\mu} \qquad (9)$$

dp : particle diameter (m)
ρ : fluid density (kg m^{-3})
For $Re < 1$, a laminar flow, the following equation is used:

$$\frac{\Delta P}{H} = 180 \frac{(1-\varepsilon_0)^2}{\varepsilon_0^3} \frac{1}{d_p^2} \mu U_0 \qquad (10)$$

ε_0 : bed porosity (dimensionless)
   In a turbulent flow, $Re > 1$, different research workers (Carman, Kozeny and Ergun) have extended the equation with a first term due to viscous forces (skin friction) and a second term, obtained for high flow rate by dimension analysis:

$$\frac{\Delta P}{H} = 4.17 \frac{(1-\varepsilon_0)^2}{\varepsilon_0^3} S^2 \mu U_0 + 0.29 \frac{(1-\varepsilon_0)}{\varepsilon_0^3} S\rho U_0^2 \qquad (11)$$

S : external specific area (m^{-1}). For a sphere $S = 6/d_p$
   It is difficult to approach the value of d_p or S for particles coated with a biofilm. However, this equation gives good agreement ($\pm$ 10 %) between calculated and experimental data.

Comiti-Renaud model

More recently, Comiti and Renaud (1989) have proposed an equation with a similar shape to the previous relations but with values for tortuosity (τ) and dynamic surface area (a_{vd}) in contact with the fluid:

$$\frac{\Delta P}{H} = 2\tau^2 \frac{(1-\varepsilon_0)^2}{\varepsilon_0^3} a_{vd}^2 \mu U_0 + 0.0968 \frac{(1-\varepsilon_0)}{\varepsilon_0^3} \tau^3 a_{vd} \rho U_0^2 \qquad (12)$$

   This equation is very useful to compare the different particles coated with biofilm. Thus, the determination of the real surface in contact with the fluid (a_{vd}) gives important information in terms of mass transfer. For biofilm-coated spherical particles, the ratio between a_{vd} and the specific surface area (S) is found to range between 1.5 to 5. The tortuosity (τ) is close to 1.5 for packing material like sand or activated carbon grains. For a fixed bed column packed with particles and biofilm, this value ranges from 2 to 5.

Breakthrough curves

General approach

Fixed beds are generally used in water treatment. Water is applied directly to one end and forced through the packing adsorbent by gravity or pressure. The pollutants present in the water are removed by transfer onto the adsorbent. The region of the bed where the adsorption takes place is called the mass transfer zone, adsorption zone or adsorption wave. As a function of time, for a constant inlet flow, the saturated zone moves through the contactor and approaches the end of the bed. Then, the effluent concentration equals the influent concentration and no more removal occurs. This phenomenon is termed breakthrough. An illustration is given in Figure 5.

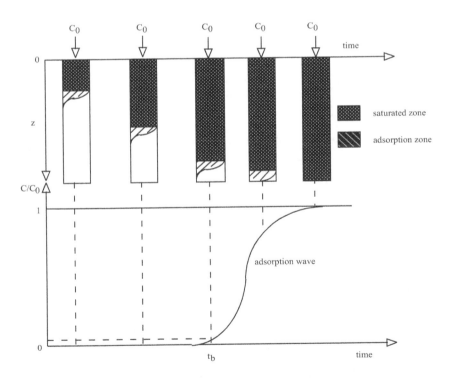

Figure 5. Schematic breakthrough curve and column saturation.

Utilization of breakthrough curves

Some information can be extracted from the breakthrough curve. The breakthrough time is determined by reporting the ratio $C_b/C_0 = 0.05$ or 0.1, i.e., when the pollutant outlet concentration is between 5 to 10 % of the outlet concentration. This percentage is a function of the desired water quality.

The total amount of solute removed ($Q_{max\ removed}$) from the feed stream upon complete saturation is given by the area above the effluent curve (C vs. t, Fig. 5), that is:

$$Q_{max\ removed} = Q\int_0^\infty ((C_0 - C)\,dt) = \varepsilon SHC_0 + (1 - \varepsilon)\, SHq_0 \qquad (13)$$

q_0:   adsorption capacity in equilibrium with C_0 (mg g^{-1})
The solute removed at $t = t_b$ is given approximately by:

$$Q_{tb\ removed} = Q(C_0 - C)t_b \qquad (14)$$

An example is given in Figure 6. Lanthanum is removed by *Pseudomonas aeruginosa* trapped in a gel (Texier *et al.* 1999, 2000, 2002). From equations 13 and 14, the data presented in Table 5 are determined and can be used to design processes.

Table 5. Design parameters obtained from breakthrough curves.

C_0 (mmol L^{-1})	t_b (min)	Q_{max} (μmol g^{-1})	Q_{tb} (mg g^{-1})
2	84	208	23
4	50	247	29
6	39	342	36

Modeling the breakthrough curves

Many models, either more or less sophisticated, are available in the literature (Ruthven 1984, Tien 1994). In this paragraph, we give three classic models useful for describing the breakthrough curves or some important operating and design data. For all the models, the assumptions are the following:

- the system is in a steady state, i.e. the flow and inlet concentrations are constant,
- there is no chemical or biological reaction, only a mass transfer occurs,
- the temperature is constant.

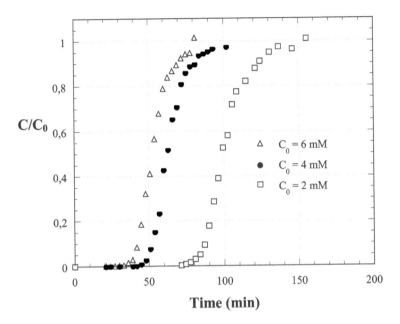

Figure 6. Breakthrough curves from a fixed bed biosorption experiment; lanthanum removal onto *Pseudomonas aeruginosa*. $U_0 = 0.76$ m h^{-1} - $Z = 300$ mm - $500 < d_p < 1,000$ µm (Adapted from Texier *et al.* 1999, 2000, 2002).

Bohart Adams model

This model is based on two kinetic equations of transfer from the fluid phase and accumulation in the inner porous volume of the material. A simple equation is obtained giving the breakthrough time (t_b) as a function of the operating conditions:

$$t_b = \frac{N_0}{C_0 U_0}\left[Z - \frac{U_0}{kN_0}\,\mathrm{Ln}\left(\frac{C_0}{C_b} - 1\right)\right] \tag{15}$$

or

$$t_b = \frac{N_0}{C_0 U_0}\left(\quad \right) \tag{16}$$

where

t_b	:	breakthrough time (h)
k	:	adsorption kinetic constant (Lg^{-1}min^{-1})
C_0	:	inlet concentration (mg L^{-1})
U_0	:	velocity in the empty bed (m h^{-1})

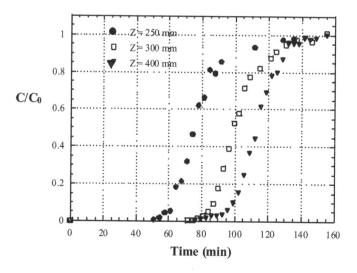

Figure 7. Breakthrough curves of lanthanum adsorbed onto *Pseudomonas aeruginosa* trapped in a polyacrylamide gel ($C_0 = 2$ mol L^{-1}, $U_0=0.76$ m h^{-1}, $500 < d_p < 1000$ μm, pH = 5).

N_0 :   adsorption capacity (mg L^{-1})
Z :   filter length (m)
Z_0 :   adsorption zone (m)

The two parameters (N_0 and Z_0 (or k)) are experimentally determined. In order to illustrate the utilization of this approach, the results are presented in Figure 7 and Table 6 (Texier *et al.* 2002). These lab experiments were performed with *Pseudomonas aeruginosa* trapped in a polyacrylamide gel adsorbing lanthanide ions at different operating conditions. From this example, some conclusions can be proposed:
-   the biosorption capacities decrease with the water velocity in the column. The mass transfer zone (Z_0) is found to be < 2 mm for $U_0 = 0.76$ m h^{-1} and 144 mm $U_0 = 2.29$ m h^{-1},
-   the size has no real influence ($125 < d_p < 1,000$ μm),
-   better capacities are obtained at higher initial concentrations,
-   the adsorption capacities are proportional to the bed depth, although the influence of this parameter is weak.

These results are in agreement with Volesky and Prasetyo (1994) who showed that this sorption model was useful for the determination of the key design parameters.

Table 6. Estimation of the characteristic biosorbent process parameters for lanthanum adsorbed onto *Pseudomonas aeruginosa* trapped in a polyacrylamide gel (adapted from Texier *et al.* 2002).

U_0 $(m h^{-1})$	C_0 $(mmol\ L^{-1})$	Z (mm)	d_p (mm)	t_p (min)	Q_{max} $(mmol\ g^{-1})$	Q_{tp} $(mg\ g^{-1})$	N_0 $(mg\ g^{-1})$	K $(Lg^{-1}min^{-1})$
0.23	2	250	500-1000	228	205	23	23	0.2
0.54	2	250	500-1000	81	199	22	23	0.3
0.76	2	250	500-1000	60	197	22	19	0.7
0.76	2	300	500-1000	84	208	23	21	0.8
0.76	2	400	500-1000	96	217	19	19	0.5
0.99	2	250	500-1000	52	171	22	21	1.2
1.38	2	250	500-1000	31	152	16	15	1.6
2.29	2	250	500-1000	12	126	7	15	1.9
0.76	2	300	250-500	102	222	30	23	0.4
0.76	2	300	125-250	90	206	27	23	0.2
0.76	4	300	500-1000	50	247	29	25	0.6
0.76	6	300	500-1000	39	342	36	33	0.4

Mass transfer model

The relations used for this model are:
- — a mass balance between the aqueous phase and the solid phase,
- — a mass transfer equation assuming a linear driving force approximation,
- — the Freundlich equation (equation 6).

An equation describing the breakthrough curves is found:

$$C(t) = \left[\frac{C_n^0}{1 + Ae^{-rt}} \right]^{\frac{1}{n-1}}$$ (17)

where

n :   Freundlich equation parameter

$C(t)$:   concentration at time t $(mg\ L^{-1})$

C_0 :   initial concentration $(mg\ L^{-1})$

A, r :   equation parameters determined experimentally

This approach has been successfully applied to pilot unit adsorption in a large number of studies (Clark 1987).

Homogeneous Surface Diffusion Model (HSDM) and Equilibrium Column Model (ECM) equations

Crittenden and co-workers (1976, 1978, 1980) developed a model based on the surface diffusion of adsorbate. Numerous applications have been performed (Montgomery 1985). In a fixed bed, the following assumptions are made:

— there is no radial dispersion; the concentration gradients exist only in the axial direction,

— plug flow exists within the bed,

— surface diffusion (kinetics limiting the mass transfer) is much greater than pore diffusion thus the contribution of pore diffusion is neglected. The adsorbent has a homogeneous surface and the diffusion flux is described by Fick's law $\left(J = D_s \dfrac{dC}{dx} \right)$.

— a linear driving force relation describes the external mass transfer from the liquid to the external surface of the solid,

— the Freundlich equation gives the adsorption equilibria between the solid and liquid phases.

An exhaustive development of this model has been presented in previous publications (Montgomery 1985). Table 7 summarizes the different equations required to describe the mechanisms.

The set of equations cannot be directly solved analytically. Solutions may be obtained using orthogonal collocation techniques. The partial differential equations are reduced to differential equations that are integrated. Computer software and calculus methodologies are described in some adsorption books and journals (Tien 1994, Basmadjian 1997, Thomas and Crittenden 1998).

Kratochvil and Volesky (2000) proposed a heavy metal ion mixture model. The assumptions are a constant feed composition, isotherm operations, uniform packing materials, homogeneity of the bed and no precipitation in the bed. The equations integrate the description of ion exchange reactions, the molar balance for sorbing species, the axial diffusion and a mass transfer equation. They applied this model to a mixture of copper and cadmium onto a packed bed of *Sargassum* algal biosorbent in the calcium form. An example of a classical breakthrough curves is presented in Figure 8.

Table 7. Homogeneous Surface Diffusion Model (HSDM) equations.

Purpose	Equation
Solid phase mass balance	$\dfrac{\partial q}{\partial t} = \dfrac{D_s}{r^2} \dfrac{\partial}{\partial r}\left[r^2 \dfrac{\partial q}{\partial r}\right]$
Initial condition	$q = 0 \ (0 \geq r \geq R, t = 0)$
Boundary conditions	$\dfrac{\partial q}{\partial t} = 0 \ (r = 0, t \geq 0)$
	$\dfrac{\partial q}{\partial t} = \dfrac{k_f}{\rho_a D_s \varphi}\ C(t) - C_s(t)$
Liquid phase mass balance	$U \dfrac{\partial C}{\partial z} = \dfrac{\partial C}{\partial t} + \dfrac{3 k_f (1 - \varepsilon)}{R \varphi \varepsilon}(C . C_s)$
Initial condition	$C = 0 \ (0 \geq z \ 0 \geq H, t < \tau)$
Boundary condition	$C = C_0 (t) \ (z = 0, t \geq 0)$
Freundlich isotherm equation	$q = KC^{1/n}$

where

k_f	:	external mass transfer coefficient (s^{-1})
D_s	:	surface diffusion coefficient (m^2 s^{-1})
R	:	particle radius (m)
φ	:	sphericity (dimensionless)
r	:	radial length of spherical shell (m)
z	:	axial direction (m)
ρ_a	:	adsorbent density (kg m^{-3})

A neural network

A new approach for the modeling of breakthrough curves is to use a statistical tool: neural networks. These are an association of several neurons (Fig. 9) connected together to make a network. This kind of approach has been applied to the adsorption of organics onto activated carbon fibers (Faur-Brasquet and Le Cloirec 2001, 2003) and lanthanide ion removal onto immobilized *Pseudomonas aeruginosa* (Texier *et al.* 2002). In this study, several architectures of neural network were tested, as shown in Figure 10, in order to model the breakthrough curves.

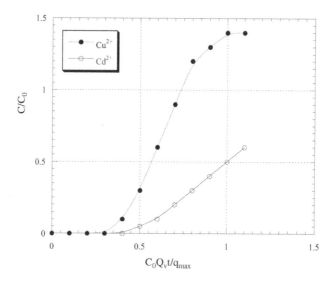

Figure 8. Breakthrough curves for multicomponent biosorption onto a biosorbent (adapted from Kratochvil and Volesky 2000).

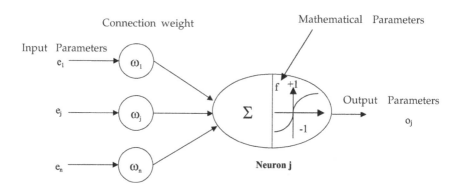

Figure 9. Presentation of a specific neuron.

It appears that the prediction ability is satisfactory for the first part of the curve ($C/C_0 < 0.25$) when the metal ion begins to be released from the column. The choice of the input parameters and the neuron network architecture is important for the prediction of experimental data. Considering that the most interesting part of the breakthrough curve to

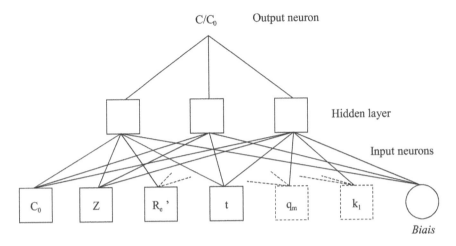

Figure 10. Neural network architectures used for modeling the breakthrough curves of lanthanide ions in the biosorption column system.

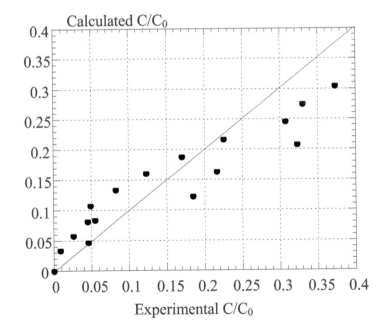

Figure 11. Agreement between experimental and predicted C/C_0 values with a neural network (C_0, Z, Re, and t) applied to the lanthanum breakthrough curve (C_0 = 2 - 6 mM, U_0 = 0.76 - 2.29 m h^{-1}, Z = 250 - 420 mm).

evaluate column performance is the first one that corresponds to the metal ion release, a comparison between the experimental and the calculated data (Fig. 11) partly illustrates the feasibility of using neural networks for biosorption. However, continued investigations are required to extend the prediction ability of such a numerical approach.

Soil and solid waste remediation

A variety of both lithotrophic and organotrophic microorganisms are known to mediate the mobilization of various elements from solids, mostly by the formation of inorganic and organic acids. These mechanisms of metal solubilization by microorganisms are currently named bioleaching. The mechanisms of metal species transformation by an industrial process are close to rapid natural biogeochemical cycles.

Bioleaching has been used since prehistoric times while the Greeks and Romans probably extracted copper from mine water more then 2000 years ago. However, it has been known for only about 50 years that bacteria are mainly responsible for the enrichment of metals in water from ore deposits and mines. Bioleaching is a simple and effective method currently used for metal extraction of gold, copper and uranium from low-grade ores. Solid industrial waste materials such as fly ash, sludges, or dust might also be microbially treated to recover metals for their re-use in metal-manufacturing industries (Krebs *et al.* 1997). Metal recovery from sulfide minerals is based on the activity of chemolithotrophic bacteria, mainly *Thiobacillus ferrooxidans* and *Thiobacillus thiooxidans*, which convert insoluble metal sulfides into soluble metal sulfates producing sulfuric acid.

Table 8. Biological leaching compared with chemical leaching (adapted from Krebs *et al.* 1997).

Advantages	Disadvantages
Leaching compounds naturally produced in situ	Long reaction times
Elevated concentration of leaching around the metal-containing particles	For field treatment, climate influence
Microbial selectivity depending on strain used	Heavy metal toxicity to microorganisms Saline concentration toxicity to microarganisms
Increase in leaching efficiency	pH variations
Excretion of surfactants	
Low energy demand	
No emission of gaseous pollutants	

Non-sulfide ores and solid industrial waste or minerals can be treated by heterotrophic bacteria and fungi. In these cases, metal extraction is due to the production of organic acids, chelating or complexing compounds excreted into their environment (Bosecker 1997). For example, *Penicillium oxalicum* produces oxalate, *Pseudomonas putida* citrate and gluconate, and *Rhizopus* sp. lactate, fumarate or gluconate. Another example is given by *Aspergillus niger* leaching metal from fly ash generated by a municipal waste incineration plant (Bosshard *et al.* 1996). In addition, the use of microorganisms is also feasible for detoxification applications to reduce environmental pollution. Metal-contaminated soils and sediments have been microbiologically treated using various *Thiobacillus* species (Gadd and White 1993, Atlas 1995).

Currently, the main techniques employed are heap, dump and *in situ* leaching. Tank leaching is practiced for the treatment of refractory gold ores. Several leaching processes of metals from ores have been patented (for references see Krebs *et al.* 1997, Brombacher *et al.* 1997). Furthermore, biohydrometallurgical processing of fly ash poses serious problems, especially at higher pulp densities, because of the high content of toxic metals and the saline and strongly alkaline (pH > 10) environment. Krebs and co-workers (1997) proposed a comparison between bioleaching techniques and chemical leaching. Some comments are given in Table 8.

Bioleaching mechanism approach

At the present time, bioleaching processes are generally based on the activity of *Thiobacillus ferrooxidans, Leptospirillum ferrooxidans* and *Thiobacillus thiooxidans*. These bacterial species convert heavy metal sulfides via biochemical oxidation reactions into water-soluble metal sulfates. The metals can be released from sulfide minerals by direct or indirect bacterial leaching (Ehrlich 1996). The bacterial strains involved are chemolithoautotrophic for *Thiobacillus* species, and strict chemolitho-autotrophic in the case of *Leptospirillum* (Sand *et al.* 1992).

Direct bacterial leaching

Direct bacterial leaching needs physical contact between the micro-organism cell and the mineral sulfide surface. The oxidation to sulfate takes place via several enzymatically-catalyzed steps. In order to consider the mechanisms, an example of iron sulfide oxidation and solubilization is presented. In this process, pyrite is oxidized to iron(III) sulfate according to the following reactions:

$$4\,FeS_2 + 14O_2 + 4H_2O \xrightarrow{\text{Bacteria}} 4FeSO_4 + 4H_2SO_4 \qquad (a)$$

$$4FeSO_2 + O_2 + 2H_2SO_4 \xrightarrow{\text{Bacteria}} 2Fe_2(SO_4)_3 + 2H_2O \qquad (b)$$

The direct bacterial oxidation of pyrite is summarized by an overall reaction:

$$4FeS_4 + 15O_2 + 2H_2O \xrightarrow{\text{Bacteria}} 2Fe_2(SO_4)_3 + 2H_2SO_4 \qquad (c)$$

These processes are aerobic and produce high quantities of sulfuric acid, which is involved in the dissolution of other minerals potentially present in the ores. Torma (1977) has shown that the following non-iron metal sulfides can be oxidized by *T. ferrooxidans* in direct interaction: covellite (CuS), chalcocite (Cu_2S), sphalerite (ZnS), galena (PbS), molybdenite (MoS_2), stibnite (Sb_2S_3), cobaltite (CoS) and millerite (NiS).

The mechanisms of attachment and the metal solubilization take place on specific sites of crystal imperfection, and the metal solubilization is due to electrochemical interactions (Mustin *et al.* 1993).

Indirect bacterial leaching

In indirect bioleaching, the bacteria generate a lixiviant, which chemically oxidizes the sulfide mineral. For example, in an acid solution containing ferric iron, metal sulfide solubilization can be described according to the following simplified reaction:

$$MeS + Fe_2(SO_4)_3 \longrightarrow MeSO_4 + 2FeSO_4 + S^0 \qquad (d)$$

where MeS is a metal sulfide.

To keep enough iron in solution, the chemical oxidation of metal sulfides must occur in an acid environment below pH 5.0. The ferrous iron arising in this reaction can be reoxidized to ferric iron by *T. ferrooxidans* or *L. ferrooxidans* and, as such, can take part in the oxidation process again. In this kind of leaching, the bacteria do not need to be in direct contact with the mineral surface. They have only a catalytic function. Effectively, the reoxidation of ferrous iron is a very slow reaction without the presence of bacteria. In the range of pH 2-3, bacterial oxidation of ferrous iron is about 10^5-10^6 times faster than the chemical reaction (Lacey and Lawson 1970). The sulfur arising simultaneously (Equation d) may be oxidized to sulfuric acid by *T. ferrooxidans* but oxidation by *T. thiooxidans*, which frequently occurs together, is much faster:

$$2S^0 + 3O_2 + 2H_2O \xrightarrow{\text{Bacteria}} 2H_2SO_4 \qquad (e)$$

In this case, the role of *T. thiooxidans* in bioleaching is to create favorable acid conditions for the growth of ferrous iron-oxidizing bacteria.

A well-known example of an indirect bioleaching process is the extraction of uranium from ores, when insoluble tetravalent uranium is oxidized to the water-soluble hexavalent uranium (equation f). The lixiviant may be generated by the oxidation of pyrite (§ equation c), which is very often associated with uranium ore (Cerda *et al.* 1993).

$$U^{IV}O_2 + Fe_2(SO_4)_3 \longrightarrow U^{IV}O_2SO_4 + 2FeSO_4 \qquad (f)$$

Leaching processes

The bioleaching of minerals is a simple and effective technology for the processing of sulfide ores and is used on an industrial scale mainly for the recovery of copper and uranium. For example, more than 25 % of the copper produced in the United States of America, and 15 % of the world production, is produced by bioleaching (Agate 1996). A typical process is represented in Figure 12. The size of the dumps varies considerably and the amount of ore may be in the range of several hundred thousand tons. The top of the dump is sprinkled continuously or flooded temporarily. Depending on the ore composition, the lixiviant may be water, acidified water or acid ferric sulfate solution produced by other leaching operations on the same mining site. Before recirculation, the leachate flows through an oxidation basin, in which the bacteria and ferric iron are regenerated.

Underground leaching (Fig. 12) is usually done in abandoned mines. Galleries are flooded and the water collected in deeper galleries is then pumped to a processing plant at the surface. The best known application of this procedure is at the Stanrock uranium mine at Elliot Lake in Ontario, Canada. The production is about 50 t of uranium oxide per year (Rawlings and Silver 1995). Moreover, ore deposits that cannot be mined by conventional methods due to their low grade or small quantity, can be leached *in situ*. In these cases, the system requires sufficient permeability of the ore-body and impermeability of the gangue rock.

The effectiveness of leaching depends largely on the development of the microorganisms and on the chemical and mineralogical composition of the ore to be leached. The maximum yield of metal extraction is achieved for the optimum growth conditions of the bacteria inducing the production of a large amount of leaching solution. Many operating factors are required such as nutrients (inorganic compounds for chemolithoautotrophic organisms), oxygen and carbon dioxide, pH (optimum pH range between 2.0 and 2.5), temperature (with an optimum close to 30°C) and chemical composition of the mineral substrate.

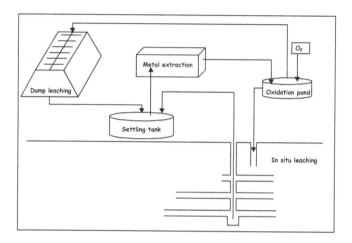

Figure 12. Flow sheet of a dump and *in situ* leaching process (adapted from Sand *et al.* 1993, Rawlings and Silver 1995).

Future developments

White and co-workers (1998) proposed a new approach for the bioreme-diation of soil contaminated with toxic metals. The microbially-catalyzed reactions, which occur in the natural sulfur cycle, were integrated in a microbiological process to remove toxic metals from contaminated soils or solid wastes. Bioleaching, using sulfuric acid produced by sulfur-oxidizing bacteria, was followed by precipitation of the leachate metals as insoluble sulfides by sulfate-reducing bacteria in anaerobic conditions.

Conclusions and trends

In this chapter, different processes using microorganisms to remove heavy metals present in water, wastewater, waste and soil have been presented:

— the continuously stirred processes for metal ion removal in wastewater by microorganisms coating particles. In this case, the equilibrium data obtained with isotherm curves show a good adsorption of several metal ions onto biofilm.

— the fixed bed reactors packed with microorganisms (biofilm, entrapped bacterial materials) are efficient to obtain a sorption (adsorption and/or ion exchange) of cations. The pressure drop is calculated with classical equations (Darcy's law, Ergun's equation or by new approaches). Some design data are obtained with the breakthrough curves. Different models have been described to simulate these curves. A statistical tool (neural networks) has been applied and good correlation has been found between experimental and calculated values.

— bioleaching was also defined and discussed in terms of mechanisms and processes. Some applications for metal extraction were presented.

The interdisciplinary nature of research and development of applications poses quite a challenge. The mechanisms of heavy metal ion removal are not well known. We need a better understanding to approach the engineering of batch or continuous reactors in order to propose this kind of technology for water and wastewater treatments or bioleaching.

Acknowledgments

A part of paragraph 3 of this chapter was previously presented at the summer school BIO-IMED II: Biofilm in Industry, Medicine and Environmental Biotechnology: the Technology, Galway, Ireland, August 9th-14th, 2003.

REFERENCES

Abdelouas, A., W. Lutze, and H.E. Nuttall. 1999. Uranium: Mineralogy, geochemistry and the environment. *Rev. Mineral.* 38: 433-473.

Agate, A.D. 1996. Recent advances in microbial mining. *W. J. Microbiol. Biotechnol.* 12: 487-495.

Aksu, Z., T. Kustal, N. Songul Gun, N. Haciosmanoglu, and M. Gholaminejad. 1991, Investigation of biosorption of Cu(II), Ni(II) and Cr (VI) ions to activated sludge bacteria. *Environ. Technol.* 12: 915-921.

Aksu, Z., Y. Sag, and T. Kustal. 1992. The biosorption of copper (II) by C. *vulgaris* and Z. *ramigera. Environ. Technol.* 13: 579-586.

Andrès, Y., H.J. McCordick, and J.C. Hubert. 1991. Complex of mycobactin from *Mycobacterium smegmatis* with scandium, yttrium and lanthanum *Biol. Metals* 4: 207-210.

Andrès, Y., H.J. McCordick, and J.C. Hubert. 1993. Adsorption of several actinide (Th, U) and lanthanide (La, Eu, Yb) ions by *Mycobacterium smegmatis. Appl. Microbiol. Biotechnol.* 39: 413-417.

Andrès Y., G. Thouand, M. Boualam, and M. Mergeay. 2000. Factors influencing the biosorption of gadolinium by microorganisms and sand. *App. Microbiol. Biotechnol* 54: 262-267.

Appanna, V.D., L.G. Gazso, J. Huang, and St. M. Pierre., 1996. A microbial model for cesium containment. *Microbios* 86: 121-126.

Arica, M.Y., C. Arpa, A. Ergene, G. Bayramoglu, and O. Genc. 2003. Ca-alginate as a support for Pb(II) and Zn(II) biosorption with immobilized *Phanerochaete chrysosporidium. Carbohydr. Polymers* 542: 167-174.

Atkinson, B.W., F. Bux, and H.C. Kasan 1998. Considerations for application of biosorption technology to remediate metal-contaminated industrial effluents. *Water SA.* 24: 129-35.

Atlas, R. M. 1995. Bioremediation. *Chem. Eng. News.* 73: 32-42.

Avery, S.V. 1995. Review, Microbial interactions with cesium - implications for biotechnology. *J. Chem. Tech. Biotechnol.* 62: 3-16.

Baléo, J.N., B. Bourges, Ph. Courcoux, C. Faur-Brasquet, and P. Le Cloirec. 2003. *Méthodologie Expérimentale,* Tec & Doc, Lavoisier, Paris, France.

Basmadjian, D. 1997. *The Little Adsorption Handbook,* CRC Press, Boca Raton, Florida.

Beolchini, F., F. Pagnanelli, L. Tora, and F. Veglio. 2003. Biosorption of copper by *Sphaerotilus natans* immobilized in polysulfone matrix: equilibrium and kinetic analysis. *Hydrometallurgy* 70: 101-112.

Beveridge, T.J., and R. G.E. Murray. 1980. Sites of metal deposition in the cell wall of *Bacillus subtilis. J. Bacteriol.* 141: 876-887.

Beveridge, T.J., and W.S. Fyfe. 1985. Metal fixation by bacterial cell walls. *Can. J. Earth Sci.* 22: 1893-1898.

Birch, L., and R. Bachofen. 1990. Complexing agents from microorganisms. *Experientia* 46: 827-834.

Bouby, M., I. Billard, and H.J. McCordyck. 1999. Complexation of UO_2^{2+} by the siderophore mycobactin S in ethanol. *Czech. J. Phys.* 49: 769-772.

Bosecker K., 1997. Bioleaching: metal solubilization by microorganisms. *FEMS Microbiology. Rev.* 20: 605-617.

Bosshard P.P., R. Bachofen, and H. Brandl. 1996. Metal leaching of fly ash from municipal waste incineration by *Aspergillus niger. Environ. Sci. Technol.* 30: 3066-3070.

Brainard, J.R., B.A. Strietelmeier, P.H. Smith, P.J. Langston-Unkefer, M.E. Barr, and R.R. Ryan 1992. Actinide binding and solubilization by microbial siderophores. *Radiochim. Acta* 58/59: 357-363.

Brasquet, C., and P. Le Cloirec. 1997. Adsorption onto activated carbon fibers: applications to water and air treatment. *Carbon* 35: 1307-1313.

Brasquet, C., and P. Le Cloirec. 2000. Pressure drop through textile fabrics - experimental data modeling using classical models and neural networks. *Chem. Eng. Sci.* 55: 2767-2778.

Brombacher, C., R. Bachofen, and H. Brandl. 1997. Biohydrometallurgical processing of solids: a patent review. *Appl Microbiol. Biotechnol.* 48: 577-587.

Brown, P., J.I. Atly, D. Parrish, S. Gill, and E. Graham. 2000. Evaluation of the adsorptive capacity of peanut hull pellets for heavy metals in solution. *Adv. Environ. Res.* 4: 19-29.

Cabral, J.P.S. 1992. Selective binding of metal ions to *Pseudomonas syringae* cells. *Microbios* 71: 47-53.

Cassidy, M.B., H. Lee, and J.T. Trevors. 1996. Environmental applications of immobilized microbial cells: a review. *J. Ind. Microbiol.* 16: 79-101.

Cerda, J., S., Gonzalez, J.M. Rios, and T. Quintana. 1993. Uranium concentrates bioproduction in Spain: A case study. *FEMS Microbiol. Rev.* 11: 253-260.

Chang, J.S., and J. Hong. 1994. Biosorption of mercury by the inactivated cells of *Pseudomonas aeruginosa* PU21. *Biotechnol. Bioeng.* 44: 999-1006.

Chang, J.S., R. Law, and C. Chang. 1997. Biosorption of lead, copper and cadmium by biomass of *Pseudomonas aeruginosa* PU21. *Water Res.* 31: 1651-1658.

Clark, R.M. 1987. Evaluating the cost and performance of field scale granular activated carbon systems. *Environ. Sci. Technol.* 21: 574-581.

Comiti, J., and M. Renaud. 1989. A new model for determining mean structure parameters of fixed beds from pressure drop measurement: application to beds packed with parallelepipedal particles. *Chem. Engn. Sci.* 44: 1539-1545.

Coulson, J.M., J.F. Richardson, J.R. Backhurst, and J.F. Harker. 1991. *Chemical Engineering*, 4th ed., Vol. 2, Butterworth Heinemann, Oxford, UK.

Crittenden, J.C. 1976. *Mathematical modeling of adsorber dynamics: single and multicomponents*. PhD Thesis, University of Michigan, Ann Arbor.

Crittenden, J.C., and W.J. Weber. Jr. 1978. Model for design of multicomponent adsorption system. *J. Environ. Eng.* 104: 185-192.

Crittenden., J.C., B.W.C. Wong, W.E. Thacker, V.L. Someyink, and R.L. Hinrichs. 1980. Mathematical modeling of sequential loading in fixed bed adsorbers. *J. Water Pollut. Control Fed.* 52: 2780-2795.

Daughney, C.J., D.A. Fowle, and D. Fortin. 2001. The effect of growth phase on proton and metal adsorption by *Bacillus subtilis*. *Geochim. Cosmochim. Acta* 65: 1025-1035.

Deckwer, W.D., F.U. Becker, S. Ledakowicz, and I. Wagner-Döbler. 2004. Microbial removal of ionic mercury in a three-phase fluidized bed reactor. *Environ. Sci. Technol.* 38: 1858-1865.

Deneux-Mustin, S., J. Rouiller, S. Durecu, C. Munier-Lamy, and J. Berthelin. 1994. Détermination de la capacité de fixation des métaux par les biomasses microbiennes des sols, des eaux et des sédiments : intérêt de la méthode du titrage potentiométrique. *C. R. Acad. Sci. Paris* 319: 1057-1062.

Doyle, R.J., T.H. Matthews, and U.N. Streips. 1980. Chemical basis for selectivity of metal ions by the *Bacillus subtilis* cell wall. *J. Bacteriol.* 143: 471-480.

Drake, L.R., C.E. Hensman, S. Lin, G.D. Rayson, and P.J. Jackson. 1997. Characterization of metal ion binding sites on *Datura innoxia* by using lanthanide ion probe spectroscopy. *Appl. Spectrosc.* 51: 1476-1483.

Eccles, H. 1999. Treatment of metal-contaminated wastes: Why select a biological process? *TIBTECH* 17: 462-465.

Ehrlich, H.L. 1996. Geomicrobiology of sulfur. Pages 508-559 in *Geomicrobiology*, 3rd ed., H.L. Ehrlich, ed., Marcel Dekker, Inc. New York.

Faur-Brasquet, C., and P. Le Cloirec. 2001. Neural network modeling of organic removal by activated carbon cloths. *J. Environ. Engn.* 127: 889-894.

Faur-Brasquet, C., and P. Le Cloirec. 2003. Modelling of the flow behavior of activated carbon cloth using a neural network approach. *Chem. Eng. Process.* 42: 645-652.

Fein, J.B., C.J. Daughney, N. Yee, and Y. Davis. 1997. A chemical equilibrium model for metal adsorption onto bacterial surfaces. *Geochim. Cosmoschim. Acta* 61: 3319-3328.

Flemming, H.C., and J. Wingender. 2003. The crucial role of extracellular polymeric substances in biofilms, Pages 178-203 in *Biofilm in Wastewater Treatment*, S. Wuertz, P. Bishop and P. Wilderer, eds., IWA Publishing, London, UK.

Fogarty, R.V., and J.M. Tobin. 1996. Fungal melanins and their interactions with metals. *Enzyme Microbial Technol.* 19: 311-317.

Fourest, E., and J.C., Roux. 1992. Heavy metal adsorption by fungal mycelial by-products: mechanisms and influence of pH. *Appl. Microbiol. Biotechnol.* 37: 399-403.

Friis, N., and P. Myers-Keith. 1986. Biosorption of uranium and thorium by *Streptomyces longwoodensis*. *Biotechnol. Bioeng.* 28: 21-28.

Gadd, G.M., and L. De Rome. 1988. Biosorption of copper by fungal melanin. *Appl. Microbiol. Biotechnol.* 29: 610-617.

Gadd, G.M. 1990. Heavy metal accumulation by bacteria and other micro-organisms. *Experientia* 46: 834-840.

Gadd, G.M. 1993. Microbial formation and transformation of organometallic and organometaloid compounds. *FEMS Microbiol. Rev.* 11: 297-316.

Gadd, G.M., and C. White. 1993. Microbial treatment of metal pollution-a working biotechnology ? *TIBTECH* 11: 353-359.

Galun, M., E. Galun, B.Z. Siegel, P. Keller, H. Lehr, and S.M. Siegel. 1987. Removal of metal ions from aqueous solution by *Penicillium* biomass: Kinetics and uptake parameters. *Water Air Soil Pollution* 33: 359-371.

Gardea-Torresdey, J, M. Hejazi, K. Tiemann, J.G. Parsons, M. Duarte-Gardea, and J. Henning. 2001. Use of hop (*Humulus lupulus*) agricultural by-products for the reduction of aqueous lead (II) environmental health hazards. J Hazard. Mater. 2784: 1-18.

Gourdon, R., S. Bhende, E. Rus, and S.S. Sofer. 1990. Comparison of cadmium biosorption by gram-positive and gram-negative bacteria from activated sludge. *Biotechnol. Lett.* 12: 839-843.

Goyal, N., S.C. Jain, and U.C. Banerjee. 2003. Comparative studies on the microbial adsorption of heavy metals. *Adv. Env. Res.* 7: 311-319.

Holan, Z.R., B. Volesky, and I. Prasetyo. 1993. Biosorption of cadmium by biomass of marine algae. *Biotechnol. Bioeng.* 41: 819-825.

Holan, Z.R., and Volesky. B. 1994. Biosorption of lead and nickel by biomass of marine algae. *Biotechnol. Bioeng.* 43: 1001-1009.

Huang, J.P., C.P. Huang, and A.L. Morehart. 1990. The removal of Cu(II) from dilute aqueous solution by *Saccharomyces cerevisiae. Water Res.* 24: 433-439.

Jilek, R., H. Prochazka, K. Stamberg, J. Katzer, and P. Nemec. P. 1975. Some properties and development of cultivated biosorbent. *Rudy* 23: 282-286.

Kapoor, A., and T. Viraraghavan. 1995. Fungal biosorption - An alternative treatment option for heavy metal bearing wastewaters: A review. *Biores. Technol.* 53: 195-206.

Kiefer, E., L. Sigg, and P. Schosseler. 1997. Chemical and spectroscopic characterization of algae surfaces. *Environ. Sci. Technol.* 31: 759-764.

Kratochvil, D., and B. Volesky. 2000. Multicomponent biosorption in fixed beds, *Water Res.* 34: 3186-3196.

Krebs, W., C. Brombacher, P.P. Bosshard, R. Bachofen, and H. Brandl., 1997. Microbial recovery of metals from solids. *FEMS Microbiol. Rev.* 20: 605-617.

Kuyucak, N., and B. Volesky. 1988. Biosorbents for recovery of metals from industrial solutions. *Biotechnol. Lett.* 10: 37-142.

Lacey D.T., and F. Lawson. 1970. Kinetics of the liquid phase oxidation of acid ferrous sulfate by the bacterium *Thiobacillus ferrooxidans. Biotechnol. Bioeng.* 12: 29-50.

Le Cloirec, P. 2002. Adsorption in water and wastewater treatments. *Handbook of Porous Solids*, F. Schüth, K.S.W. Sing and J. Weitkamp Editors, Wiley-VCH, Weinheim, Germany, Vol. 5, Ch. 6.7.

Ledin, M., K. Pedersen, and B. Allard. 1997. Effects of pH and ionic strength on the adsorption of Cs, Sr, Zn, Cd and Hg by *Pseudomonas putida. Water Air Soil Pollut.* 93: 367-381.

Levenspiel, O. 1979. *The Chemical Reactor Minibook.* OSU Book, Corvallis, Oregon.

Liu, D.H.F., B.G. Liptack, and P.A. Bouis. 1997. *Environmental Engineer's Handbook*, 2nd ed., Lewis Publishers, Boca Raton, Florida.

Lloyd J.R. 2003. Microbial Reduction of metals and radionuclides. FEMS Microbiol. Rev. 27: 411-425.

Lovley, D.R. 1993. Dissimilatory metal reduction. *Annu. Rev. Microbiol.* 47: 263-290.

Lovley, D.R. 1995. Bioremediation of organic and metal contaminants with dissimilatory metal reduction. *J. Ind. Microbiol.* 14: 85-93.

Lovley, D.R., E.J.P., Phillips, Y.A. Gorby, and E.R. Landa. 1991. Microbial reduction of uranium. *Nature* 350: 413-416.

Manahan, S.E. 1997. *Environmental Science and Technology*, Lewis Publishers, Boca Raton, Florida.

Mav Kai, S., Anolves Y., Routavon G., Grambow B. 2003. Complexation studies of europium (iii) on *Bacillus subtilis*: Fixation sites and biosorption modeling. *J. Coll. Interf. Sci.* 262: 351-361.

Marshall, W.E., L.H. Wartelle, D.E. Boler, and C.A. Toles. 2000. Metal ion adsorption by soybean hulls modified with citric acid: A comparative study. *Environ Technol.* 21: 601-607.

Mattuschka, B., and G. Straube. 1993. Biosorption of metals by waste biomass. *J. Chem. Technol. Biotechnol.* 58: 57-63.

Mersmann, A., G. Schneider, and H. Voit. 1990. Selection and design of aerobic bioreactor. *Chem. Eng. Tech.* 13: 357-370.

McEldowney, S., and M. Fletcher. 1986. Effect of growth conditions and surface characteristics of aquatic bacteria on their attachment to solid surfaces. *J. Gen. Microbiol* 132: 513-523.

Montgomery, J.M. 1985. *Water Treatment, Principles & Design*, John Wiley & Sons, New York.

Morley, G.F., and G.M. Gadd. 1995. Sorption of toxic metals by fungi and clay minerals. *Mycol. Res.* 9: 1429-1438.

Mullen, M.D., D.C. Wolf, F.G. Ferris, T.J. Beveridge, C.A. Flemming, and G.W. Bailey. 1989. Bacterial sorption of heavy metals. *Appl. Environ. Microbiol.* 55: 3143-3149.

Mustin C., Ph. de Donato, J. Berthelin, and Ph. Marion. 1993. Surface sulphur as promoting agent of pyrite leaching by *Thiobacillus ferrooxidans*. *FEMS Microbiol. Rev.* 11: 71-77.

Nourbakhsh M., Y. Sag, B. Ozer, Z. Aksu, T. Kustal, A.M. Caglar., *et al.* 1994. A comparative study of various biosorbents for removal of Chromium (VI) ions from industrial wastewaters. *Process Biochem.* 29: 1-5.

Ogale, S.S., and D.N. Deobagkar. 1988. A high molecular weight plasmid of *Zymomonas mobilis* harbours genes for $HgCl_2$ resistance. *Biotechnol. Lett.* 10: 43-48.

Pagnanelli, F., F. Beolchini, A. Esposito, L. Toro, and F. Veglio. 2003a. Mechanistic modeling of heavy metal biosorption in batch and membrane reactor system. *Hydrometallurgy* 71: 201-208.

Pagnanelli, F., A. Esposito, L. Toro, and F. Veglio 2003b. Metal speciation and pH effect on Pb, Cu, Zn and Cd biosorption onto *Sphaerotilus natans:* Langmuir-type empirical model. *Water Res.* 37: 627-633.

Penumarti, N., and G.K. Khuller. 1983. Subcellular distribution of mannophosphoinositides in *Mycobacterium smegmatis* during growth. *Experientia* 39: 882-884.

Rawlings, D.E., and S. Silver. 1995. Mining with microbes. *Biotechnology* 13: 773-778.

Reddad, Z., C. Gérente, Y. Andrès, and P. Le Cloirec. 2002a. Adsorption of several metal ions onto a low cost biosorbent: Kinetic and equilibrium studies. *Environ. Sci. Technol.* 36: 2067-2073.

Reddad, Z., C. Gérente, Y. Andrès, and P. Le Cloirec. 2002b. Modeling of single and competitive metal adsorption onto a natural polysaccharide. *Environ. Sci. Technol.* 36: 2242-2248.

Ruthven, D.M. 1984. *Principles of Adsorption and Adsorption Processes*, John Wiley & Sons, New York.

Remacle, J. 1990. The cell wall and metal binding. Page 83-92 in *Biosorption of Heavy Metals*, B. Volesky, ed., CRC Press, Boca Raton, Florida.

Sag, Y., and T. Kutsal. 1995. Biosorption of heavy metals by *Zooglea ramigera*: Use of adsorption isotherms and a comparison of biosorption characteristics. *Biochem. Eng. J.* 60: 181-188.

Sakagushi, T., and A. Nakajima. 1991. Accumulation of heavy metal such as uranium and thorium by microorganisms, *Mineral Bioprocessing*, The Minerals, Metals and Materials Soc.

Salt, D.E., R.C. Prince, A.J.M. Baker, I. Raskin, and I.J. Pickering. 1999. Zinc ligands in the metal hyperaccumulator *Thlaspi caerulescens* as determined using X-ray absorption spectroscopy. *Environ. Sci. Technol.* 33: 713-717.

Sand, W., K. Rohde, B. Sobotke, and C. Zenneck. 1992. *Evaluation of Leptospirillum ferrooxidans* for leaching. *Appl. Environ. Microbiol.* 58: 85-426.

Sand, W., R. Hallmann, K. Rohde, B. Sobotke, and S. Wentzien. 1993. Controlled microbiological in-situ stope leaching of a sulphidic ore. *Appl. Microbiol. Biotechnol.* 40: 421-426.

Sarret, G., A. Manceau, L. Spadini, J.-C., Roux, J.-L., Hazemann, Y. Soldo, L. Eybert-Bérard, and J.-J. Menthonnex. 1998. Structural determination of Zn and Pb binding sites in *Penicillium chrysogenum* cell walls by EXAFS spectroscopy. *Environ. Sci. Technol.* 32: 1648-1655.

Savvaidis, I., M.N. Hughes, and R.K. Poole. 1992. Differential pulse polarography: a method of directly measuring uptake of metal ions by live bacteria without separation of biomass and medium. *FEMS Microbiol. Lett.* 92: 181-186.

Schiewer, S., and M.H. Wong. 2000. Ionic strength effects in biosorption of metals by marine algae. *Chemosphere* 41: 271-282.

Schweiger, A. 1991. *Angew. Chem.* 103: 223-250.

Scott, J.A., S.J. Palmer. 1988. Cadmium biosorption by bacterial exopolysaccharide. *Biotechnol. Lett.* 10: 21-24.

Shumate II, S.E., and G.W. Strandberg. 1985. Accumulation of metals by microbial cells. pages 235-247 in *Comprehensive Biotechnology*, Vol. 4, M. Moo-Young, Pergamon Press, New York.

Sitting, M. 1981. *Handbook of Toxic and Hazardous Chemicals*, Noyes Publications, Park Ridge, New Jersey.

Solaris, P., A.I., Zouboulis, K.A. Matis, and G.A. Stalidis. 1996. Removal of toxic metals by biosorption onto nonliving sewage sludge. *Separation Sci. Technol.* 31: 1075-1092.

Strandberg, G.W., S.E. Shumate II, and J.R., Parrott. Jr. 1981. Microbial cells as biosorbents of heavy metals: accumulation of uranium by *Saccharomyces cerevisiae* and *Pseudomonas aeruginosa*. *Appl. Environ. Microbiol.* 41: 237-245.

Texier, A.C., Y. Andrès and P. Le Cloirec. 1997. Selective biosorption of lanthanide ions by *Mycobacterium smeymatis*. Environ. Technol. 18: 835-841.

Texier, A.C., Y. Andrès, and P. Le Cloirec. 1999. Selective biosorption of lanthanide (La, Eu, Yb) ions by *Pseudomonas aeruginosa Environ. Sci. Technol.* 33: 489-495.

Texier, A.C., Andrès, Y., M. Illemassene, and P. Le Cloirec. 2000. Characterization of lanthanide ions binding sites in the cell wall of *P. aeruginosa. Environ. Sci. Technol.* 34: 610-615.

Texier, A.C., Y. Andrès, BC. Faur-Brasquet, and P. Le Cloirec. 2002. Fixed bed study for lanthanide (La, Eu, Yb) ions removal from aqueous solutions by immobilized *Pseudomonas aeruginosa:* experimental data and modelization, *Chemosphere* 47: 333-342.

Thomas, W.J., and B. Crittenden. 1998. *Adsorption Technology and Design,* Butterworth Heinemann, Oxford, UK.

Tiemann, K.J., J.L. Gardea-Torresdey, G. Gamez, K. Dokken, S. Sias, M.W. Renner, and L.R. Furenlid. 1999. Use of X-ray absorption spectroscopy and esterification to investigate Cr (III) and Ni (II) ligands in alfalfa biomass. *Environ. Sci. Technol.* 33: 150-154.

Tien, C. 1994. *Adsorption Calculations and Modeling,* Butterworth - Heinemann, Boston, Massachusetts.

Torma, A.E. 1977. The role of *Thiobacillus ferrooxidans* in hydrometallurgical processes. Pages 1-37 in *Advances in Biochemical Engineering,* vol. 6, T.K. Ghose, A. Fiechter and N. Blakebrough, eds., Springer-Verlag, Heidelberg.

Trussel, R.R., and M. Chang. 1999. Review of flow through porous media as applied to head loss in water filters. *J. Environ. Eng.* 125: 998-1006.

Tsezos, M., and B. Volesky. 1982a. The mechanism of uranium biosorption by *Rhizopus arrhizus. Biotechnol. Bioeng.* 24: 385-401.

Tsezos, M., and B. Volesky. 1982b. The mechanism thorium biosorption by *Rhizopus arrhizus. Biotechnol. Bioeng.* 24: 955-969.

Van der Wal, A., W. Norde, A.J.B. Zehnder, and J. Lyklema. 1997. Determination of the total charge in the cell walls of gram-positive bacteria. *Colloids and Surfaces B: Biointerfaces* 9: 81-100.

Veglio, F., and F. Beolchini. 1997. Removal of metals by biosorption: a review. *Hydrometallurgy* 44: 301-316.

Veglio, F., A. Esposito, and A.P. Reverberi. 2003. Standardisation of heavy metal biosorption tests: equilibrium and modeling study. *Process Biochem.* 38: 953-961.

Volesky, B. 1990. Removal and recovery of heavy metals by biosorption. Pages 7-44 in *Biosorption of Heavy Metals,* B. Volesly, ed., CRC Press, Boca Raton, Florida.

Volesky, B. 1992. Removal of heavy metals by biosorption. *Harnessing Biotechnology for the 21st Century*, M.R. Ladisch and A. Bose, eds., Am. Chem. Soc., Washington, DC.

Volesky, B., H. May, and Z.R. Holan. 1993. Cadmium biosorption by *Saccharomyces cerevisiae*. *Biotechnol. Bioeng.* 41: 826-829.

Volesky, B. 1994. Advances in biosorption of metals: Selection of biomass types. *FEMS Microbiol. Rev.* 14: 291-302.

Volesky, B., and I. Prasetyo. 1994. Cadmium removal in a biosorption column. *Biotechnol. Bioeng.* 43: 1010-1015.

Wartelle, L.H., and W.E. Marshall. 2000. Citric acid modified agricultural by-products as copper ion adsorbents. *Adv. Environ. Res.* 4: 1-7.

White C., A.K. Sharma, and G. Gadd. 1998. An integrated microbial process for the bioremediation of soil contaminated with toxic metals. *Nature Biotechnol.* 16: 572-575.

Wobus, A., F. Kloep, K. Röske, and I. Röske. 2003. Influence of population structure on the performance of biofilm reactor. Pages 232-259 in *Biofilm in Wastewater Treatment*, S. Wuertz, P. Bishop and P. Wilderer, eds., IWA Publishing, London, UK.

Zouboulis A.I., K.A. Matis, M. Loukidou, and F. Sebesta. 2003. Metal biosorption by PAN-immobilized fungal biomass in simulated wastewaters. *Colloids and Surface A: Physiochem. Eng. Aspects* 212: 185-195.

Guidance for the Bioremediation of Oil-Contaminated Wetlands, Marshes, and Marine Shorelines

[1]Albert D. Venosa and [2]Xueqing Zhu

[1]U. S. Environmental Protection Agency, 26 W. Martin Luther King Drive, Cincinnati, OH 45268, USA
[2]Department of Civil and Environmental Engineering, University of Cincinnati Cincinnati, OH 45221, USA

Introduction

In the fall of 2001, EPA completed publishing a comprehensive guidance document on the bioremediation of marine shorelines and freshwater wetlands (Zhu *et al*. 2001). Two years later, EPA followed up with a second guidance document devoted exclusively to salt marshes (Zhu *et al*. 2004). This chapter summarizes both documents by incorporating their salient features in one concise report so that readers do not need to refer to the main documents to extract information important to them. If more detailed explanations are desired, one can always refer back to the original documents.

Marine shorelines are important public and ecological resources that serve as a home to a variety of wildlife and provide public recreation. Marine oil spills, particularly large scale spill accidents, have posed great threats and cause extensive damage to the marine coastal environments. For example, the spill of 37,000 metric tons (11 million gallons) of North Slope crude oil into Prince William Sound, Alaska, from the Exxon Valdez in 1989 led to the mortality of thousands of seabirds and marine mammals, a significant reduction in population of many intertidal and subtidal organisms, and many long-term environmental impacts (Spies *et al*. 1996). In 1996, the Sea Empress released approximately 72,000 tons of Forties crude oil and 360 tons of heavy fuel oil at Milford Haven in South Wales and posed a considerable threat to local fisheries, wildlife, and tourism (Edwards and White 1999, Harris 1997).

Compared to marine oil spills, inland oil spills have received much less attention. However, freshwater spills are very common, with more than 2000 oil spills, on average, taking place each year in the inland waters of the continental United States (Owens *et al.* 1993). Although freshwater spills tend to be of a smaller volume than their marine counterparts (Stalcup *et al.* 1997), they have a greater potential to endanger public health and the environment because they often occur within populated areas and may directly contaminate surface water and groundwater supplies.

Catastrophic accidents have increased public awareness about the risks involved in the storage and transportation of oil and oil products and have prompted more stringent regulations, such as the enactment of the 1990 Oil Pollution Act by Congress (OPA90). However, because oil is so widely used, despite all the precautions, it is almost certain that oil spills and leakage will continue to occur. Thus, it is essential that we have effective countermeasures to deal with the problem.

Coastal wetlands are influenced by tidal action. They provide natural barriers to shoreline erosion, habitats for a wide range of wildlife including endangered species, and key sources of organic materials and nutrients for marine communities (Mitsch and Gosselink 2000). Coastal wetlands may be classified into tidal salt marshes, tidal fresh water marshes, and mangrove swamps (Mitsch and Gosselink 2000).

In the early 1990s, it was estimated that the total area of coastal wetlands in the United States was approximately 3.2 million ha (32,000 km^2), with about 1.9 million ha or 60 percent of the total coastal wetlands as salt marshes and 0.5 million ha as mangrove swamps (Mitsch and Gosselink 2000). Coastal wetlands are no longer viewed as intertidal wastelands, and their ecological and economic values have been increasingly recognized.

The threat of crude oil contamination to coastal wetlands is particularly high in certain parts of the U.S., such as the Gulf of Mexico, where oil exploration, production, transportation, and refineries are extensive (Lin and Mendelssohn 1998). Oil and gas extraction activities in coastal marshes along the Gulf of Mexico have been one of the leading causes of wetland loss in the 1970s (Mitsch and Gosselink 2000). Despite more stringent environmental regulations, the risk of an oil spill affecting these ecosystems is still high because of extensive coastal oil production, refining, and transportation.

Although conventional methods, such as physical removal, are the first response option, they rarely achieve complete cleanup of oil spills. According to the Office of Technology Assessment (OTA 1990), current mechanical methods typically recover no more than 10-15% of the oil after a major spill. Bioremediation has emerged as a highly promising secondary treatment option for oil removal since its application after the 1989 Exxon Valdez spill (Bragg *et al.* 1994, Prince *et al.* 1994). Bioremediation has been defined as "the act of adding materials to contaminated environments to cause an acceleration of the natural biodegradation processes" (OTA 1991). This technology is based on the premise that a large percentage of oil components are readily biodegradable in nature (Atlas 1981, 1984, Prince 1993). The success of oil spill bioremediation depends on our ability to establish and maintain conditions that favor enhanced oil biodegradation rates in the contaminated environment. There are two main approaches to oil spill bioremediation:

- Bioaugmentation, in which known oil-degrading bacterial cultures are added to supplement the existing microbial population, and

- Biostimulation, in which the growth of indigenous oil degraders is stimulated by the addition of nutrients or other growth-limiting substrates, and/or by alterations in environmental conditions (e.g., surf-washing, oxygen addition by plant growth, etc.).

Both laboratory studies and field tests have shown that bioremediation, biostimulation in particular, may enhance the rate and extent of oil biodegradation on contaminated shorelines (Prince 1993, Swannell *et al.* 1996). Recent field studies have also demonstrated that addition of hydrocarbon degrading microorganisms will not enhance oil degradation more than simple nutrient addition (Lee *et al.* 1997a, Venosa *et al.* 1996, Zhu *et al.* 2001). Bioremediation has several advantages over conventional technologies. First the application of bioremediation is relatively inexpensive. For example, during the cleanup of the Exxon Valdez spill, the cost of bioremediating 120 km of shoreline was less than one day's costs for physical washing (Atlas 1995). Bioremediation is also a more environmentally benign technology since it involves the eventual degradation of oil to mineral products (such as carbon dioxide and water), while physical and chemical methods typically transfer the contaminant from one environmental compartment to another. Since it is based on natural processes and is less intrusive and disruptive to the contaminated site, this "green technology" may also be more acceptable to the general public.

Bioremediation also has its limitations. Bioremediation involves highly heterogeneous and complex processes. The success of oil bioremediation depends on having the appropriate microorganisms in place under suitable environmental conditions. Its operational use can be limited by the composition of the oil spilled. Bioremediation is also a relatively slow process, requiring weeks to months to take effect, which may not be feasible when immediate cleanup is demanded. Concerns also arise about potential adverse effects associated with the application of bioremediation agents. These include the toxicity of bioremediation agents themselves and metabolic by-products of oil degradation and possible eutrophic effects associated with nutrient enrichment (Swannell *et al.* 1996). Bioremediation has been proven to be a cost-effective treatment tool, if used properly, in cleaning certain oil-contaminated environments. Few detrimental treatment effects have been observed in actual field operations.

Currently, one of the major challenges in the application of oil bioremediation is the lack of guidelines regarding when and how to use this technology. Although extensive research has been conducted on oil bioremediation during the last decade, most existing studies have concentrated on either evaluating the feasibility of bioremediation for dealing with oil contamination, or testing favored products and methods (Mearns 1997). Only a limited number of pilot-scale and field trials, which may provide the most convincing demonstrations of this technology, have been carried out. To make matters worse, many field tests have not been properly designed, well controlled, or correctly analyzed, leading to skepticism and confusion among the user community (Venosa 1998). The need exists for a detailed and workable set of guidelines for the application of this technology for oil spill responders that answers questions such as when to use bioremediation, what bioremediation agents should be used, how to apply them, and how to monitor and evaluate the results. Scientific data for the support of an operational guidelines document has recently been provided from laboratory studies and field trials carried out by EPA, University of Cincinnati, and Fisheries and Oceans Canada.

Biostimulation (Nutrient Amendment)

Nutrient addition has been proven to be an effective strategy to enhance oil biodegradation in various marine shorelines. Theoretically, approximately 150 mg of nitrogen and 30 mg phosphorus are consumed in the conversion of 1,000 mg of hydrocarbon to cell material (Rosenberg and Ron 1996). Therefore, a commonly used strategy has been to add nutrients at concentrations that approach a stoichiometric ratio of C:N:P of 100:5:1. Recently, the potential application of resource-ratio theory in hydrocarbon biodegradation was discussed (Head and Swannell 1999, Smith *et al.* 1998). This theory suggests that manipulating the N:P ratio may result in the enrichment of different microbial populations, and the optional N:P ratio can be different for degradation of different compounds (such as hydrocarbons mixed in with other biogenic compounds in soil). However, the practical use of these ratio-based theories remains a challenge. Particularly, in marine shorelines, maintaining a certain nutrient ratio is impossible because of the dynamic washout of nutrients resulting from the action of tides and waves. A more practical approach is to maintain the concentrations of the limiting nutrient or nutrients within the pore water at an optimal range (Bragg *et al.* 1994, Venosa *et al.* 1996). Commonly used nutrients include water soluble nutrients, solid slow-release nutrients, and oleophilic fertilizers. Each type of nutrient has its advantages and limitations.

Water soluble nutrients. Commonly used water soluble nutrient products include mineral nutrient salts (e.g., KNO_3, $NaNO_3$, NH_3NO_3, K_2HPO_4, $MgNH_4PO_4$), and many commercial inorganic fertilizers (e.g., the 23:2 N:P garden fertilizer used in the Exxon Valdez case). They are usually applied in the field through the spraying of nutrient solutions or spreading of dry granules. This approach has been effective in enhancing oil biodegradation in many field trials (Swannell *et al.* 1996, Venosa *et al.* 1996). Compared to other types of nutrients, water soluble nutrients are more readily available and easier to manipulate to maintain target nutrient concentrations in interstitial pore water. Another advantage of this type of nutrient over organic fertilizers is that the use of inorganic nutrients eliminates the possible competition of carbon sources. The field study by Lee *et al.* (1995b) indicated that although organic fertilizers had a greater effect on total heterotrophic microbial growth and activity, the inorganic nutrients were much more effective in stimulating crude oil degradation.

However, water soluble nutrients also have several potential disadvantages. First, they are more likely to be washed away by the actions of tides and waves. A field study in Maine demonstrated that water soluble nutrients might be washed out within a single tidal cycle on high-energy beaches (Wrenn *et al.* 1997a). Second, inorganic nutrients, ammonia in particular, should be added carefully to avoid reaching toxic levels. Existing field trials, however, have not observed acute toxicity to sensitive species resulting from the addition of excess water soluble nutrients (Mearns *et al.* 1997, Prince *et al.* 1994). Third, water soluble nutrients may have to be added more frequently than slow release nutrients or organic nutrients, resulting in more labor-intensive, costly, and physically intrusive applications.

Granular nutrients (slow-release). Many attempts have been made to design nutrient delivery systems that overcome the washout problems characteristic of intertidal environments (Prince 1993). Use of slow release fertilizers is one of the approaches used to provide continuous sources of nutrients to oil contaminated areas. Slow release fertilizers are normally in solid forms that consist of inorganic nutrients coated with hydrophobic materials like paraffin or vegetable oils. This approach may also cost less than adding water-soluble nutrients due to less frequent applications.

Olivieri *et al.* (1976) found that the biodegradation of a crude oil was considerably enhanced by addition of a paraffin coated $MgNH_4PO_4$. Another slow-release fertilizer, Customblen (vegetable oil coated calcium phosphate, ammonium phosphate, and ammonium nitrate), performed well on some of the shorelines of Prince William Sound, particularly in combination with an oleophilic fertilizer (Atlas 1995, Pritchard *et al.* 1992, Swannell *et al.* 1996). Lee and Trembley (1993) also showed that oil biodegradation rates increased with the use of a slow-release fertilizer (sulfur-coated urea) compared to water soluble fertilizers.

However, the major challenge for this technology is control of the release rates so that optimal nutrient concentrations can be maintained in the pore water over long time periods. For example, if the nutrients are released too quickly, they will be subject to rapid washout and will not act as a long-term source. On the other hand, if they are released too slowly, the concentration will never build up to a level that is sufficient to support rapid biodegradation rates, and the resulting stimulation will be less effective than it could be.

Oleophilic nutrients. Another approach to overcome the problem of water soluble nutrients being rapidly washed out was to utilize oleophilic organic nutrients (Atlas and Bartha 1973, Ladousse and Tramier 1991). The rationale for this strategy is that since oil biodegradation mainly occurs at the oil-water interface and since oleophilic fertilizers are able to adhere to oil and provide nutrients at the oil-water interface, enhanced biodegradation should result without the need to increase nutrient concentrations in the bulk pore water.

Variable results have also been produced regarding the persistence of oleophilic fertilizers. Some studies showed that Inipol EAP 22, an oleophilic fertilizer, could persist in a sandy beach for a long time under simulated tide and wave actions (Santas and Santas 2000, Swannell *et al.* 1995). Others found that Inipol EAP 22 was rapidly washed out before becoming available to hydrocarbon-degrading bacteria (Lee and Levy 1987, Safferman 1991). Another disadvantage with oleophilic fertilizers is that they contain organic carbon, which may be biodegraded by microorganisms in preference to petroleum hydrocarbons (Lee *et al.* 1995b, Swannell *et al.* 1996), and may also result in undesirable anoxic conditions (Lee *et al.* 1995a, Sveum and Ramstad 1995).

In summary, the effectiveness of these various types of nutrients will depend on the characteristics of the contaminated environment. Slow-release fertilizers may be an ideal nutrient source if the nutrient release rates are well controlled. Water-soluble fertilizers are likely more cost-effective in low-energy and fine-grained shorelines where water transport is limited. And oleophilic fertilizers may be more suitable for use in high-energy and coarse-grained beaches or rocky outcroppings.

Bioaugmentation (Microbial Amendments)

The rationale for adding microbial cultures to an oil-contaminated site includes the contention that indigenous microbial populations may not be capable of degrading the wide range of substrates that are present in complex mixtures such as petroleum and that seeding may reduce the lag period before bioremediation begins (Forsyth *et al.* 1995, Leahy and Colwell 1990). For this approach to be successful in the field, the seed microorganisms must be able to degrade most petroleum components, maintain genetic stability and viability during storage, survive in foreign and hostile environments, effectively compete with indigenous microorganisms already adapted to the environmental conditions of the site, and move through the pores of the sediment to the contaminants (Atlas 1977, Goldstein *et al.* 1985).

Many vendors of bioremediation products claim their product aids the oil biodegradation process. The U.S. EPA has compiled a list of bioremediation agents (USEPA 2000) as part of the National Oil and Hazardous Substances Pollution Contingency Plan (NCP) Product Schedule, which is required by the Clean Water Act, the Oil Pollution Act of 1990, and the National Contingency Plan for a product to be used as an oil spill countermeasure. However, even though the addition of microorganisms may be able to enhance oil biodegradation in the laboratory, its effectiveness has never been convincingly demonstrated in the field (Zhu *et al.* 2004). In fact, field studies have indicated that bioaugmentation is not effective in enhancing oil biodegradation in marine shorelines, and nutrient addition or biostimulation alone had a greater effect on oil biodegradation than the microbial seeding (Jobson *et al.* 1974, Lee and Levy 1987, Lee *et al.* 1997b, Venosa *et al.* 1996). The failure of bioaugmentation in the field may be attributed to the fact that the carrying capacity of most environments is likely determined by factors that are not affected by an exogenous source of microorganisms (such as predation by protozoans, the oil surface area, or scouring of attached biomass by wave activity), and that added bacteria seem to compete poorly with the indigenous population (Tagger *et al.* 1983, Lee and Levy 1989, Venosa *et al.* 1992). Therefore, it is unlikely that exogenously added microorganisms will persist in a contaminated beach even when they are added in high numbers.

Fortunately, oil-degrading microorganisms are ubiquitous in the environment, and they can increase rapidly by many orders of magnitude after being exposed to crude oil (Atlas 1981, Lee and Levy 1987, Pritchard and Costa 1991). Therefore, in most environments, there is usually no need to add hydrocarbon degraders.

Guidelines

Typically, bioremediation is used as a polishing step after conventional mechanical cleanup options have been applied, although it could also be used as a primary response strategy if the spilled oil does not exist as free product and if the contaminated area is remote enough not to require immediate cleanup or not accessible by mechanical tools. However, one of the major challenges in the application of oil bioremediation is lack of guidelines regarding the selection and use of this technology. Although extensive research has been conducted on oil bioremediation in the last decade, most existing studies have been concentrated on either evaluating the feasibility of bioremediation for dealing with oil contamination or testing favored products and methods (Mearns 1997). Only a few limited operational guidelines for bioremediation in marine shorelines have been proposed (Lee 1995, Lee and Merlin 1999, Swannell *et al.* 1996). As a result of recent field studies (Lee *et al.* 1997b, Venosa *et al.* 1996), we now know that there is usually little need to add hydrocarbon-degrading microorganisms because this approach has been shown not to enhance oil degradation more than simple nutrient addition. Therefore, the guidelines that have been developed for oil bioremediation are confined to using biostimulation strategies, mainly nutrient addition, to accomplish the cleanup.

A general procedure or plan for the selection and application of bioremediation technology is illustrated in Figure 1. The major steps in a bioremediation selection and response plan include:

(1) Pre-treatment assessment – This step involves the determination of whether bioremediation is a viable option based on the type of oil that has been spilled, its concentration, the presence of hydrocarbon-degrading microorganisms, concentrations of background nutrients, the type of shoreline that has been impacted, and other environmental factors (pH, temperature, presence of oxygen, remoteness of the site, logistics, etc.).

(2) Design of treatment and monitoring plan – After the decision is made to use bioremediation, further assessments and planning are needed prior to the application. This involves selection of the rate-limiting treatment agents (e.g., nutrients), determination of application strategies for the rate-limiting agents, and design of sampling and monitoring plans.

(3) Assessment and termination of treatment – After the treatment is implemented according to the plan, assessment of treatment efficacy and determination of appropriate treatment endpoints are performed based on chemical, toxicological, and ecological analysis.

The overall flow diagram describing the steps one should follow in deciding whether and how to bioremediate an oil-contaminated site is shown below (Zhu *et al.* 2001):

The major steps in the above diagram are described in more detail below.

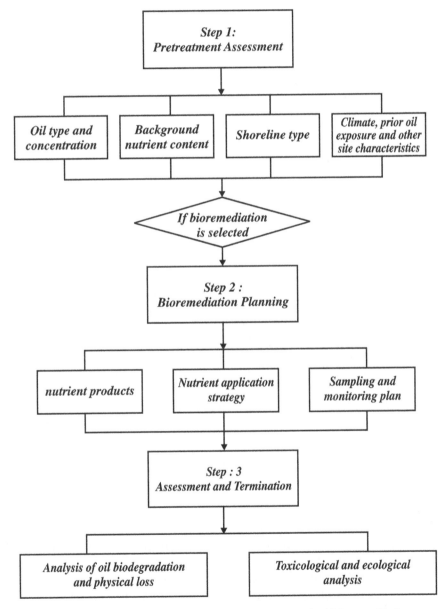

Figure 1. Procedures tor the selection and application of oil bioremediation.

Pre-Treatment Assessment

Pretreatment assessment involves some preliminary investigations to assess whether bioremediation is a viable option and to define the rate-determining process, which include the evaluation of (1) oil types and concentrations, (2) background nutrient content, (3) shoreline types, and (4) other environmental factors such as the prevalent climate and prior oil exposures.

Oil type. The degradation rate for the same oil components may vary significantly for different oils. It has been found that the rate and extent of biodegradation of biodegradable components (e.g., n-alkanes) decreases with the increase of non-biodegradable fractions (e.g., resins and asphaltenes) (Uraizee *et al.* 1998, Westlake *et al.* 1974). Therefore, the heavier crude oils are likely to be less biodegradable than lighter crude oils. McMillen *et al.* (1995) investigated the biodegradability of 17 crude oils with API gravity ranging from 14° to 45°. They concluded that crude oil with greater than 30° API gravity can be considered readily biodegradable, and those with less than 20° API gravity (heavier oils) are slow to biodegrade. Similar results were obtained by other researchers (Hoff *et al.* 1995, Sugiura *et al.* 1997). Wang and Bartha (1990) also investigated the effects of bioremediation on residues of fuel spills in soils. The results showed that the treatability by bioremediation for the fuel residues are in the order of jet fuel > heating oil > diesel oil. However, more work is still required to classify crude oils and refined products with respect to their theoretical amenability to cleanup by bioremediation. Field experience has suggested that oils that have been subjected to substantial biodegradation might not be amenable to bioremediation due to the accumulation of polar components in the oils (Bragg *et al.* 1994, Oudet *et al.* 1998).

Oil concentration. For sites contaminated with oils at low concentrations, biodegradation is also less likely to be limited by nutrients or oxygen. Therefore, bioremediation may not be effective in enhancing biodegradation in these cases. Natural attenuation may be a more viable option. High concentrations of hydrocarbons may cause inhibition of biodegradation due to toxic effects, although the inhibitory concentration varies with oil composition. Therefore, there should be an optimum oil concentration range for bioremediation applications, below which degradation is not easily stimulated, and above which inhibition occurs. However, this concentration range, particularly the maximum concentration of oil amenable to bioremediation, has not been well quantified. Field experiences in Prince William Sound, Alaska, showed that less than 15g oil/kg sediments could be treated using bioremediation (Swannell *et al.* 1996). Xu *et al.* (2001) recently investigated the effect of oil concentration in a microcosm study using weathered Alaska North Slope crude oil. The results showed that crude oil concentrations as high as 80 g oil/kg dry sand were still amenable to biodegradation. Favorable oil concentrations for bioremediation are also related to background conditions, such as shoreline types, which will be discussed later.

Background nutrient content. Assessment of background nutrient concentrations is critical in determining whether bioremediation should be considered a viable option, whether natural attenuation should be considered, and/or which nutrient (nitrogen or phosphorus) should be added for oil bioremediation. In marine environments, nutrients are generally limiting due to the naturally low nitrogen and phosphorus concentrations in seawater (Floodgate 1984). Nutrient content is more variable in freshwater systems and is normally abundant in freshwater wetlands (Cooney 1984, Mitsch and Gosselink 1993). However, background nutrients also depend on other site characteristics such as local industrial and domestic effluents and agricultural runoff.

Recent field studies indicate that natural nutrient concentrations in some marine shorelines may be high enough to sustain rapid intrinsic rates of biodegradation without human intervention (Oudet et al. 1998, Venosa et al. 1996) The field trial in Delaware (Venosa et al. 1996) showed that although biostimulation with inorganic mineral nutrients significantly accelerated the rate of hydrocarbon biodegradation, the increase in biodegradation rate over the intrinsic rate (i.e., slightly greater than twofold for the alkanes and 50% for the PAHs) would not be high enough to warrant a recommendation to actively initiate a major, perhaps costly, bio-remediation action in the event of a large crude oil spill in that area. The investigators observed that maintenance of a threshold nitrogen concentration of 3-6 mg N/L in the interstitial pore water was stimulatory for hydrocarbon biodegradation.

A similar conclusion was also reached in a field trial to evaluate the influence of a slow-release fertilizer on the biodegradation rate of crude oil spilled on interstitial sediments of an estuarine environment in the Bay of Brest, France (Oudet et al. 1998). Due to the high background levels of N and P at the study site, no significant difference in biodegradation rates was detected following nutrient addition. It was proposed that bioremediation by nutrient enrichment would be of limited use if background interstitial pore water levels of N exceed 1.4 mg/L, which is close to the finding from the aforementioned Delaware study.

The recommendation is that, in the event of a catastrophic oil spill impacting a shoreline, one of the first tasks in pretreatment assessment is to measure the natural nutrient concentrations within the interstitial water in that environment. If they are high enough, further investigation is required to determine whether such a nutrient loading is typical for that area and season (i.e., determine the impact of chronic runoff from nearby agricultural practice and local industrial and domestic effluents). The decision to use bioremediation by addition of nutrients should be based on how high the natural levels are relative to the optimal or threshold nutrient concentrations.

Types of shorelines. The characteristics of impacted site play an important role in the decision to use bioremediation. Preliminary investigation involves the assessment of the need for bioremediation based on wave and tidal energy, the sediment characteristics, and geomorphology of the shoreline.

Shoreline energy and hydrology. Oil may be removed rather rapidly under high wave and tide influence. In high-energy environments, bioremediation products are also more difficult to apply successfully since they may be washed out rapidly. High wave energy will also scour degrading microorganisms attached to the sediment particles, and diminish the net oil biodegradation rate that can be achieved. A tracer study conducted in Scarborough, Maine, demonstrated that washout rate of nutrients from the bioremediation zone will be strongly affected by the wave activity of the contaminated beach (Wrenn *et al.* 1997). However, washout due to tidal activity alone in the absence of significant wave action is relatively slow, and nutrients will probably remain in contact with oiled beach material long enough to effectively stimulate oil biodegradation on such beaches.

However, many of the same characteristics that make low-energy beaches favorable for bioremediation cleanup from a nutrient persistence perspective might make other conditions unfavorable with respect to other important factors. For example, availability of oxygen is more favorable on high-energy beaches than on low-energy beaches. Aeration mechanisms for near-surface coastal sediments involve exchange of oxygenated surface water with oxygen-depleted pore water by wave-induced pumping and tidal pumping. For low energy beaches, tidal pumping is the only likely aeration mechanism, and as a result, the surface sediments are more likely to be anoxic than are similar depths on high-energy beaches (Brown and McLachlan 1990). The probability of moisture (or water activity) limitation is also higher on low-energy beaches, because wave run up provides water to supratidal sediments on high-energy beaches during neap tides. Therefore, it is essential to thoroughly characterize the factors that are likely to be rate limiting on each contaminated site before deciding and designing a bioremediation response strategy.

Shoreline substrate. Although successful bioremediation application and field trials have been carried out on cobble, medium sand, fine sand,

and some salt marsh shorelines (Bragg *et al.* 1994, Lee and Levy 1991, Swannell *et al.* 1999a, Venosa *et al.* 1996), different shoreline substrates or sediment types will affect the feasibility and strategies of using bioremediation. In a 7-month field study, Lee and Levy (1991) compared the bioremediation of a waxy crude oil on a sandy beach and a salt marsh shoreline at two oil concentrations, 3% (v/v) and 0.3% (v/v) to beach sand and salt marsh sediments. At the lower oil concentrations within the sand beach, oil biodegradation proceeded equally rapidly in both the fertilized plot and the unfertilized control. However, at the higher oil concentrations on the sandy beach, oil biodegradation rates were enhanced by nutrient addition. In contrast, addition of nutrients to the salt marsh sediments containing the lower (0.3%) oil concentration resulted in enhanced rates of biodegradation. This additional need for nutrients at the lower oil concentrations is consistent with the notion that nutrient demands within a salt marsh environment are higher, due to the size of the microbial population within an organic-carbon rich environment. At the higher oil concentration (3%) within the salt marsh sediments, insignificant rates of oil degradation were reported following fertilization. The results clearly demonstrated that the success of bioremediation depends on the characteristics of the shoreline and the factors that limit biodegradation. On the sandy beach, nutrients are likely the limiting factor; however, on a salt marsh, oxygen availability is the key limitation. Similar results have been obtained in the field study conducted in a freshwater wetland (Garcia-Blanco *et al.* 2001, Venosa *et al.* 2002), which also indicated that oxygen availability was likely a major rate-limiting factor in the wetland environments.

Summary of pretreatment assessment. In summary, the following pretreatment assessments should be conducted to determine whether bioremediation is a viable option in response to a spill incident:

- Determine whether the spilled oil is potentially biodegradable.
- Determine whether the nutrient content at the impacted area is likely to be an important limiting factor by measuring the background nutrient concentrations within the interstitial water in that environment.
- Determine whether the shoreline characteristics are favorable for using bioremediation-high-energy rocky beaches and some low energy shorelines such as some wetlands are considered not likely to be very amenable to nutrient addition.

Selection of Nutrient Products

After bioremediation is determined to be a potentially effective cleanup option based on the preliminary investigations, further assessments and planning are needed before applying it. The first task is to select appropriate nutrient products through both screening tests and assessments based on characteristics of the contaminated site.

Nutrient selection based on efficacy and toxicity. To assist response personnel in the selection and use of spill bioremediation agents, it is useful to have some simple, standard methods for screening the performance and toxicity of bioremediation products as they become available (Blenkinsopp *et al.* 1995, Haines *et al.* 1999, Lepo and Cripe 1998a). EPA uses a tiered approach (NETAC 1993, Thomas *et al.* 1995), which provides empirical evidence through the use of laboratory shake flask treatability studies to estimate a product's effectiveness in accelerating biodegradation of weathered crude oil. It also provides information on the relative changes in aliphatic and aromatic oil constituent concentrations over time. Conducting treatability tests using batch or flow-through micro- or mesocosms is another commonly used approach.

Field studies provide the most convincing demonstration of the effectiveness of oil bioremediation because laboratory studies simply cannot simulate real world conditions such as spatial heterogeneity, climate change, and mass transfer limitations. Since conducting a field study just to determine that a product might work is unrealistic and economically burdensome, the practical approach in selection of nutrient products for the bioremediation of an oil spill would be through laboratory tests, microcosm tests in particular, as well as evaluations based on existing field study results in similar environmental conditions.

Effect of nutrient type may also depend on the properties of shoreline substrates. Jackson and Pardue (1999) found that addition of ammonia as compared to nitrate appeared to more effectively simulate degradation of crude oil in salt marsh soils in a microcosm study. The ammonia requirement was only 20% of the concentration of nitrate to achieve the same increase in degradation. The authors concluded that ammonia was less likely to be lost from the microcosms by washout due to its higher adsorptive capacity to sediment organic matter. However, in a microcosm study using sandy sediments, it was found that there were no significant differences in the nutrient washout rates or the abilities of ammonium and nitrate to support oil biodegradation. These results suggest that although adsorption may be an important difference between ammonium and nitrate in sediments with high cation-exchange capacities (CECs), such as marsh sediments, it is unlikely to be significant in sediments with low CECs, such as sand.

Determination of the Optimal Nutrient Loading and Application Strategy

After the initial selection of nutrient products that meet the requirements of efficacy and safety, the next step is to determine the proper nutrient loading and the best nutrient application strategies. Major considerations in this task include the determination of optimal nutrient concentration, frequency of addition, and methods of addition. Finally, selection of appropriate nutrient products should also be conducted in conjunction with this process.

Concentration of nutrients needed for optimal biostimulation. Since oil biodegradation largely takes place at the interface between oil and water, the effectiveness of biostimulation depends on the nutrient concentration in the interstitial pore water of oily sediments (Bragg *et al.* 1994, Venosa *et al.* 1996). The nutrient concentration should be maintained at a high enough level to support maximum oil biodegradation based on the kinetics of nutrient consumption. Higher concentrations will not only provide no added benefit but also may lead to potentially detrimental ecological and toxicological impacts.

Studies on optimal nutrient concentrations have been conducted both in the laboratory and in the field. Boufadel *et al.* (1999) investigated the optimal nitrate concentration for alkane biodegradation in continuous flow beach microcosms using heptadecane as a model alkane immobilized onto sand particles at a loading of 2 g heptadecane/kg sand. They determined that a continuous supply of approximately 2.5 mg N/L supported maximum heptadecane biodegradation rates. Du *et al.* (1999) also investigated the optimal nitrogen concentration for oil biodegradation using weathered Alaska North Slope crude oil in the same microcosms with an oil loading of 5 g/kg sand. The results showed that nitrate concentrations below approximately 10 mg N/L limited the rate of oil biodegradation. The higher nutrient requirement was attributed to the more complex substrate (crude oil) compared to the pure heptadecane of Boufadel *et al.* (1999). The more complex substrate (crude oil) of Du *et al.* (1999) also likely selected a different population of degraders than those that grew on the pure heptadecane (Boufadel *et al.* 1999), which might have contributed to the different growth rate characteristics observed.

Ahn (1999) further studied the effect of nitrate concentrations under tidal flow conditions instead of continuous flow. He used the same beach microcosms as Du *et al.* (1999) filled with sand loaded with weathered Alaska North Slope crude oil at 5 g/kg sand. A nutrient solution with nitrate concentrations ranging from 6.25 to 400 mg N/L was supplied semi-diurnally to simulate tidal flow. The results indicated that the optimum nitrate concentration for maximum oil biodegradation rate was over 25 mg N/L. Some laboratory studies have reported that greater than 100 mg N/L was required to stimulate maximum biodegradation rates (Atlas and Bartha 1992, Reisfeld *et al.* 1972), but this observation probably reflects a stoichiometric rather than a kinetic requirement, since these experiments were conducted in closed batch reactors.

Compared to the results from laboratory studies, nutrient concentrations that supported high oil biodegradation rates were found to be lower in field studies. For example, the field tests that were conducted after the Exxon Valdez oil spill in Prince William Sound, Alaska, showed that the rate of oil biodegradation was accelerated by average interstitial nitrogen concentrations of about 1.5 mg N/L (Bragg *et al.* 1994). A similar result was obtained from the study conducted in the Bay of Brest, France, in which nitrogen was not a limiting factor when the interstitial pore water concentrations exceeded 1.4 mg N/L (Oudet *et al.* 1998). The Delaware field trial also showed that the background nitrate concentration (0.8 mg N/L) was sufficient to support fairly rapid natural (but not maximal) rates of alkane and PAH biodegradation (Venosa *et al.* 1996). Increasing the average nitrate concentration in the bioremediation zone of the experimental plots to between 3 and 6 mg N/L resulted in a moderate increase in the oil biodegradation rate.

Observations from the referenced field studies suggest that concentrations of approximately 1 to 2 mg/L of available nitrogen in the interstitial pore water is sufficient to meet the minimum nutrient requirement of the oil degrading microorganisms for the approximately 6-hour exposure time to the contaminated substrate during a tidal cycle. However, laboratory microcosm results as well as the Delaware field study suggest that higher concentrations of nitrogen can lead to accelerated hydrocarbon biodegradation rates. Since the minimum nitrogen concentration needed to satisfy the nitrogen demand in a tidal cycle is 1 to 2 mg N/L, and since concentrations of nitrogen in pore water that lead closer to maximum rates of biodegradation can be several-fold to as much as an order of magnitude higher, it is recommended that biostimulation of oil impacted beaches should occur when nitrogen concentrations of at least 2 to as much as to 5-10 mg N/L are maintained in the pore water with the decision on higher concentrations to be based on a broader analysis of cost, environmental impact, and practicality.

The frequency of nutrient addition to maintain the optimal concentration in the interstitial pore water mainly depends on shoreline types or nutrient washout rates. On marine shorelines, contamination of coastal areas by oil from offshore spills usually occurs in the intertidal zone where the washout of dissolved nutrients can be extremely rapid.

Oleophilic and slow-release formulations have been developed to maintain nutrients in contact with the oil, but most of these rely on dissolution of the nutrients into the aqueous phase before they can be used by hydrocarbon degraders (Safferman 1991). Therefore, understanding the transport of dissolved compounds in intertidal environments is critical in designing nutrient addition strategies, no matter what type of fertilizer is used.

The Maine field study on nutrient hydrodynamics demonstrated that during spring tide, nutrients could be completely removed from a high-energy beach within a single tidal cycle. It may take more than two weeks to achieve the same degree of washout from a low-energy beach. Washout during the neap tide can be much slower because the bioremediation zone will be only partially covered by water in this period. Since nutrients may be completely washed out from high-energy beaches within a few days, and remain in low energy beaches for several weeks, the optimal frequency of nutrient application should be based on observations of the prevalent tidal and wave conditions in the bioremediation zone. For example, a daily nutrient application may be needed for a high-energy beach during spring tide. But weekly or monthly additions may be sufficient for low-energy beaches when the nutrients are applied during neap tide. Nutrient sampling, particularly in beach pore water, must also be coordinated with nutrient application to ensure that the nutrients become distributed throughout the contaminated area and that target concentrations are being achieved. The frequency of nutrient addition should be adjusted based on the nutrient monitoring results.

Methods of nutrient addition. Nutrient application methods should be determined based on the characteristics of the contaminated environment, physical nature of the selected nutrients, and the cost of the application. Shoreline energy and geometry are important factors in determination of nutrient application methods. The tracer study in Maine (Wrenn *et al.* 1997) suggested that surface application of nutrients may be ineffective on high-energy beaches because wave action in the upper intertidal zone may cause nutrients from the surface layers of the beach to be diluted directly into the water column, resulting in their immediate loss from the bioremediation zone. Daily application of water-soluble nutrients onto the beach surface at low tide could be a feasible approach (Venosa *et al.* 1996), although this method is highly labor-intensive. Nutrients that are released from slow-release or oleophilic formulations will probably behave similarly to water-soluble nutrients with respect to nutrient washout. Formulations with good long-term release characteristics probably will never achieve optimal nutrient concentrations in environments with high washout rates. Therefore, they will not be effective on high-energy beaches unless the release rate is designed to be high enough to achieve adequate nutrient concentrations while the tide is out.

Compared to high-energy shorelines, application of nutrients on low-energy beaches is much less problematic. Since washout due to tidal activity alone is relatively slow, surface application of nutrients is an effective and economical bioremediation strategy on low-energy beaches.

Dry granular fertilizers may be slow-release (e.g., Customblen in Alaska) or water soluble, solid granules (e.g., prilled ammonium nitrate). Granular fertilizers are easier and more flexible to apply using commercially available whirlybird-type hand spreaders. Although this type of fertilizer is also subject to washout by wave and tidal action, dry granular fertilizers are probably the most cost-effective way to control nutrient concentrations. Liquid oleophilic nutrients are also relatively easy to apply by using hand-held or backpack sprayers. This type of fertilizer is significantly more expensive than granular fertilizers. Water-soluble nutrient solutions are normally delivered to the beach by a sprinkler system after dissolving nutrient salts in a local water source. Although this type of nutrient may be easier to manipulate to maintain target concentrations in interstitial pore water, its application may require more complicated equipment such as large mixing tanks, pumps, and sprinklers. Also, use of sprinklers in a seawater environment is problematic since saltwater causes clogging of the nozzles, requiring frequent maintenance.

Based on current experiences and understandings, application of dry granular fertilizer to the impact zone at low tide is probably the most cost-effective way to control nutrient concentrations.

Sampling and Monitoring Plan. To properly evaluate the progress of bioremediation, a comprehensive and statistically valid sampling and monitoring plan must be developed before the application of bioremediation. The sampling and monitoring plan should include important efficiency and toxicity variables, environmental conditions, and sampling strategies.

Important variables to be monitored in an oil bioremediation project include limiting factors for oil biodegradation (e.g., interstitial nutrient and dissolved oxygen concentrations), evidence of oil biodegradation (e.g., concentrations of oil and its components), environmental effects (e.g., ecotoxicity levels), and other water quality variables (e.g., temperature and pH). A monitoring plan for a full-scale bioremediation application should include as a minimum those measurements as critical variables.

Since oil biodegradation in the field is usually limited by availability of nutrients, nutrient analysis, particularly the nutrient concentrations in the pore water, is one of the most important measurements in developing proper nutrient addition strategies and assessing the effect of oil bioremediation. The frequency of nutrient sampling must be coordinated with nutrient application, making certain that the treatment is reaching and penetrating the impact zone, target concentrations of nutrients are being achieved, and toxic nutrient levels are not being reached. Otherwise, nutrient application strategies should be adjusted accordingly. The location from which nutrient samples are collected is also important. Recent research on solute transport in the intertidal zone has shown that nutrients may remain in the beach subsurface for much longer time periods than in the bioremediation zone (Wrenn *et al.* 1997). Nutrient concentration profiles along the depth of the oil-contaminated region may be monitored by using multi-port sample wells or sand samples collected from the oil-contaminated region.

The oil sampling depth should be determined based on the preliminary survey of oil distribution. It can be established by determining the maximum depth of oil penetration, then adding a safety factor, which will be chosen based on the observed variation in oiled depth, to ensure that the samples will encompass the entire oiled depth throughout the project. The safety factor will be modified if observations during the bioremediation application suggest that the depth of oil penetration has changed.

The success of oil bioremediation will be judged by its ability to reduce the concentration and environmental impact of oil in the field. To effectively monitor biodegradation under highly heterogeneous conditions, it is necessary that concentrations of specific analytes (i.e., target alkanes and PAHs) within the oil be measured using chromatographic techniques (e.g., GC/MS) and that they be reported relative to a conservative biomarker such as hopane. However, from an operational perspective, more rapid and less costly analytical procedures are also needed to satisfy regulators and responders on a more real time, continual basis. Existing TPH technologies are generally not reliable and have little biological significance. TLC-FID seems to be a promising screening tool for monitoring oil biodegradation (Stephens *et al.* 1999).

In addition to monitoring the treatment efficacy for oil degradation, the bioremediation monitoring plan should also incorporate reliable ecotoxicological endpoints to document treatment effectiveness for toxicity reduction. Commonly used ecotoxicity monitoring techniques, such as the Microtox® assay and an invertebrate survival bioassay, may provide an operational endpoint indicator for bioremediation activities on the basis of toxicity reduction (Lee *et al.* 1995a).

Statistical considerations. To ensure that monitored results reflect the reality in a highly heterogeneous environment, it is important that a bioremediation sampling plan be designed according to valid statistical principles that include randomization, replication, and representative controls. A random sampling plan should be used to minimize bias and to evaluate treatment effects and their variance within the bioremediation zone. For samples with a high degree of spatial heterogeneity, which will be the case for most oil spill sites, stratified sampling strategies might be used. For example, the sampling field on a marine shoreline may be divided into a number of sectors or quadrants based on the homogeneity of geomorphology within each sector (e.g., upper and lower intertidal zones), and independent samples should be taken in each sector according to the rule of proportionality (e.g., taking more samples in more heavily oiled sites).

Although economic factors could be restrictive, efforts should be made to ensure that an adequate number of samples be taken to achieve a given accuracy and confidence. Power analysis should be used to assist in the determination of sample replications required in a monitoring plan. For example, if oil distribution and shoreline characteristics are highly heterogeneous, variance will be high, thus requiring more replicates to detect significant treatment effects. If background nutrients are high, treatment differences will be low, and more replicates will also be required. By comparing three shoreline assessment designs used for the Exxon Valdez oil spill, Gilfillan et al. (1999) also proposed several strategies to increase power (i.e., the probability that significant differences between two or more treatments are detected when indeed they exist). One of the approaches to increase power is to select sampling sites from only the most heavily oiled locations. This strategy may not be feasible for assessing the oil degradation within the whole bioremediation zone, although it may be useful for evaluating the effect of bioremediation on ecological recovery since the ecological injury most likely occurs at the heavily oiled locations.

A control area normally refers to a set-aside untreated site, which has similar physical and biological conditions as the treated site. Although on-scene coordinators prefer not to leave oiled sites untreated, it is difficult to assess the true impact of a treatment without control or set- aside areas (Hoff and Shigenaka 1999). When selecting control areas, one must consider not only the similarity of the conditions but also the effect of sand and nutrient exchanges between the treated and untreated areas.

Bioremediation Strategies in Freshwater and Saltwater Wetlands

Although the same decision-making and planning principles that were described above for bioremediation of marine shorelines should also apply to wetland environments, the feasible bioremediation strategies are likely to be different due to the distinct characteristics of wetlands. The potential effectiveness of different amendments is based on the findings of the St. Lawrence River field study (Garcia-Blanco *et al.* 2001a, Venosa *et al.* 2002, Lee *et al.* 2001) and the Dartmouth, Nova Scotia, study.

Unlike other types of marine shorelines (e.g., sandy beaches), the most important limitation for cleanup of an oil-contaminated marine wetland is oxygen availability. Wetland sediments often become anoxic a few mm to cm below the soil surface. When substantial penetration of spilled oil into anoxic sediments has taken place, available evidence suggests that bio-stimulation with nutrient addition has limited potential for enhancing oil biodegradation, and it would likely be best simply to leave it alone and not risk further damage to the environment by trampling and the associated bioremediation activities. Therefore, the evaluation of oil penetration and oxygen availability is probably the most important pre-treatment assess-ment for determining whether bioremediation is a viable option.

Nutrient amendment. Since nutrients could be limited in wetland sediments during the growing season in particular, addition of nutrients would seem to have some potential for enhancing oil biodegradation in such an environment. However, the results from the St. Lawrence River freshwater wetland field study showed that no significant enhancement was observed in terms of the oil biodegradation following biostimulation through addition of nutrients (either ammonium or nitrate). After 21 weeks, reduction of target parent and alkyl- substituted PAHs averaged 32% in all treatments. Reduction of target alkanes was of similar magnitude. The removal of PAHs in nutrient-amended plots was only slightly better than natural attenuation after 64 weeks of treatment. Oil analysis from the top 2 cm sediment samples showed that the plots amended with ammonium nitrate and with *Scirpus pungens* plants cut back demonstrated a significant enhancement in target hydrocarbon reduction over natural attenuation as well as all other treatments. This suggests that biostimulation may be effective only in the top layer of the soil, where aerobic conditions are greater, and when hydrocarbon-degrading microorganisms do not have to compete for nutrients with the growing wetland plants.

Coastal marshes are generally considered high-nutrient wetlands. However, inorganic bioavailable nutrient concentrations in salt marsh sediments may exhibit a strong seasonal pattern with a concentration peak usually during the summer months probably due to a high mineralization rate at a higher temperature. The available nutrient levels could also be elevated as a result of runoff, fire, and death of plants. If these events are sporadic, biostimulation may still be appropriate when the nutrient levels fall below threshold concentrations.

Only a few studies have been reported on the optimal nutrient concentration in salt marsh environments. In a microcosm study using salt marsh sediment slurry, Jackson and Pardue (1999) found that oil degradation rates could be increased with increasing concentrations of ammonia in the range of 10 – 670 mg N/L, with most of the consistent rate increases occurring between 100 – 670 mg N/L. They further proposed a critical nitrogen concentration range of 10 – 20 mg N/L. Harris *et al.* (1999) examined the nutrient dynamics during natural recovery of an oil-contaminated brackish marsh and found that there was an inter-dependency between the natural nutrient levels and the extent of oil degradation when the background nitrogen concentration in pore water declined from 40 mg N/L to 5 mg N/L. Evidence from bioremediation field studies also suggested that concentrations of approximately 5 to 10 mg/L of available nitrogen in the interstitial pore water is sufficient to meet the minimum nutrient requirement of the oil degrading microorganisms (Mills *et al.* 1997). As mentioned earlier, the threshold concentration range for optimal hydrocarbon biodegradation on marine shorelines is around 2 to 10 mg N/L based on field experiences on sandy beaches (Bragg *et al.* 1994, Venosa *et al.* 1996) and in an estuarine environment (Oudet *et al.* 1998). The apparent higher threshold nitrogen concentrations in salt marshes are mainly due to the lack of information with respect to oil biodegradation under lower nitrogen concentrations, since all the existing field studies were conducted in salt marshes with background nitrogen concentrations of at least 5 mg N/L (Harris *et al.* 1999, Mills *et al.* 1997, Shin *et al.* 1999). Therefore, it is reasonable to recommend, as for other types of shorelines, that biostimulation of oil impacted salt marshes should occur when nitrogen concentrations of at least 2 to as much as to 5-10 mg N/L are maintained in the pore water with the decision on higher concentrations to be based on a broader analysis of cost, environmental impact, and practicality. In practice, a safety factor should be used to achieve target concentrations, which will depend on anticipated nutrient washout rates, selected nutrient types, and application methods. The safety factor used in salt marsh environments may generally be smaller than that used in higher energy beaches due to the reduced degree of nutrient washout expected in salt marshes. However, the factors that lead to higher nutrient losses in wetland environments may also be important, such as sediment adsorption, plant uptake, and denitrification (if applicable).

As far as frequency of nutrient application is concerned, weekly to monthly additions may be sufficient for biostimulation of salt marshes when the nutrients are applied during neap tide. It is even possible that only one nutrient dose is required for the bioremediation of some coastal marshes. A study on the nutrient dynamics in an oil contaminated brackish marsh showed that it took more than one year for nutrient concentrations to decrease to background levels after being naturally elevated by flooding and perturbations due to the spill (Harris *et al.* 1999). However, this may not be truly indicative of nutrient application dynamics, since exogenous nutrients were not added in this case. Nutrient sampling, particularly in sediment pore water, must be coordinated with nutrient application to ensure that the nutrients become distributed throughout the contaminated area and that target concentrations are being achieved. The frequency of nutrient addition should be adjusted based on the nutrient monitoring results.

Oxygen amendment. Oxygen is the most likely factor limiting oil biodegradation in freshwater wetland environments. An appropriate technology for increasing the oxygen concentration in such environments, other than reliance on the wetland plants themselves to pump oxygen down to the rhizosphere through the root system, has yet to be developed. Existing oxygen amendment technologies developed in terrestrial environments, such as tilling, forced aeration, and chemical methods are not likely to be cost-effective for bioremediation of freshwater wetlands since they often involve expensive and overly intrusive practices that do more harm than good.

During the St. Lawrence River field trial (Garcia-Blanco *et al.* 2001, Venosa *et al.* 2002), after the first nutrient and oil applications, the top 1-2 cm surface soil in all plots was manually raked using cast iron rakes. This was done to minimize loss of oil from the plots due to tidal action and to uniformly incorporate the nutrients and the oil into the soil. However, the oil analysis results suggested that the tilling of surface soil might have slowed the overall oil biodegradation rates by enhancing oil penetration deep into the anaerobic sediments. Based on these observations, surface tilling will not be an effective strategy for increasing the oxygen concentration in freshwater wetlands.

Phytoremediation. Phytoremediation is emerging as a potentially viable technology for cleanup of soils contaminated with petroleum hydrocarbons (Frick *et al.* 1999). However, this technique has not been used as a wetland oil spill countermeasure. Only limited studies have been carried out on the effectiveness of phytoremediation in enhancing oil degradation in marine shorelines and freshwater wetlands. Lin and Mendelssohn (1998) found in a greenhouse study that application of fertilizers in conjunction with the presence of transplants (*S. alterniflora* and *S. patens*) significantly enhanced oil degradation in a coastal wetland environment. In the case of freshwater wetlands, the St. Lawrence River study suggested that although application of fertilizers in conjunction with the presence of a wetland plant (*Scirpus pungens*) may not significantly enhance oil degradation, it could enhance habitat recovery through the stimulation of vigorous vegetative growth and reduction of sediment toxicity and oil bioavailability (Lee *et al.* 2001a). The effectiveness of oil phytoremediation in freshwater wetland environments still requires further study.

Natural attenuation. Natural attenuation has been defined as the reliance on natural processes without human intervention to achieve site-specific remedial objectives (USEPA 1999b). Monitoring is still required to determine how effective the natural cleanup is progressing. Previous research on wetlands, both freshwater and saltwater, have shown that oxygen may be the limiting factor determining the rate of self-purification. For example, the St. Lawrence River Study demonstrated that the availability of oxygen, not nutrients, was likely the limiting factor for oil biodegradation in freshwater wetlands if subsurface penetration has taken place. However, no feasible technique is currently available for increasing oxygen concentration under such an environment. As a result of this study, natural attenuation has been recommended as the most cost-effective strategy for oil spill cleanup in freshwater wetlands when the oil concentration is not high enough (e.g., less than 30 g/kg soil; Longpre *et al.* 1999) to destroy wetland vegetation. However, this recommendation should be tempered if little penetration has occurred. In the latter case, when all the oil contamination is located on the surface, biostimulation might be an appropriate remedy.

Conclusions and Recommendations

The overall conclusions are as follows. First, with respect to marine sandy shorelines, natural attenuation may be appropriate if background nutrient concentrations were high enough that intrinsic biodegradation would take place at close to the expected maximum rate. The Delaware study proved this clearly. Certainly in nutrient-limited places like Prince William Sound, Alaska, nutrient addition should accelerate cleanup rates many-fold. However, the decision to use the natural attenuation approach may be tempered by the need to protect a certain habitat or vital resource from the impact of oil. For example, using Delaware as the model, every spring season, horseshoe crabs migrate to the shoreline of Delaware for their annual mating season. Millions of eggs are laid and buried a few mm below the surface of the sand. Migrating birds making their way from South America to Arctic Canada fly by this area and feed upon these eggs to provide energy to continue their long flight. If an oil spill occurred in February or March, it would certainly be appropriate to institute bioremediation to accelerate the disappearance of the oil prior to the horseshoe crab mating season despite the expected high natural attenuation rate. So, even in the case where background nutrients are high enough to support rapid biodegradation, addition of more nutrients would help protect such a vital resource. If the spill occurred during the summer, and no vital natural resources were threatened by the spill, then reliance on natural attenuation might be the wisest course of action. Of course, removal of free product and high concentrations of oil should still be conducted by conventional means even if a no bioremediation action is warranted by the circumstances.

With respect to freshwater wetlands and salt marshes, data reviewed demonstrated that, if significant penetration of oil takes place into the subsurface, biodegradation would take place very slowly and ineffectively. This is because of the anaerobic conditions that quickly occur in these types of saturated environments, and anaerobic biodegradation of petroleum oils is much slower and less complete than under aerobic conditions. One of the objectives of the St. Lawrence River experimental design was to determine the amenability of wetlands to biodegradation when oil has penetrated into the sediment. The oil was artificially raked into the sediment to mimic such an occurrence. Consequently, no significant treatment effects were observed because all the nutrients in the world would not stimulate biodegradation if oxygen were the primary limiting material. If penetration did not take place beyond a few mm, then bioremediation might be an appropriate cleanup technology, since more oxygen would be available near the surface. It is clear that whatever oxygen gets transported to the root zone by the plants is only sufficient to support plant growth and insufficient to support the rhizosphere microorganisms to degrade contaminating oil. In the salt marsh study conducted in Nova Scotia, the oil was not raked into the subsurface, and substantial biodegradation took place since the oil was exposed to more highly aerobic conditions. Thus, data generated from both wetland studies point to the same overall conclusions in regard to the need to bioremediate a wetland environment. Oxygen availability is the key, and if aerobic conditions prevail in all parts of the impact zone, then nutrient availability becomes the critical variable. If sufficient nutrients are already available, natural attenuation might be the appropriate action to take.

However, if ecosystem restoration is the primary goal rather than oil cleanup, the data strongly suggest that nutrient addition would accelerate and greatly enhance restoration of the site. Abundant plant growth took place in the nutrient-treated plots despite the lack of oil disappearance from the extra nutrients. Furthermore, the stimulation lasted more than one growing season even though nutrients were never added after the first year. Clearly, the plants took up and stored the extra nitrogen for use in subsequent growing seasons, so restoration of the site was abundantly evident in a few short months.

Thus, in conclusion, the decision to bioremediate a site is dependent on cleanup, restoration, and habitat protection objectives and whatever factors that are present that would have an impact on success. Responders must take into consideration the oxygen and nutrient balance at the site. If the circumstances are such that no amount of nutrients will accelerate biodegradation, then the decision should be made on the need to accelerate oil disappearance to protect a vital living resource or simply to speed up restoration of the ecosystem. If there is no immediate need to protect a vital resource or restore the ecosystem, then natural attenuation may be the appropriate response action. These decisions are clearly influenced by the circumstances of the spill.

REFERENCE

Ahn, C.H. 1999. The characteristics of crude oil biodegradation in sand columns under tidal cycles. M. S. Thesis, University of Cincinnati, Ohio.

Atlas, R.M. 1977. Stimulated petroleum biodegradation. *Crit. Rev. Microbiol.* 5: 371-386.

Atlas, R.M. 1981. Microbial degradation of petroleum hydrocarbons: An environmental perspective. *Microbiol. Rev.* 45: 180-209.

Atlas, R.M. 1984. *Petroleum Microbiology*, Macmillan Publishing Company, New York.

Atlas, R.M. 1995. Bioremediation of petroleum pollutants. International Biodeterioration and Biodegradation, 317-327.

Atlas, R.M., and R. Bartha, 1973. Stimulated biodegradation of oil slicks using oleophilic fertilizers. *Environ. Sci. Technol.* 7: 538-541.

Atlas, R.M. and R. Bartha. 1992. Hydrocarbon biodegradation and oil spill bioremediation. *Adv. Microb. Ecol.* 12: 287-338.

Blenkinsopp, S., G. Sergy, Z. Wang, M.F. Fingas, J. Foght, and D.W.S. Westlake. 1995. Oil spill bioremediation agents: Canadian efficiency test protocols. *Proceedings of 1995 International Oil Spill Conference*, American Petroleum Institute, Washington, D.C.

Boufadel, M.C., P. Reeser, M.T. Suidan, B.A. Wrenn, J. Cheng, X. Du, T.L. Huang, and A.D. Venosa. 1999. Optimal nitrate concentration for the biodegradation of n-heptadecane in a variably-saturated sand column. *Environ. Technol.* 20: 191-199.

Bragg. J.R., R.C. Prince, E.J. Harner, and R.M. Atlas. 1994. Effectiveness of bioremediation for the Exxon Valdez oil spill. *Nature* 368: 413-418.

Brown, A.C., and A. McLachan. 1990. *Ecology of Sandy Shores*, Elsevier, New York.

Cooney, J.J. 1984. The fate of petroleum pollutants in freshwater ecosystems. Pages 355-398 in *Petroleum Microbiology*, R.M. Atlas, ed., Macmillan Publishing Company, New York.

Du, X., P. Reeser, M.T. Suidan, T.L. Huang, M. Moteleb, M.C. Boufadel, and A.D. Venosa. 1999. Optimal nitrate concentration supporting maximum crude oil biodegradation in microcosms. *Proceedings of 1999 International Oil Spill Conference*. American Petroleum Institute, Washington, D.C.

Edwards, R., and I. White. 1999. The Sea Empress oil spill: environmental impact and recovery. *Proceedings of 1999 International Oil Spill Conference*. American Petroleum Institute, Washington, D.C.

Floodgate, G. 1984. The fate of petroleum in marine ecosystems. Page 355-398 in *Petroleum Microbiology*, R.M. Atlas, ed., Macmillan Publishing Company, New York.

Forsyth, J.V., Y.M. Tsao, and R.D. Blem. 1995. Bioremediation: when is augmentetion needed ? Pages 1-14 in Bioaugmentation for site Remediation, R.E. Hinchee et al., eds., Battelle Press, Columbus, Ohio.

Frick, C.M., J.J. Germida, and R.E. Farrell. 1999. Assessment of phytoremediation as an *in-situ* technique for cleaning oil-contaminated sites. *Proceedings of the Phytoremediation Technical Seminar*, Environment Canada, Ottawa.

Garcia-Blanco, S., M. Motelab, A.D. Venosa, M.T. Suidan, K. Lee, and D.W. King. 2001. Restoration of the oil-contaminated Saint Lawrence River shoreline: Bioremediation and phytoremediation. *Proceedings of 2001 International Oil Spill Conference*. American Petroleum Institute, Washington, D.C.

Gilfillan, E.S., E.J. Harner. J.E. O'Reilly, D.S. Page, and W.A. Burns. 1999. A comparison of shoreline assessment study designs used for Exxon Valdez oil spill. *Mar. Pollut. Bull.* 38: 380.

Goldstein. R.M., L.M. Mallory, and M. Alexander. 1985. Reasons for possible failure of inoculation to enhance biodegradation. *Appl. Environ. Microbiol.* 50: 977-983.

Haines J.R., E.L. Holder, K.M. Miller, and A.D. Venosa. 1999. Laboratory assessment of bioremediation products under freshwater conditions. *Proceedings of 1999 International Oil Spill Conference*. American Petroleum Institute, Washington, D.C.

Harris, B.C., J.S. Bonner, R.L. Autenrieth. 1999. Nutrient dynamics in marsh sediments contaminated by an oil spill following a flood. *Environ. Technol.* 20: 795-810.

Harris, C. 1997. The Sea Empress incident: overview and response at sea. *Proceedings of 1997 International Oil Spill Conference*. American Petroleum Institute, Washington, D.C.

Head, I.M., and R.P.J. Swannell. 1999. Bioremediation of petroleum hydrocarbon contaminants in marine habitats. *Curr. Opin. Biotechnol.* 10: 234-239.

Hoff, R., and G. Sergy, C. Henry, S. Blennkinsopp, and P. Roberts. 1995. Evaluating biodegradation potential of various oils. *Proceedings of the 18th*

Arctic and Marine Oilspill Program (AMOP) Technical Seminar, Environment Canada, Ottawa.

Hoff, R., and G. Shigenaka. 1999. Lessons from ten years of post-Exxon Valdez monitoring on intertidal shorelines. *Proceedings of 1999 International Oil Spill Conference*. American Petroleum Institute, Washington, D.C.

Jackson, W.A., and J.H. Pardue. 1999. Potential for enhancement of biodegradation of crude oil in Louisiana salt marshes using nutrient amendments. *Water Air Soil Pollut.* 109: 343-355.

Jobson. A.M., F.D. Cook, and D.W.S. Westlake. 1974. Effect of amendments on the microbial utilization of oil applied to soil. *Appl. Microbiol.* 27: 166-171.

Ladousse, A., and B. Tramier. 1991 . Results of 12 years of research in spilled oil bioremediation: Inipol EAP 22, *Proceedings of 1991 Oil Spill Conference*. American Petroleum Institute Washington, D.C.

Leahy, J.G. and R.R. Colwell. 1990. Microbial degradation of hydrocarbons in the environment. *Microbiol. Rev.* 53: 305-315.

Lee, K. 1995. Bioremediation studies in low-energy shoreline environments. *Proceedings of Second International Oil Spill Research and Development Forum.* International Marine Organization, London, U.K.

Lee, K., and E.M. Levy. 1987. Enhanced biodegradation of a light crude oil in sandy beaches. *Proceedings of 1987 Oil Spill Conference.* American Petroleum Institute, Washington, D.C.

Lee, K., and E.M. Levy. 1989. Enhancement of the natural biodegradation of condensate and crude oil on beaches of Atlantic Canada. *Proceedings of 1989 Oil Spill Conference.* American Petroleum Institute., Washington, D.C.

Lee, K., K.G. Doe, L.E.J. Lee, , M.T. Suidan, and A.D. Venosa. 2001. Remediation of an oil contaminated experimental freshwater wetland: Habitat recovery and toxicity reduction. *Proceedings of the 2001 International Oil Spill Conference.* American Petroleum Institute, Washington, D.C.

Lee, K. and E.M. Levy. 1991. Bioremediation: waxy crude oils stranded on low-energy shorelines. *Proc. 1991 Internat. Oil Spill Conf., Amer. Petroleum Institute, Washington, D.C.* pp. 541-547.

Lee, K., T. Lunel, P. Wood, R. Swannell, and P. Stoffyn-Egli. 1997a. Shoreline cleanup by acceleration of clay-oil flocculation processes. *Proceedings of 1997 International Oil Spill Conference.* American Petroleum Institute, Washington, D.C.

Lee. K., and F.X. Merlin. 1999. Bioremediation of oil on shoreline environments: development of techniques and guidelines. *Pure Appl. Chem.* 71: 161-171.

Lee, K., and G.H. Trembley. 1993. Bioremediation: application of slow-release fertilizers on low energy shorelines. *Proceedings of the 1993 Oil Spill Conference,* American Petroleum Institute, Washington, D.C.

Lee, K., R. Siron, and G.H. Tremblay. 1995a . Effectiveness of bioremediation in reducing toxiciy in oiled intertidal sediments. Pages 117-127 in *Microbial*

Processes for Bioremediation, eds., R.E. Hinchee *et al.* Battelle Press, Columbus, Ohio. yes.

Lee, K., G.H. Tremblay, J. Gauthier, S.E. Cobanli and M. Griffin. 1997b. Bioaugmentation and biostimulation: a paradox between laboratory and field results. *Proceedings of 1997 International Oil Spill Conference.* American Petroleum Institute, Washington, D.C.

Lee, K., G.H. Tremblay, and S.E. Cobanli. 1995b. Bioremediation of oiled beach sediments: Assessment of inorganic and organic fertilizers. *Proceedings of 1995 Oil Spill Conference.* American Petroleum Institute, Washington, D.C.

Lepo, J.E., and C.R. Cripe. 1998a. Development and application of protocols for evaluation of oil spill bioremediation. U.S. EPA, Gulf Breeze Environmental Research Laboratory, EPA/600/S-97/007.

Lin, Q., and I.A. Mendelssohn. 1998. The combined effects of phytoremediation and biostimulation in enhancing habitat restoration and oil degradation of petroleum contaminated wetlands. *Ecol. Engineering* 10: 263-274.

Longpre, D., K. Lee, V. Jarry, A. Jaouich, A.D. Venosa, and M.T. Suidan. 1999. The response of *Scirpus pungens* to crude oil contaminated sediments. *Proceedings of the Phytoremediation Technical Seminar,* Environment Canada, Ottawa.

McMillen, S.J., A.G. Requejo, G.N. Young, P.S. Davis, P.D. Cook, J.M. Kerr, and N.R. Gray. 1995. Bioremediation potential of crude oil spilled on soil. Pages 91-99 in *Microbial Processes for Bioremediation*, R.E. Hinchee, F.J. Brockman, C.M. Vogel *et al.*, eds., Battelle Press, Columbus, Ohio.

Means, W.J. 1997. Cleaning oiled shores: putting bioremediation to the test. *Spill Sci. Technol. Bull.* 4: 209-217.

Mills, M.A., J.S. Bonner, M.A. Simon, T.J. McDonald and R.L. Antenrieth. 1997. Bioremediation of a controlled oil release in a wetland. Proc. 20th Arctic and Marine Oil Spill Program Technical Science, Env. Canada Ottowa, pp. 606-616.

Mitsch. W.J., and J.G. Gosselink. 1993. *Wetlands,* Van Nostrand Reinhold, New York.

Mitsch, W.J., and J.G. Gosselink. 2000. *Wetlands,* John Wiley and Sons, Inc., New York.

National Environmental Technology Application Corporation (NETAC). 1993. Evaluation Methods Manual: Oil Spill Response Bioremediation Agents. University of Pittsburgh Applied Research Center, Pittsburgh, Pennsylvania.

Office of Technology Assessment. 1990. Coping With An Oiled Sea: An Analysis of Oil Spill Response Technologies, OTA-BP-O-63, Washington, D.C.

Office of Technology Assessment. 1991 . Bioremediation of Marine Oil Spills: An Analysis of Oil Spill Response Technologies, OTA-BP-O-70, Washington, D.C.

Olivieri, R., P. Bacchin, A. Robertiello, N. Odde, L. Degen, and A. Tonolo. 1976. Microbial degradation of oil spills enhanced by a slow release fertilizer. *Appl. Environ. Microbiol.* 31: 629-634.

Oudet J., F.X. Merlin, and P. Pinvidic. l998. Weathering rates of oil components in a bioremediation experiment in estuarine sediments. *Mar. Environ. Res.* 45: 113-125.

Owens, E.H., E. Taylor, R. Marty, and D.I. Little. 1993. An inland oil spill response manual to minimize adverse environmental impacts. *Proceedings of 1993 International Oil Spill Conference.* American Petroleum Institute, Washington, D.C.

Prince, R.C. 1993. Petroleum spill bioremediation in marine environments. *Critical Rev. Microbiol.* 19: 217-242.

Prince, R.C., D.L. Elmendorf, J.R. Lute, C.S. Hsu, C.E. Haith, J.D. Senius, G.J. Dechert, G.S. Douglas, and E.L. Butler.1994. 17a(H), 21b(H)-Hopane as a conserved internal marker for estimating the biodegradation of crude oil. *Environ. Sci. Technol.* 28: 142-145.

Pritchard, P.H., and C.F. Costa. 1991. EPA's Alaska oil spill bioremediation project. *Environ. Sci. Technol.* 25: 372-379.

Pritchard, P.H, J.G. Mueller, J.C. Rogers, F.V. Kremer, and J.A. Glaser. 1992. Oil spill bioremediation: experiences, lessons and results from the Exxon Valdez oil spill Alaska. *Biodegradation* 3: 109-132.

Reisfeld, A., E. Rosenberg, and D. Gutnick. 1972. Microbial degradation of crude oil: factors affecting the dispersion in sea water by mixed and pure cultures. *Appl. Microbiol.* 24: 363-368.

Rosenberg, E., and E.Z. Ron. 1996. Bioremediation of petroleum contamination. Page 100-124 in *Bioremediation: Principles and Applications*, R.L. Crawford and D.L. Crawford, eds., Cambridge University Press, U.K.

Safferman, S.I. 1991. Selection of nutrients to enhance biodegradation for the remediation of oil spilled on beaches. *Proceedings of 1991 International Oil Spill Conference.* American Petroleum Institute, Washington, D.C.

Santas, R., and P. Santas. 2000. Effects of wave action on the biodegradation of crude oil saturated hydrocarbons. *Mar. Polluti. Bull.* 40: 434-439.

Shin, W.S., P.T. Tate, W.A. Jackson, and J.H. Pardue. 1999. Bioremediation of an experimental oil spill in a salt marsh. Page 33-55 in *Wetland and Remediation: An International conference.* J.L. Means and R.E. Hinchee eds., Battelle Press, Columbus, Ohio.

Smith, V.H., D.W. Graham., and D.D. Cleland. 1998. Application of resource ratio theory to hydrocarbon degradation, *Environ. Sci. Technol.* 32: 3386-3395.

Spies, R.B., S.D. Rice, D.A. Wolfe, B.A. Wright. 1996. The effect of the Exxon Valdez oil spill on Alaskan coastal environment. *Proceedings of the 1993 Exxon Valdez Oil Spill Symposium*, American Fisheries Society, Bethesda, Maryland.

Stalcup, D., G. Yoshioka, E. Mantus, and B. Kaiman, 1997. Characteristics of oil spills: inland versus coastal. *Proceedings of 1997 International Oil Spill Conference.* American Petroleum Institute, Washington, D.C.

Stephens, F.L., J.S. Bonner, and R.L. Autenrieth. 1999. TLC/FID analysis of compositional hydrocarbon changes associated with bioremediation. *Proceedings of 1999 International Oil Spill Conference.* American Petroleum Institute, Washington, D.C.

Sugiura, K., M. Ishihara, T. Shimauchi, and S. Harayama. 1997. Physicochemical properties and biodegradability of crude oil. *Environ. Sci. Technol.* 31: 45-51.

Sveum, P., and S. Ramstad. 1995. Bioremediation of oil contaminated shorelines with organic and inorganic nutrients. Page 201-217 in *Applied Bioremediation of Petroleum Hydrocarbons,* R.E. Hinchee *et al.,* eds., Battelle Press, Columbus, Ohio.

Swannell, R.P.J., B.C. Croft, A.L. Grant, and K. Lee. 1995. Evaluation of bioremediation agent in beach microcosms. *Spill Sci. Technol. Bull.* 2: 151-159.

Swannell, R.P.J., K. Lee, and M. Mcdonagh. 1996. Field evaluations of marine oil spill bioremediation. *Microbiol. Rev.* 60: 342-365.

Swannell, R.P.J., D. Mitchell, D.M. Jones, S.P. Petch, I.M. Head, A. Willis, K. Lee, and J.E.Lepo. 1999, Bioremediation of oil-contaminated fine sediments. Proc. 1999 International Oil Spill Conf., Amer. Petroleum Institute, Inst. Washington, D.C., pp. 751-756.

Tagger, S., A. Bianchi, M. Julliard, J. Le Petit, and B. Roux. 1983. Effect of microbial seeding of crude oil in seawater. *Mar. Biol.* 78: 13-21.

Thomas, G., R. Nadeau, and J. Ryabik. 1995. Increasing readiness to use bioremediation response to oil spills. *Proceedings of Second International Oil Spill Research and Development Forum.* International Marine Organization, London, U.K.

USEPA 1999. Monitored Natural Attenuation of Petroleum Hydrocarbons, EPA 600-F-98-021, Office of Research and Development, U.S. Environmental Protection Agency, Washington, D.C.

USEPA 2000. NCP Product Schedule, http://www.epa.gov/oilspill. USEPA 2000. NCP Product Schedule and Notebook http://www.epa.gov/oilspill/ncp/ncp_index.htm

Uraizee, F.A. A.D. Venosa, and M.T. Suidan. 1998. A model for diffusion controlled bioavailability of crude oil components. *Biodegradation* 8: 287-296.

Venosa, A.D. 1998. Oil spill bioremediation of coastal shorelines: a critique. Pages 259-301 in *Bioremediation: Principles and Practice. Vol. III. Bioremediation Technologies,* S.K. Sikder and R.I. Irvine, eds., Technomic Publishing, Lancaster, Pennsylvania.

Venosa, A.D., J.R. Haines, and D.M. Allen. 1992. Efficacy of commercial inocula in enhancing biodegradation of crude oil contaminating a Prince William Sound beach. *J. Ind. Microbiol.* 10: 1-11.

Venosa, A.D., K. Lee, M.T. Suidan, S. Garcia-Blanco, S. Cobanli, M. Moteleb, J.R. Haines, G. Tremblay, and M. Hazelewood 2002. Bioremediation and biorestoration of a crude oil-contaminated freshwater wetland on the St. Lawrence River. *Bioremediation* 6: 261-281.

Venosa, A.D., M.T. Suidan, B.A. Wrenn, K.L. Strohmeirer, J.R. Haines, B.L. Eberhart, D.W. King, and E. Holder. 1996. Bioremediation of experimental oil spill on the shoreline of Delaware Bay. *Environ. Sci. Technol.* 30: 1764-1775.

Wang, X., and R. Bartha. 1990. Effects of bioremediation on residues: activity and toxicity in soil contaminated by fuel spills. *Soil Biol. Biochem.* 22: 501-506.

Westlake, D.W.S., A. Jobson, R. Phillipee, and F.D. Cook. 1974. Biodegradability and crude oil composition. *Can. J. Microbiol.* 20: 915-928.

Wrenn, B.A., M.T. Suidan, K.L. Strohmeier, B.L. Eberhart, G.J. Wilson, and A.D. Venosa. 1997. Nutrient transport during bioremediation of contaminated beaches: Evaluation with lithium as a conservative tracer. *Water Res.* 31: 515-524.

Xu, Y., M.T. Suidan, S. Garcia-Blanco, and A.D. Venosa. 2001. Biodegradation of crude oil at high oil concentration in microcosms, *Proceedings of the 6th International In-Site and On-Site Bioremediation Symposium*, Battelle Press, Columbus, Ohio.

Zhu, X., A.D. Venosa, M.T. Suidan, and K. Lee. 2001. Guidelines for the bioremediation of marine shorelines and freshwater wetlands. {HYPERLINK " http://www.epa.gov/oilspill/pdfs/bioremed.pdf}.

Zhu, X., A.D. Venosa, M.T. Suidan, and K. Lee. 2004. Guidelines for the bioremediation of oil-contaminated salt marshes. EPA/600/R-04/074. {Hyperlink : "http://www.epa.gov/oilspill/pdfs saltmarshbiormd.pdf"}.

Bioremediation of Petroleum Contamination

Ismail M.K. Saadoun[1] and Ziad Deeb Al-Ghzawi[2]

[1]Department of Applied Biological Sciences, College of Science and Arts,
[2]Department of Civil Engineering, College of Engineering,
Jordan University of Science and Technology, Irbid-22110, Jordan
E-mail: isaadoun@just.edu.jo

Introduction

As landfills have become more and more scarce and cost prohibitive, interest in biological methods to treat organic wastes, and in particular petroleum contamination, has increased and received more attention. Petroleum fuel spills which resulted from damage, stress, and corrosion of pipelines, transportation accidents, leakage of storage tanks and various other industrial and mining activities are classified as hazardous waste (Bartha and Bossert 1984) and are considered as the most frequent organic pollutants of terrestrial and aquatic ecosystems (Bossert *et al.* 1984, Margesin and Schinnur 1997). It is estimated that 1.7-6.8 million tonnes of oil, with a best estimate of 3.2 million tonnes per annum, are released from all sources into the environment. The majority of this is not due to the oil industry and tanker operations, which only account for approximately 14% of the input, but to other industrial and general shipping activities (ITOPF 1990). Estimates suggest that there are between 100,000 and 300,000 tanks leaking petroleum or petroleum-based products in the USA (Mesarch and Nies 1997, Lee and Gongaware 1997). The petroleum leaks are of particular interest as petroleum can contain up to 20% benzene, toluene, ethylbenzene and xylene (BTEX), and these are on the hazardous list. The BTEX compounds, although not miscible with water, are mobile and can contaminate the groundwater (Bossert and Compeau 1995), which is recognized as a serious and widespread environmental problem. The Nawrus spill in 1984, during the Iran/Iraq War, resulted in an unknown but massive quantity of oil being spilled (Watt 1994b). Following the Gulf War in 1991, estimates between four and eight million barrels (1,000 tonnes = 7,500 barrels) were released into the Arabian Gulf and in the Kuwaiti Desert making this the largest oil spill in history (Purvis 1999). The size of this spill is brought into perspective when it is compared to other major spills around the world such as the *Amoco Cadiz* off the coast of Brittany (France), spilling 200,000 tonnes (1.5 million barrels), or the *Torrey Canyon, Braer, Sea Empress* and the super tanker *Breaf* off the coast of Shetland (UK) in 1993 with a maximum spill of 84,000 tonnes (607,300 barrels), or the *Exxon Valdez* in Prince William Sound, Alaska (US), which was approximately 36,224 tonnes (261,904 barrels) (Watt 1994a), as well as other spills in Texas, Rhode Island and Delaware Bay (Atlas 1991).

Terrestrial spills are also clear as the outcome of the Gulf War in 1991 and formation of the oil lakes in the Kuwaiti Desert, as well as the failure of the Continental Pipeline near Crosswicks, New Jersey, that resulted in the spill of approximately 1.9 million liters of kerosene that inundated 1.5 hectares of agricultural land (Dibble and Bartha 1979). The spills in gasoline stations due to leakage may be small but continuous and prolonged. However, the vast majority of spills are small (i.e., less than 7 tonnes) and data on numbers and amounts is incomplete. Over 80% of recorded oil spills are less than 1,000 tonnes (7,500 barrels), and only 5% of recorded spills are greater than 10,000 tonnes. An accepted average sample size of an oil spill is about 700 tonnes (5,061 barrels) (ITOPF 1990). The number of large spills (>700 tonnes) has decreased significantly during the last 20 years (Table 1). The average number of large spills per year during the 1990s was about a third of that witnessed during the 1970s. Table 2 shows a brief summary of 20 selected major oil spills since 1967.

Bioremediation is an important option for restoration of oil-polluted environments. Technology and approaches of this process will be presented in this manuscript.

Table 1. Number of spills over 7 tonnes (http://www.itopf.com/stats.html).

Year	7-700 tonnes	>700 tonnes	Quantity Spilt × 10^3 tonnes
1970-1974	189	125	1114
1975-1979	342	117	2012
1980-1984	221	41	570
1985-1989	124	48	513
1990-1994	165	48	907
1995-1999	108	25	194
2000-2002	46	9	101

Table 2. Selected major oil spills (http://www.itopf.com/stats.html).

Shipname	Year	Location	Spill (10^3) tonnes
Torrey Canyon	1967	Scilly Isles, UK	119
Sea Star	1972	Gulf of Oman	115
Jakob Maersk	1975	Oporto, Portugal	88
Urquiola	1976	La Coruna, Spain	100
Hawaiian Patriot	1977	300 nautical miles off Honolulu	95
Amoco Cadiz	1978	off Brittany, France	223
Atlantic Empress	1979	off Tobago, West Indies	287
Independenta	1979	Bosphorus, Turkey	95
Irenes Serenade	1980	Navarino Bay, Greece	100
Castillo de Bellver	1983	off Saldanha Bay, South Africa	252
Odyssey	1988	700 nautical. miles off Nova Scotia, Canada	132
Khark 5	1989	120 nautical. miles off Atlantic coast of Morocco	80
Exxon Valdez	1989	Prince William Sound, Alaska, USA	37
ABT Summer	1991	700 nautical miles off Angola	260
Haven	1991	Genoa, Italy	144
Aegean Sea	1992	La Coruna, Spain	74
Katina P.	1992	off Maputo, Mozambique	72
Braer	1993	Shetland Islands, UK	85
Sea Empress	1996	Milford Haven, UK	72
Prestige	2002	Off the Spanish coast	77

Crude oil

Crude oil is an extremely complex and variable mixture of organic compounds which consist mainly of hydrocarbons in addition to heterocyclic compounds that contain sulphur, nitrogen and oxygen, and some heavy metals. The different hydrocarbons that make up crude oil come in a wide range of molecular weight compounds, from the gas methane to the high molecular weight tars and bitumens, and of molecular structure: straight and branched chains, single or condensed rings and aromatic rings. The two major groups of aromatic hydrocarbons are monocyclic, such as benzene, toluene, ethylbenzene and xylene (BTEX), and the polycyclic aromatic hydrocarbons (PAHs) such as naphthalene, anthracene and phenanthrene.

Factors affecting the biodegradation of petroleum hydrocarbons

To understand the different technologies applied in bioremediation of petroleum contamination, it is necessary to be introduced to the physicochemical, hydrological and microbiological factors that control bioremediation of the contaminant. Therefore, this section outlines the different factors affecting the biodegradation of the petroleum hydrocarbons.

Reports on the microbial ecology of hydrocarbon degradation and how both environmental and biological factors could determine the rate at which and extent to which hydrocarbons are removed from the environment by biodegradation have been published (Leahy and Colwell 1990, Venosa and Zuh 2003).

Numerous factors are known to affect both the kinetics and the extent of hydrocarbon removal from the environment. These include the following:

Chemical Composition and Hydrocarbon Concentration

The asphaltenes (phenols, fatty acids, ketones, esters and porphyrins), the aromatics, the resins (pyridines, quinolines, carbazoles, sulfoxides, and amides) and the saturates are the classes of petroleum hydrocarbons (PHCs) (Colwell 1977). Susceptibility of hydrocarbons to microbial degradation has been shown to be in the following order: n-alkanes > branched alkanes > low-molecular-weight aromatics > cyclic alkenes (Perry 1984). Alkanes are usually the easiest hydrocarbons to be degraded by their conversion to alcohol via mixed function oxygenase activity (Singer and Finnerty 1984). The simpler aliphatics and monocyclic aromatics are readily degradable, but more complex compounds such as PAHs are not easily degraded and may persist for some time. The persistence will be increased if the compound is also toxic or its breakdown products are toxic to the soil microflora. For example, phenol and hydroquinone are the major products of benzene oxidation with the ability of hydroquinone to exert a toxic effect as accumulated concentrations inhibit the degradation of other pollutants (Burback and Perry 1993). The order of degradation mentioned above is not universal, however; naphthalene and alkylaromatics are extensively degraded in water sediments prior to hexadecane and n-alkane, respectively (Cooney *et al.* 1985, Jones *et al.* 1983). Fedorak and Westlake (1981) have reported a more rapid attack of aromatic hydrocarbons during the degradation of crude oil by marine microbial populations from a pristine site and a commercial harbor.

High-molecular-weight aromatics, resins, and asphaltenes have been shown to feature a slow rate of biodegradation (Jobson *et al.* 1972, Walker and Colwell 1976). Oils with a high proportion of low molecular weight material are known as 'light oils' and flow easily, while 'heavy oils' are the reverse. The more complex and less soluble oil components will be degraded much more slowly than the lighter oils. In the case of the oil tanker *Braer*, carrying light crude oil, the oil was dispersed in a matter of hours (Scragg 1999).

High concentrations of hydrocarbons in water means heavy undispersed oil slicks causing a limited supply of nutrients and oxygen, and thus resulting in the inhibition of biodegradation. Protection of oil from dispersion by wind and wave action in beaches, harbors, small lakes and ponds explains the presence of high concentrations of hydrocarbons in these places and the accompanied negative effects on biodegradation. The lowest rates of degradation of crude oil were observed in protected bays, while the highest rates happened in the areas of greatest wave action (Rashid 1974). Oil sludge contaminating the soil at high concentrations also inhibits microorganisms in their action (Dibble and Bartha 1979). Recently, Tjah and Autai (2003) found maximal degradation of Nigerian light crude oil occurred in soil contaminated at a 10% (v/w) concentration. However, minimal degradation was noted in soil contaminated with 40% (v/w). This indicates that the quantity of crude oil spilled in soil influences the rate and total extent of disappearance of the soil in the environment.

Physical State

The physical state of petroleum hydrocarbons has a marked effect on their biodegradation. Crude oil in aquatic systems, usually does not mix with seawater, and therefore, floats on the surface, allowing the volatilization of the 12 carbons or less components. The rate of dispersion of the floating oil will depend on the action of waves which in turn is dependent on the weather. Crude oil with a high proportion of 'light oils' flows easily and will be dispersed in a short time. As a result of wind and wave action, oil-in-water or water-in-oil (mousse) emulsions may form (Cooney 1984), which in turn increase the surface area of the oil and thus its availability for microbial attack. However, a low surface-to-volume ratio as a result of formation of large masses (plates) of mousse or large aggregates of weathered and undegraded oil (tarballs) inhibits biodegradation because these plates and tarballs restrict the access of microorganisms (Davis and Gibbs 1975, Colwell *et al.* 1978). Providenti *et al.* (1995) reported that one of the factors that limits biodegradation of oil pollutants in the environment is their limited availability to microorganisms. Availability of the compound for degradation within the soil plays a crucial factor in the determination of the rate of hydrocarbon degradation. Soil, freshwater lakes and marine hydrocarbon-utilizing bacteria have been demonstrated to synthesize and release biosurfactants which, greatly enhance their effectiveness in handling or uptake of hydrocarbons (Broderick and Cooney 1982, Jobson *et al.* 1974, Singer and Finnerty 1984). Therefore, to overcome this problem surfactants have been added to contaminated soils and sea water to improve access to the hydrocarbons (Mihelcic *et al.* 1993, NRC 1989), with different chemical dispersant formulations having been studied as means of increasing the surface area and hence enhancing breakdown of hydrocarbon pollutants. The chemical formulation of the dispersant (i.e., its concentration and the dispersant/oil application ratio) have been shown to determine its effectiveness in enhancing the biodegradation of oil slicks (Leahy and Colwell 1990). However, some sources indicated that not all dispersants enhance biodegradation (Mulkin-Phillips and Stewart 1974, Robichaux and Myrick 1972).

The soil structure, its porosity and composition, and the solubility of the compound itself will affect availability. For example, a consortium of pre-isolated oil-degrading bacteria in association with three species of plants effectively remediated contaminated silt-loam soil more than silt, loam and sand loam with an average 80% reduction of total petroleum hydrocarbon (Ghosh and Syed 2001). The effect of three different soil matrices, namely Texas sand, Baccto topsoil, and Hyponex topsoil on California crude oil (5% wt) bioremediation kinetics was studied by Huesemann and Moore (1994). Their results showed that soil type has a significant effect on commulative oxygen consumption kinetics with the highest values in Hyponex topsoil, less in Baccto topsoil, and least in Texas sand. They hypothesized that the addition of crude oil to soil could cause both an increase in bacterial numbers and a change in bacterial ecology resulting in enhanced biodegradation of the inherent soil organic matter compared to the crude oil-deficient control.

Soil particle size distribution also affects microbial growth, so that a soil with an open structure will encourage aeration and thus the rate of degradation will be affected likewise (Scragg 1999). In addition to that, infiltration of oil into the soil would prevent evaporative losses of volatile hydrocarbons, which can be toxic to microorganisms (Leahy and Colwell 1990). Particulate matter can reduce, by absorption, the effective toxicity of the components of oil, but absorption and adsorption of hydrocarbons to humic substances probably contribute to the formation of persistent residues (Leahy and Colwell 1990).

Physical Factors

Temperature

Temperature has a considerable influence on petroleum biodegradation by its effect on the composition of the microbial community and its rate of hydrocarbon metabolism, and on the physical nature and chemical composition of the oil (Atlas 1981). Some small alkanes components of petroleum oil are more soluble at 0 °C than at 25 °C (Polak and Lu 1973), and elevated temperatures can influence nonbiological losses, mainly by evaporation. In some cases the decrease in evaporation of toxic components at lower temperatures was associated with inhibited degradation (Floodgate 1984). Atlas and Bartha (1992) found that the optimum temperature for biodegradation of mineral oil hydrocarbons under temperate climates is in the range of 20-30 °C. Most mesophilic bacteria on the other hand perform best at about 35 °C. Even though temperatures in the range of 30-40 °C maximally increase the rates of hydrocarbon metabolism (Leahy and Colwell 1990). Also, a fast rate of crude oil degradation in oil-contaminated sites in Tiruchirappali, India, was reported a tropical climate prevailing there during most of the year (Raghavan and Vivekanandan 1999).

At low temperatures, the rate of biodegradation of oil is reduced as a result of the decreased rate of enzymatic activities, or the "Q_{10}" (the change in enzyme activity caused by a 10 °C rise) effect (Atlas and Bartha 1972, Gibbs et al. 1975). Negligible degradation of oil was exhibited in the Arctic marine ice (Atlas et al. 1978) and in the frozen tundra soil (Atlas et al. 1976). However, Huddleston and Cresswell (1976) reported that petroleum biodegradation in soils at temperatures as low as -1.1 °C went on as long as the soil solution remained in its liquid form. Nevertheles, cold climates may select for lower temperature indigenous microorganisms with high biodegradation activities (Colwell et al. 1978, Margesin and Schinner 1997, Pritchard et al. 1992, ZoBell 1973); and a considerable potential for oil bioremediation in Alpine soils with a significant enhancement by biostimulation or inorganic supply was reported by Margesin (2000). Biodegradation of petroleum hydrocarbons in frozen Arctic soil has been reported by Rike and his colleagues (2003), who conducted an in situ study at a hydrocarbon contaminated-Arctic site. They concluded that 0°C is not the ultimate limit for in situ biodegradation of hydrocarbons by cold adapted microorganisms and that biodegradation can proceed with the same activity at subzero temperatures during the winter at the studied Arctic site.

Pressure

The importance of pressure is confined to the deep-ocean environment where the oil that reaches there will be degraded very slowly by microbial populations. Thus, certain recalcitrant fractions of the oil could persist for decades (Colwell and Walker 1977). Schwarz *et al.* (1974, 1975) monitored the degradation of hydrocarbons by a mixed culture of deep-sea sediment bacteria under 1 atm and 495 or 500 atm at 4 °C. After a 40-week high-pressure incubation, 94% of the hexadecane was degraded, the same amount that occurred after 8 weeks at 1 atm (Schwarz *et al.* 1975).

Moisture

Bacteria rely upon the surrounding water film when they exchange materials with the surrounding medium through the cell membrane. At soil saturation, however, all pore spaces are filled with water. At a 10% moisture level in soil the osmotic and matrix forces may reduce metabolic activity to marginal levels. Soil moisture levels in the range of 20-80% of saturation generally allow suitable biodegradation to take place (Bossert and Bartha 1984), while 100% saturation inhibits aerobic biodegradation because of lack of oxygen.

Chemical factors

Oxygen

In most petroleum-contaminated soils, sediments, and water, oxygen usually is the limiting requirement for hydrocarbon biodegradation (Hinchee and Ong 1992, Miller *et al.* 1991) because the bioremediation methods for reclamation of these contaminated sites is mainly based on aerobic processes. Bacteria and fungi in their breaking down of aliphatic, cyclic and aromatic hydrocarbons involve oxygenase enzymes (Singer and Finnerty 1984, Perry 1984, Cerniglla 1984), for which molecular oxygen is required (Atlas 1984). The availability of oxygen in soils, sediments, and aquifers is often limiting and dependent on the type of soil and whether the soil is waterlogged (Atlas 1991a). Oxygen concentration has been identified as the rate-limiting variable in the biodegradation of petroleum hydrocarbons in soil (von Wedle *et al.* 1988) and of gasoline in groundwater (Jamison *et al.* 1975).

Anaerobic hydrocarbon degradation has been shown to occur at very slow rates (Bailey *et al.* 1973, Boopathy 2003, Jamison *et al.* 1975, Ward and Brock 1978, Ward *et al.* 1980) and its ecological significance appears to be minor (Atlas 1981, Bossert and Bartha 1984, Cooney 1984, Floodgate 1984,

Ward *et al.* 1980). However, several studies have shown that anaerobic hydrocarbon metabolism may be an important process in certain conditions (Head and Swannell 1999). Furthermore, the biodegradation of some aromatic hydrocarbons such as BTEX compounds, has been clearly demonstrated to occur under a variety of anaerobic conditions (Krumholz *et al.* 1996, Leahy and Colwell 1990). Anoxic biodegradation has shown that the BTEX family of compounds, except benzene, can be mineralized or transformed cometabolically (Flyvbjerg *et al.* 1991) under denitrifying conditions. Arcangeli and Arvin (1994) investigated the biodegradation of BTEX compounds in a biofilm system under nitrate-reducing conditions and they confirmed that nitrate can be used to enhance *in situ* TEX biodegradation of a contaminated aquifer. These results suggested that denitrifying bacteria can utilize toluene, ethylbenzene and xylene as sources of carbon. Also, experiments on the degree of the microbial degradation of organic pollutants in a landfill leachate in iron reducing aquifer zones specifically to degrade toluene, have been done, with complete degradation occurring in 70-100 days at a rate of 3.4-4.2 µg/(L day) of this hydrocarbon (Albrechtsen 1994).

pH

While the pH of the marine environment is characterized by being uniform, steady, and alkaline, the pH values of various soils vary over a wide range. In soils and poorly buffered treatment situations, organic acids and mineral acids from the various metabolic processes can significantly lower the pH. The overall biodegradation rate of hydrocarbons is generally higher under slightly alkaline conditions. So appropriate monitoring and adjustments should be made to keep such systems in the pH range of 7.0-7.5. The pH of the soil is an important factor for anthracene and pyrene degradation activity of introduced bacteria (*Sphingomonas paucimobilis* BA 2 and strain BP 9). A shift of the pH from 5.2 to 7.0 enhanced anthracene degradation by *S. paucimobilis* strain BA 2. However, a pH of 5.2 did not lead to total inhibition of activity (Kästner *et al.* 1998).

Salinity

Few studies have dealt with the effect of salinity on microbial degradation of oil. Ward and Brock in 1978 showed that rates of hydrocarbon metabolism decreased with increasing salinity (33-284 g/L) as a result of a general reduction in microbial metabolic rates. Also, Diaz *et al.* (2000) found that the biodegradation of crude oil was greatest at lower salinities and decreased at salinities more than twice that of normal seawater. The use of sea water instead of fresh water in remediation of hydrocarbon contaminated desert soil blocked the hydrocarbon attenuation effect (Radwan *et al.* 2000). However, Shlaris (1989) reported a general positive correlation between salinity and rates of mineralization of phenanthrene and naphthalene in estuarine sediments. In another study, Mille *et al.* (1991) noted that the amount of oil degraded initially increased as the salt concentration increased to a level of 0.4 mol/L (23.3 g/L) of NaCl and thereafter decreased with increasing salt concentration.

Water activity (a_w)

Leahy and Colwell (1990) in their review of microbial degradation of hydrocarbons in the environment suggested that hydrocarbon biodegradation in terrestrial ecosystems may be limited by the water available (a_w ranges from 0.0 to 0.99) for microbial growth and metabolism. Optimal rates of biodegradation of oil sludge in soil have been reported at 30/90% water saturation (Dibble and Bartha 1979). In contrast to the terrestrial environment, water activity in the aquatic environment is stable at 0.98 (Bossert and Bartha 1984) and may limit hydrocarbon biodegradation of tarballs deposited on beaches (Atlas 1981).

Nutrients

Spilled oil contains low concentrations of inorganic nutrients. Thus the C/N or C/P ratios are high and often limit microbial growth (Atlas 1981, Cooney 1984). If these ratios are adjusted by the addition of nitrogen and phosphorus in the form of oleophilic fertilizers (e.g., Inipol EAP22), biodegradation of the spilled oil will be enhanced (Atlas 1991). The release of nutrients from these products that contain substantial amounts of nitrogen, phosphorus, and other limiting compounds is slow. Thus the nutrient retention time is increased in contrast to water-soluble fertilizers which, have a restricted retention time. Oleophilic fertilizers are essential in environments with high water exchange or if water transport is limited, and proved to be more effective than water-soluble fertilizers when the spilled oil resided in the intertidal zone (Halmø *et al.* 1985, Halmø and Sveum 1987, Sendstad 1980, Sendstad *et al.* 1982, 1984). The effect of different nutrient combinations (C/N/P) on biodegradation of oil deposited on shorelines has been investigated by Sveum *et al.* (1994) by monitoring the total number of bacteria, the metabolically active bacteria, and oil degradation. Such treatment appeared to result in an increased degradation of oil, compared to non-treated crude oil or crude oil treated with Inipol EAP22 (Sveum *et al.* 1994).

Several investigators observed increased rates of biodegradation of crude oil or gasoline in soil and groundwater when inorganic fertilizer amendment was used (Dibble and Bartha 1979, Jamison *et al.* 1975, Jobson *et al.* 1974, Margesin 2000, Verstraete *et al.* 1976). Others (Lehtomaki and Niemela 1975) reported contradictory results which were postulated to be due to heterogenous and complex soil composition plus some other factors such as nitrogen reserves in soil and the presence of nitrogen-fixing bacteria (Bossert and Bartha 1984). Other forms of fertilizers organic carbons (glucose/peptone) were used to fertilize oily desert soil, which resulted in a dramatic increase in the number of hydrocarbon utilizing microorganisms and enhanced attenuation of hydrocarbons (Radwan *et al.* 2000).

Biological Factors

The rate of petroleum hydrocarbon biodegradation in the environment is determined by the populations of indigenous hydrocarbon degrading microorganisms, the physiological capabilities of those populations, plus other various abiotic factors that may influence the growth of the hydrocarbon-degraders (Atlas 1981, Leahy and Colwell 1990). Leahy and Colwell (1990) reviewed this subject and concluded that hydrocarbon biodegradation depends on the composition of the microbial community and its adaptive response to the presence of hydrocarbons.

Among all microorganisms, bacteria and fungi are the principal agents in hydrocarbon biodegradation, with bacteria assuming a dominant role in the marine ecosystems and fungi becoming more important in freshwater and terrestrial environments. Hydrocarbon-utilizing bacteria and fungi are readily isolated from soil, and the introduction of oil or oily wastes into soil caused appreciable increases in the numbers of both groups (Jensen 1975, Lianos and Kjoller 1976, Pinholt *et al.* 1979). In the case of algae and protozoa, on the other hand, the evidence suggests there is no ecologically significant role played by these groups in the degradation of hydrocarbons (Bossert and Bartha 1984, O'Brien and Dixon 1976).

Microbial communities with a history of being previously exposed to hydrocarbon contamination exhibit a higher potential of biodegradation than communities with no history of such exposure. The process of getting organisms to be adapted to hydrocarbon pollutants includes selective enrichment (Spain *et al.* 1980, Spain and van Veld 1983). Such treatment encourages the hydrocarbon-utilizing microorganisms and the build-up of their proportion in the heterotrophic community. The effect of adaptation or utilizing cultures adapted to pollutants is clear in the experiments of Jussara *et al.* (1999) when they showed a 42.9% reduction of the heavy fraction of light Arabian oil in sandy sediments in 28 days. Native flora achieved only 11.9% removal of these compounds. Roy (1992), and Williams and Lieberman (1992), utilizing acclimated bacteria, have also described some successful applications of microbial seeding.

Although microbial enumeration is not a direct measure of their activity in soils, it provides an indication of microbial vitality and/or biodegradative potential. In a crude petroleum oil contaminated soil, biodiversity may indicate how well the soil supports microbial growth (Bossert and Compeau 1995). This is clear in a study by Al-Gounaim and Diab (1998) where they found that the distribution of oil-degrading bacteria in the Arabian Gulf water at Kuwait ranged from 0.3-15.2 x 10^3 CFU/L at Shuwaikh Station (a commercial harbour) and 0.1-5.8 x 10^3 CFU/L at Salmiya (a relatively unpolluted control site). Their percentages among all the heterotrophic bacteria were in the range of 0.2-22.8 % in Shuwaikh water and 0.1-8.8% in Salmiya water. The ratios of CFU/L of oil-degrading bacteria obtained from Shuwaikh to those obtained from Salmiya were in the range of 1.5-57.0. In addition, the distribution of the type and the number of microorganisms at a given site may help to characterize that site with respect to the concentration and duration of the contaminant. Fresh spills and/or high levels of contaminants often kill or inhibit large sectors of the soil microbiota, whereas soils with lower levels or old contamination show greater numbers and diversity of microorganisms (Bossert and Bartha 1984, Dean-Ross 1989, Leahy and Colwell 1990, Walker and Colwell 1976). Saadoun (2002) observed that long duration contamination sites showed greater numbers of microorganisms, whereas fresh spills reduced the bacterial number in the crude oil polluted soil. The recovered bacteria from these contaminated soils mainly belonged to the genera *Pseudomonas, Enterobacter* and *Acinetobacter* (Saadoun 2002). Radwan *et al.* (1995) reported a predominance of members of the genus *Pseudomonas*, in addition to *Bacillus, Streptomyces* and *Rhodococcus*, in the various oil-polluted Kuwaiti Desert soil samples subjected to various types of management. Rahman *et al.* (2002) showed that bacteria are the most dominant flora in gasoline and diesel station soils and *Corynebacterium* was the predominant genus. The prevalence of members of the genus *Pseudomonas* in all soils tested by Saadoun (2004) confirms previous reports (Ijah and Antai 2003) about the widespread distribution of such bacteria in hydrocarbon-polluted soils and reflects their potential for use aganist these hydrocarbon contaminants, and thus to clean these polluted sites (Cork and Krueger 1991). Another way of obtaining more organisms adapted to hydrocarbon pollutants is by genetic manipulation. This would allow the transfer of degradative ability between bacteria and particularly in soil. Thus, a rapid adaptation of the bacterial population to a particular compound is promoted and the pool of hydrocarbon-catabolizing genes carrier organisms within the community is clearly enhanced. Therefore, the number of hydrocarbon utilizing organisms would be increased. These genes may also be associated with a plasmid DNA (Chakrabarty 1976) which encodes for enzymes of hydrocarbon catabolism leading to an increased frequency of plasmid-bearing microorganisms. The capability of these microorganisms to degrade hydrocarbon pollutants and their suitability to be used as seed organisms at the contaminated sites could be further manipulated by recombinant DNA technology.

Bioremediation (Definition and Technology)

Bioremediation can be defined as a natural or managed biological degradation of environmental pollution. The indigenous microorganisms normally carry out bioremediation and their activity can be enhanced by a more suitable supply of nutrients and/or by enhancing their population. Therefore, this process exploits such microorganisms and their enzymatic activities to effectively remove contaminants from contaminated sites. This process is a cost effective means of cleanup of hydrocarbon spills from contaminated sites as it involves simple procedures only and it is an environmentally friendly technology which optimizes microbial degradation activity via control of the pH, nutrient balance, aeration and mixing (Desai and Banat 1997). Also, bioremediation is a versatile alternative to physicochemical treatments (Atlas 1991a, Bartha 1986) and produces non-toxic end products such as CO_2, water and methane from petroleum hydrocarbons (PHCs) (Walter *et al.* 1997).

Among the developed and implemented technologies for remediaion of petroleum contamination (EPRI-EEI 1989, Miljoplan 1987), there are technologies that can be conducted both *in situ* (Bartha *et al.* 1990, Mathewson *et al.* 1988) and on site (API 1980, CONCAWE 1980). Both technologies are discussed in the following sections.

In Situ Bioremediation

In situ bioremediation is a very site specific techonlogy that involves establishing a hydrostatic gradient through the contaminated area by flooding it with water carrying nutrients and possibly organisms adapted to the contaminants. Water is continously circulated through the site until it is determined to be clean.

The most effective means of implementing *in situ* bioremediation depends on the hydrology of the subsurface area, the extent of the contaminated area and the nature (type) of the contamination. In general, this method is effective only when the subsurface soils are highly permeable, the soil horizon to be treated falls within a depth of 8-10 m and shallow groundwater is present at 10 m or less below ground surface. The depth of contamination plays an important role in determining whether or not an *in situ* bioremediation project should be employed. If the contamination is near the groundwater but the groundwater is not yet contaminated then it would be unwise to set up a hydrostatic system. It would be safer to excavate the contaminated soil and apply an on site method of treatment away from the groundwater.

The average time frame for an *in situ* bioremediation project can be in the order of 12-24 months depending on the levels of contamination and depth of contaminated soil. Due to the poor mixing in this system it becomes necessary to treat for long periods of time to ensure that all the pockets of contamination have been treated.

The *in situ* treatment methods of contaminated soil include the following:

1-Bioventing

This process combines an increased oxygen supply with vapour extraction. A vacuum is applied at some depth in the contaminated soil which draws air down into the soil from holes drilled around the site and sweeps out any volatile organic compounds. The development and application of venting and bioventing for *in situ* removal of petroleum from soil have been shown to remediate approximately 800 kg of hydrocarbons by venting, and approximately 572 kg by biodegradation (van Eyk 1994).

2-Biosparging

This is used to increase the biological activity in soil by increasing the O_2 supply via sparging air or oxygen into the soil. In some instances air injections are replaced by pure oxygen to increase the degradation rates. However, in view of the high cost of this treatment in addition to the limitations in the amount of dissolved oxygen available for microorganisms, hydrogen peroxide (H_2O_2) was introduced as an alternative, and it was used on a number of sites to supply more oxygen. Each liter of commercially available H_2O_2 (30%) would produce more than 100 L of O_2 (Schlegel 1977), and was more efficient in enhancing microbial activity during the bioremediation of contaminated soils and ground-waters (Brown and Norris 1994, Flathman *et al.* 1991, Lee *et al.* 1988, Lu 1994, Lu and Hwang 1992, Pardieck *et al.* 1992). The H_2O_2 put into the soil would supply ~ 0.5 mg/L of oxygen from each mg/L of H_2O_2 added, but a disadvantage comes from its dangerous toxicity to microorganismss even at low concentrations (Brown and Norris 1994, Scragg 1999).

3-Extraction

In this case the contaminants and their treatment are extracted on the surface in bioreactors.

4-Phytoremediation

The use of living green plants for the removal of contaminants and metals from soil is known as phytoremediation. Terrestrial, aquatic and wetland plants and algae can be used for the phytoremediation process under specific cases and conditions of hydrocarbon contamination (Nedunuri *et al*. 2000, Radwan *et al*. 2000, Siciliano *et al*. 2000). A database (PhytoPet©) containing information on plants with a demonstrated potential to phytoremediate or tolerate petroleum hydrocarbons was developed by Farrell *et al*. (2000) to serve as an inventory of plant species with the above mentioned potential in terrestrial and wetland environments in western Canada. One of the search results generated by this database is a list of 11 plant species capable of degrading (or assisting in the degradation of) a variety of petroleum hydrocarbons (Table 3), and which may have potential for phytoremediation efforts in western Canada.

The accidental release of oil from oil wells and broken pipelines and the vast amount of burnt and unburnt crude oil from the burning and gushing oil wells that followed the Gulf War of 1991 have driven Radwan and his colleagues to devise a feasible technology for enhancing the petroleum hydrocarbon remediation of Kuwaiti desert areas that were polluted with crude oil. Broad beans (*Vicia faba*) and lupine (*Lupine albus*) plants were tested and the results showed that *V. faba* tolerated up to 10% crude oil (sand/crude oil, w/w) (Radwan *et al*. 2000). However, *L. albus* died after three weeks of exposure to a 5% oil concentration. Also, the leaflet areas of *V. faba* and *L. albus*, were respectively reduced by 40% and 13% at a concentration of 1% of oil. Other plants, such as as Bermuda grass and Tall fescue, were also investigated for their capabilities to remediate petroleum sludge under the influence of inorganic nitrogen and phosphorus fertilizers. About a 49% reduction of TPH occurred in the first six months, but there were no significant differences between the two species and the control (unvegetated). After one year, TPH was reduced by 68, 62 and 57% by Bermuda, fescue, and control, respectively. Radwan and his colleagues (2000) concluded that the optimal remediation was obtained by fertilization that produced a C:N:P ratio of 100:2:0.2.

On Site Bioremediation

Here the contaminated soil is excavated and placed into a lined treatment cell. Thus, it is possible to sample the site in a more thorough and, therefore, representative manner. On site treatment involves land treatment or land farming, where regular tilling of the soil increases aeration and the supplement area is lined and dammed to retain any contaminants that leak out. The use of the liner is an added benefit, since the liner prevents

Table 3. Plants native to western Canada and with a demonstrated ability to phytoremediate petroleum hydrocarbons.

Common Name	Scientific Name	Family	Growth Form	Petroleum Hydrocarbons	Mechanism of Phytore- mediation
Western wheatgrass	*Agropyron smithii*	Gramineae	grass	chrysene, benzo[a] pyrene, benz[a] anthracene dibenz[a,h] anthracene	unknown
Big bluestem	*Andropogon gerardi*	Gramineae	grass grass	chrysene, benzo[a] pyrene, benz[a] ant-hracene, dibenz[a,h] anthracene	unknown
Side oats grama	*Bouteloua curtipendula*	Gramineae	------	chrysene, benzo[a] pyrene, benz [a] anthracene, dibenz [a,h] anthracene	unknown
Blue grama	*Bouteloua gracilis*	Gramineae	grass	chrysene, benzo [a] pyrene, benz [a] anthracene, dibenz [a,h] anthracene	unknown
Common buffalograss	*Buchloe dactyloides*	Gramineae	grass	naphthalene, fluorene, phenanthrene	unknown
Prairie buffalograss	(*Buchloe dactyloides* var. Prairie)	Gramineae	grass	naphthalene, fluorene, phenanthrene	unknown
Canada wild rye	*Elymus canadensis*	Gramineae	grass	chrysene, benzo [a] pyrene, benz [a] anthracene, dibenz [a,h] anthracene	unknown
Red fescue rhizosphere	*Festuca rubra* var. Arctared	Gramineae	grass	crude oil and diesel	effect (suspected)
Poplar trees	*Populus deltoides x nigra*	Salicaceae	deciduous	potential to phyto-remediate benzene, toluene, o-xylene	rhizosphere effect
Little bluestem	*Schizchyrium Scoparious* or *Andropogon scoparious*	Gramineae	grass	chrysene, benzo [a] pyrene, benz [a] anthracene, dibenz [a,h] anthracene	unknown
Indiangrass	*Sorghastrum nutans*	Gramineae	grass	chrysene, benzo [a] pyrene, benz [a] anthracene, dibenz [a,h] anthracene	unknown

migration of the contaminants and there is no possibility of contaminating the groundwater. However, excavation of the contaminated soil adds to the cost of a bioremediation project as does the liner and the landfarming equipment. In addition to these costs, it is necessary to find enough space to treat the excavated soil on site. This process allows for better control of the system by enabling the engineering firm to dictate the depth of soil well as the exposed surface area. As a consequence of the depth and exposed surface area of the soil being determined, one is able to better control the temperature, nutrient concentration, moisture content and oxygen availability.

The average time frame for an on site bioremediation project is 60-90 days, depending on the level of contamination. Bossert and Compeau (1995) reported that the average half-life for degradation of diesel fuel and heavy oil is in the order of 54 days with this type of bioremediation.

Biostimulation (Environmental Modification) Versus Bioaugmentation (Microbial Seeding)

Approaches to bioremediation include the application of microorganisms with specific enzymatic activities and/or environmental modification to permit increased rates of degradative activities by indigenous microorganisms. In most cases the organisms employed are bacteria, however, fungi and plants have also been used.

The organisms used often naturally inhabit the polluted matrix. However, they may inhabit a different environment and be used as seed organisms for their capability to degrade a specific class of substances. Dagley (1975) suggested that indigenous oil utilizing microorganisms, which have the ability to degrade organic compounds, have an important role in the disappearance of oil from soil.

There are two techniques for utilizing bacteria to degrade petroleum in the aquatic and terrestrial environments. One method, biostimulation, uses the indigenous bacteria which are stimulated to grow by introducing nutrients into the soil or water environment and thereby enhancing the biodegradation process. The other method, bioaugmentation, involves culturing the bacteria independently and then adding them to the site. Leavitt and Brown (1994) presented and compared case studies of bioremediation versus bioaugmentation for removal of crude oil contaminant. One study focused on using bioreactors to treat tank bottoms where crude oil storage had been stored and compared the indigenous organisms to known petroleum degraders. The other study demonstrated land treatment of weathered crude oil in drilling mud; one of the plots studied had only indigenous organisms, while the other utilized a commercial culture with a recommended nutrient blend. These investigators concluded that some conventional applications may not require bioaugmentation, and for some bioremediation applications biostimulation of indigenous organisms is the best choice considering cost and performance.

1-Biostimulation

This process involves the stimulation of indigenous microorganisms to degrade the contaminant. The microbial degradation of many pollutants in aquatic and soil environments is limited primarily by the availability of nutrients, such as nitrogen, phosphorus, and oxygen. The addition of nitrogen- and phosphorus-containing substrates has been shown to stimulate the indigenous microbial populations. Zucchi *et al.* (2003), while studying the hydrocarbon-degrading bacterial community in laboratory soil columns during a 72-day biostimulation treatment with a mineral nutrient and surfactant solution of an aged contamination of crude oil-polluted soil, found a 39.5% decrease of the total hydrocarbon content. The concentrations of available nitrogen and phosphorus in seawater have been reported to be severely limiting to microbial hydrocarbon degradation (Atlas and Bartha 1972, Leahy and Colwell 1990). The problem of nutrient limitations has been overcome by applying fertilizers (Atlas 1977, Dibble 1979, Jamison *et al.* 1975, Jobson *et al.* 1974, Margesin 2000, Verstraete *et al.* 1976) which, range from soluble and slow release agricultural fertilizers of varying formulations to specialized oleophilic nitrogen-and phosphorus-containing fertilizers for use in treating oil spills. The cost of fertilizer and the potential for groundwater contamination encourage more conservative application rates. Most agricultural fertilizers contain excessive phosphorus and potassium. Urea and ammonium compounds are added to such fertilizers to bring up the nitrogen levels. Laboratory experiments by Dibble and Bartha (1979) showed a C:N ratio of 60:1 and a C:P ratio of 800:1 to be optimum.

Another course of action is the addition of a second carbon source to stimulate *cometabolism* (Semprini 1997). Cometabolism occurs when an organism is using one compound for growth and gratuitously oxidizes a second compound that is resistant to being utilized as a nutrient and energy source by the primary organism, but the oxidation products are available for use by other microbial populations (Atlas and Bartha 1993). This cooxidation process was noted by Leadbetter and Foster (1958) when they observed the oxidation of ethane, propane and butane by *Pseudomonas methanica* growing on methane, the only hydrocarbon supporting growth. Beam and Perry (1974) described this phenomenon when *Mycobacterium vaccae* cometabolized cyclohexane while growing on propane. The cyclohexane is oxidized to cyclohexanol, which other bacterial populations (*Pseudomonas*) can then utilize. Therefore, such cometabolism transformation in a mixed culture or in the environment may lead to the recycling of relatively recalcitrant compounds, that do not support the growth of any microbial culture (Atlas and Bartha 1993). The study of Burback and Perry (1993) demonstrated that *M. vaccae* can catabolize a number of major groundwater pollutants to more water-soluble compounds. When toluene and benzene were present concomitantly, toluene was catabolized and benzene oxidation was delayed (Burback and Perry 1993).

2-Bioaugmentation

This process involves the introduction of preselected organisms to the site for the purpose of increasing the rate or extent, or both, of biodegradation of contaminants. It is usually done in conjunction with the development and monitoring of an ideal growth environment, in which the selected bacteria can live and work. The selected microorganisms must be carefully matched to the waste contamination present as well as the metabolites formed. Effective seed organisms are characterized by their ability to degrade most petroleum components, genetic stability, viability during storage, rapid growth following storage, a high degree of enzymatic activity and growth in the environment, ability to compete with indigenous microorganisms, nonpathogenicity and inability to produce toxic metabolites (Atlas 1991b).

Mixed cultures have been most commonly used as inocula for seeding because of the relative ease with which microorganisms with different and complementary biodegradative capabilities can be isolated (Atlas 1977). Different commercial cultures were reported to degrade petroleum hydrocarbons (Compeau *et al.* 1991, Leavitt and Brown 1994, Chhatre *et al.* 1996, Mangan 1990, Mishra *et al.* 2001, Vasudevan and Rajaram 2001). Compeau *et al.* (1991) compared two different commercial cultures to indigenous microorganisms with respect to their ability to degrade petroleum oil in soil. Neither of the cultures was capable of degrading the oil. The case studies of Leavitt and Brown (1994) evaluated the benefits of adding such bacterial cultures in terms of cost and performance to bioremediation systems. The potential of a bacterial consortium for degradation of Gulf and Bombay High crude oil was reported by Chhatre *et al.* (1996). They showed that some members of the consortium were able to enzymatically degrade 70% of the crude oil, while others effectively degraded crude oil by production of biosurfactant and rhamnolipid. The wide range of hydrocarbonclastic capabilities of the selected members of the bacterial consortium led to the degradation of both aromatic and aliphatic fractions of crude oil in 72 hours.

In a recent study by Ruberto *et al.* (2003) on the bioremediation of a hydrocarbon contaminated Antarctic soil demonstrated a 75% removal of the hydrocarbon when the contaminated soil was bioaugmented with a psychrotolerant strain (B-2-2) and that bioaugmentation improved the bioremediation efficiency.

Fungi have also been used. Lestan and Lamar (1996) used a number of fungal inocula to bioaugment soils contaminated with pentachlorophenol (PCP) which resulted in the removal of 80-90% within four weeks. A high rate trichloroethylene (TCE) transformant strain of *Methylosinus trichosporium* was selected and used in a field study to degrade TCE efficiently (Erb *et al.* 1997). Two white rot fungal species, *Irpex lacteus* and *Pleurotus ostreatus*, were used as inoculum for bioremediation of petroleum hydrocarbon-contaminated soil from a manufactured-gas-plant-area. The two fungal species were able to remove PAHs from the contaminated soil where the concentrations of phenanthrene, anthracene, fluorranthene and pyrene decreased up to 66% after a 10-week treatment (Šašek *et al.* 2003).

However, some degradative pathways can produce intermediates, which are trapped in dead-end pathways, or transform the pollutants into toxic compounds. Such a situation can be improved by the addition of a seed culture of selected or genetically engineered microoorganisms. The use of these genetically manipulated organisms to degrade a variety of pollutants has been suggested as a way to increase the rate or extent of biodegradation of pollutants. The genes encoding the enzymes of biodegradative pathways often reside on plasmids (Chakrabarty *et al.* 1973 Chakrabarty 1974). The best studied plasmid-based pathway is the toluene degradation by *Pseudomonas putida* mt-2 and the plasmid TOL (Glazer and Nikaido 1994). Kostal *et al.* (1998) reported that the ability of *Pseudomonas* C12B to utilize n-alkanes (C_9-C_{12}) and n-alkenes (C_{10} and C_{12}) of medium chain length is plasmid-encoded. These plasmids are usually transferred to other microorganisms by conjugation by which homologous regions of DNA will recombine to generate a fusion plasmid carrying the enzymes for more than one degradative pathway. For example, Chakrabarty (1974) transferred a camphor-degrading plasmid (CAM plasmid) into a bacterium carrying a plasmid with the genes for degrading octane (OCT plasmid). As a result of their homologous regions, the CAM and OCT plasmids recombined to form a fusion plasmid that encoded enzymes for both pathways. Subsequent mating with other strains can generate a bacterium that can degrade a variety of different types of hydrocarbons. Chakrabarty and his colleagues generated the first engineered microorganisms with degradative properties in the 1970s. Chakrabarty obtained the first U.S.

patent for a genetically engineered hydrocarbon-degrading pseudomonad. The engineered organism was capable of degrading a number of low molecular weight aromatic hydrocarbons, but did not degrade the higher molecular weight persistent polynuclear aromatics, and thus has not been used in the bioremediation of oil spills.

Bioremediation of Marine Oil Spills

The *Exxon Valdez* spill of almost 11 million U.S. gallons (37,000 metric tonnes) of crude oil into the water of Prince William Sound, Alaska, brought into focus the necessity for a major study of bioremediation. Then it witnessed the largest application of bioremediation technology (Pritchard 1990, Pritchard and Costa 1991). The initial approach was by physical cleanup of the spilled oil by washing shorelines with high-pressure water. Then the collected oil was removed with skimmers, followed by the application of carefully chosen fertilizers to stimulate the biodegradation of the remaining oil by the indigenous microbial populations. The spillage from the oil tanker, *Exxon Valdez*, accident provided the opportunity for in-depth studies on the efficiency of inorganic mineral nutrient application on the biological removal of oil from the rocky shore. Three different forms of nitrogen and phosphate fertilizer were investigated (Chianelli et al. 1991, Ladousse and Tramier 1991, Pritchard 1990). The first was a water-soluble fertilizer with a ratio of 23:2, nitrogen to phosphorus. The second one was a a slow-release formulation of soluble nutrients encased in a polymerized vegetable oil and marketed under the trade name Customblen™ 28-8-0 (Grace-Sierra Chemicals, Milpitas, California). It contains ammonium nitrate, calcium phosphate and ammonium phosphates with a nitrogen to phosphorus ratio of 28:3.5. The formula (Osmocote ™) was studied as a slow release fertilizer by Xu and his colleagues (2003) who investigated the effect of various dosages of such ferilizer in stimulating an indigenous microbial biomass in oil-contaminated beach sediments in Singapore. An addition of 0.8% Osmocote ™ to the sediments was sufficient to maximize metabolic activity of the biomass, and the biodegradation of C_{10}-C_{33} straight-chain alkanes. The third one was an oleophilic fertilizer designed to adhere to oil and marketed under the trade name Inipol EAP22 ™ (CECA S.A. 92062 Paris La Defense, France). It is a microemulsion of a saturated solution of urea in oleic acid, containing tri(laureth-4)-phosphate and 2-butoxyethanol. It is applied only where the oil is on the surfaces. The application rates were approximately 360 g/m^2 of Inipol EAP22 ™ plus 17 g/m^2 of Customblen™ to areas that were clean on the surface but had subsurface oil. The optimization of fertlizer concentrations for stimulating bioremediation in contaminated marine substrates is desirable for minimizing undesirable ecological impacts, particularly eutrophication from algal blooms and toxicity to fish and invertebrates. The oleophilic fertilizer gave the best results. It stimulated biodegradation to the extent that surfaces of the oil-blackened rocks on the shoreline turned white and were essentially oil-free only 10 days after treatment (Atlas 1991a). Therefore, the use of Inipol and Customblen was approved for shoreline treatment and was used as a major part of the cleanup effort. This was adopted in a joint Exxon/USEPA/Alaska Department of Environmental Conservation Monitoring Program to follow the effectiveness of the bioremediation treatment. The program succeeded in demonstrating that bioremediation was safe and effective as the rates of bioremediation increased at least three-fold (Chianelli et al. 1991, Prince et al. 1990).

Cleaning of the *Mega Borg* oil tanker spill off the Texas coast involved the application of a seed culture with a secret catalyst, produced by the Alpha Corporation, to the oil at sea (Mangan 1990), but the effectiveness of the Alpha Corporation seeding culture to stimulate biodegradation has not been verified, nor has the effectiveness of the culture been confirmed by the USEPA in laboratory tests (Fox 1991).

The large amounts of oil spilled after the events that followed the Gulf War of 1991 stimulated the interest of several researchers to focus on the problem of this petroleum contamination and how the heavy spilage of oil altered the content of the sediments. The results generated from these studies were used to assess the degree of environmental damage caused by the oil spills during the Gulf War (Al-Lihaibi and Al-Omran 1995, Al-Muzaini and Jacob 1996, Saeed et al. 1996). For example, the concentration of petroleum hydrocarbons (PHCs) in the sediments of the open area of the Arabian Gulf was reported by Al-Lihaibi and Al-Omran (1995) and found to be between 4.0 and 56.2 µg/g, with an overall average of 12.3 µg/g. Before the Gulf War, Fowler (1988) reported that the concentrations of PHCs in the sediments of the offshore area ranged from 0.1-1.5 µg/g. The levels of PAHs in the sediments from the Shuaiba industrial area of Kuwait were determined and the levels were considerably higher than those reported for samples collected from the same area prior to the Gulf War (Saeed et al. 1996). The toxic metals (V, Ni, Cr, Cd and Pb) content in the sediments of the same area was also determined by Al-Muzaini and Jacob (1996).

The choice "to do nothing" to the spilled oil in the Arabian Gulf turned out to be a beneficial choice. When polluted areas were left alone, extensive mats of cyanobacteria appeared on the floating oil layers (Al-Hasan et al. 1992). Included in those mats was an organotrophic bacterium which is capable of utilizing crude oil as a sole source of carbon and energy (Al-Hasan et al. 1992). It was believed that cyanobacteria (*Microcoleus chthonoplastes* and *Phormidium corium*) can at least initiate the biodegradation of hydrocarbons in oil by oxidizing them only to the corresponding alcohols. Other bacteria, yeasts and fungi can then consume these alcohols by oxidizing them to aldehydes, and finally to fatty acids, then degrade them further by beta oxidation to acetyl CoA which can be used for the production of cell material and energy (Al-Hasan et al. 1994). The results indicated that the biomass as well as the biliprotein content of both specils of cyanobacteria studied increased when cultures were provided with crude oil or individual n-alkanes, which suggests they would be valuable agents for bioremediation purposes. Samples from similar mats developing in oil contaminated sabkhas along the African coasts of the Gulf of Suez and in the pristine Solar Lake, Sinai, showed efficient degradation of crude oil in the light, followed by development of an intense bloom of *Phormidium* spp. and *Oscillatoria* spp. (Cohen 2002).

Watt (1994a) discussed various techniques to clean up oil pollution in the Marine Wildlife Sanctuary for the Arabian Gulf Region. Among the techniques discussed was bioremediation, which suggested an enhancment of oil degradation after the addition of nutrients. The importance of inorganic fertilizers to enhance biodegradation of spilled oil in the marine environment has been discussed in a previous section.

Bioremediation of Contaminated Soils

Degradation of oil in soil by microorganisms can be measured by a variety of strategies. To measure the potential of microorganisms to degrade hydrocarbons (HC) in soil, detection and enumeration of HC-degrading bacteria in hydrocarbon-contaminated soils was tested. The results generated from this approach usually show that contaminated soils contain more microorganisms than uncontaminated soils, but the diversity of the microorganisms is reduced (Al-Gounaim and Diab 1998, Bossert and Compeau 1995, Mesarch and Nies 1997, Saadoun 2002).

Biotreatment of oil-polluted sites involves environmental modification rather than seeding with microbial cultures. The findings of Wang and Bartha (1990) on bioremediation of residues of fuel spills in soil indicated that bioremediation treatment (fertilizer application plus tilling) can restore fuel spill contaminated soils in 4-6 weeks to a degree that can support plant cover. Wang et al. (1990) continued the work to remove PAH components of diesel oil in soil and found that bioremediation treatment almost completely eliminated PAHs in 12 weeks. A bioremediation treatment that consisted of liming, fertilization and tilling was evaluated on a laboratory scale for its effectiveness in cleaning up sand, loam and clay loam contaminated by gasoline, jet fuel, heating oil, diesel oil or bunker C (Song et al. 1990). The disappearance of hydrocarbons was maximal at 27 °C in response to bioremediation treatment.

After the Gulf War in 1991 when a huge amount of oil was released into the Kuwaiti Desert, many techniques were developed to remediate the contaminated soils. To do nothing to the oil lakes would have been hazardous to public health and to the environment. However, completely clean stones and other solid materials lifted from the oil-soaked soil in the oil lake in the Kuwaiti desert have been observed (Al-Zarban and Obuekwe 1998). Evidently oil-degrading microorganisms were attached to the surfaces that developed in crevices of stones and other solid materials (Al-Zarban and Obuekwe 1998). Phytoremediation of the contaminated soil in the lakebed has also been investigated. The initial observations of moderate to weakly contaminated areas showed that plants belonging to the family Compositae, that were growing in black, oil polluted sand, always had white clean roots (El-Nemr et al. 1995). The soil immediately adjacent to the roots was also clean, while sand nearby was still polluted. These studies showed oil-utilizing microorganisms, which are associated with the roots, take up and metabolize hydrocarbons quickly, which helps to detoxify and remediate the soil. El-Nemr and his colleagues suggested that remediation of the contaminated soil in the lakebed and under the dry conditions of Kuwait would work well in moderate and weakly contaminated areas by densely cultivating oil-polluted desert areas with selected crops that tolerate oil and whose roots are associated with oil degrading microorganisms. Heavily contaminated areas would first have to be mixed with clean sand to dilute the oil to tolerable levels for the plants to survive (El-Nemr et al. 1995). The third alternative involves several techniques that use fungi in bioremediation of the soil. The techniques include land farming, windrow composting piles and static bioventing piles. Before these techniques were applied, the soil was removed by excavation then taken to a specially designed containment area where it was screened to remove tarry material and large stones. The soil was then amended with fertilizer and a mixture of compost and wood chips to improve water-holding capacity and to provide the microorganisms with sufficient carbon and nutrients. When the soil was thoroughly mixed, the three bioremediation techniques were performed (Al-Awadhi et al. 1998a). The land farming method involved spreading the soil mixture to a thickness of 30 centimeters in four land farming plots. The plots were irrigated with fresh water from a pivot irrigation system. The soil water content was maintained in the optimal range of 8-10 %. Every soil plot was inoculated individually through the irrigation system by use of a sprinkler connected to a pump. The soil was tilled at least twice a week with a rototiller to maintain aeration and mixing (Al-Awadhi et al. 1998a). For the second bioremediation approach, eight windrow composting piles were cons-tructed of the same soil mixture as was used in land farming with the fertilizer and wood chips added (Al-Awadhi *et al.* 1998a). The soil was also inoculated with the fungus *Phanerochaete chrysosporium* by adding it to the water running through the irrigation system (Al-Awadhi *et al.* 1998c). All the piles were 1.5 meters tall, 20 meters long and 3 meters wide. The piles had perforated pipes buried inside them at different heights and spacings to supply constant water and nutrients. Once a month, the soil piles were turned using front-end loaders for mixing and aerating. One pile was covered with plastic to study the effect of increasing water retention (Al-Awadhi *et al.* 1998a, b). Finally, four static soil piles were also constructed in much the same way as the windrow piles except that the piles were fitted with perforated plastic pipes laid on the ground in the piles (Al-Awadhi *et al.* 1998a). The pipes were hooked to an air compressor that provided a continuous supply of air to the pile. The perforated pipes were also used to provide soil, fertilizer and the fungal inocculum and the same mix of soil was used (Al-Awadhi *et al.* 1998a). All sites were monitored on a monthly basis for one year (Al-Awadhi *et al.* 1998a). Soil tests were performed to analyze for oil content and other key factors like nutrient concentrations and microbial counts. In general, all treatments reduced the oil concentration compared to doing nothing or passive bioremediation which was the experimental control. The highest oil degradation rate was observed in the soil that was landfarmed where oil content was reduced by 82.5%, then the windrow piles, 74.2% and the static bioventing, 64.2%. Using large volumes of fresh water to leach out the salts also reduced soil salinity levels. Although landfarming and the windrow soil pile methods resulted in more oil degradation than soil bioventing, soil bioventing was deemed the better method to use. This conclusion was based on the high operation and maintenance costs associated with landfarming and windrow piles. The costs were high because of the amount and intensity of labor and the heavy field equipment needed for the operation. Soil bioventing also required a much smaller area for operation compared to the other two methods (Al-Awadhi *et al.* 1998a).

Oily sludge that is generated by the petroleum industry is another form of hazardous hydrocarbon waste that contaminates soil. A carrier-based hydrocarbon-degrading bacterial consortium was used for bioremediation of a 4000 m^2 plot of land that belongs to an oil refinery (Barauni, India) and was contaminated with approximately 300 tonnes of oily sludge. The application of 1 kg of such consortium/10 m^2 area and nutrients degraded 90.2% of the TPH in 120 days; however, only 16.8% of the TPH was degraded in the untreated control (Mishra *et al.* 2001). This study confirmed the value large-scale use of this type of consortium and nutrients for the treatment of land contaminated with oily sludge. Other experiments were undertaken for bioremediation of such waste contaminated soil in the presence of a bacterial consortium, inorganic nutrients, compost and a bulking agent (wheat bran). During the 90-day experimental period, the wheat bran-amended soil showed a considerable increase in the number of bacterial populations and 76% hydrocarbon removal compared to 66% in the case of the inorganic nutrients amended soil. Addition of the bacterial consortium in different amendments significantly enhanced the removal of oil from the petroleum sludge from different treatment units (Vasudevan and Rajaram 2001).

Bioremediation of Oil Contaminated Groundwater and Aquifers

Contamination of groundwater by the accidental release of petroleum hydrocarbons (PHC) is a common problem for drinking water supplies (U.S. National Research Council 1993). The crude oil spill site near Bemidji, Minnesota, is one of the better characterized sites of its kind in the world. The results generated from the Bemidji research project were the first to document the fact that the extent of crude oil contamination can be limited by natural attenuation (intrinsic bioremediation).

Biodegradation is the only process that leads to a reduction of the total mass of PHC or ideally results in complete mineralization of these contaminants, forming only CO_2, water, and biomass. *In situ* biodegradation of PHC in aquifers is considered to be a cost-effective and environmentally sound remediation method (Lee *et al.* 1988) because PHC are mineralized by naturally-occurring microorganisms (intrinsic bioremediation) (Rifai *et al.* 1995). Therefore, for effective petroleum biodegradation in such anaerobic contaminated sites, it is essential to supply oxygen and nutrients to stimulate the biodegradation of the leaked petroleum. The performance of aerobic *in situ* bioremediation in such anaerobic contaminated sites is limited due to low solubility of O_2 and its rapid consumption (Lee *et al.* 1988, Bouwer 1992). To supply more oxygen to enhance bioremediation of contaminated groundwaters, forced aeration (Jamison *et al.* 1975, 1976) and hydrogen peroxide (Flathman *et al.* 1991, Lee *et al.* 1988, Lu 1994, Pardiek *et al.* 1992) have been used. Lu (1994) used hydrogen peroxide as an alternative oxygen source that enhanced the biodegradation of benzene, propionic acid and *n*-butyric acid in a stimulated groundwater system. Lu found that the ratio of organics biodegraded to the amount of hydrogen peroxide added decreased with the increase of influent of hydrogen peroxide concentration, indicating that hydrogen peroxide was not efficiently utilized when its concentration was high. Berwanger and Barker (1988) and Wilson *et al.* (1986) have successfully remediated BTEX compounds in an anaerobic groundwater situation using enhanced *in situ* aerobic remediation.

Different methods were developed to assess the *in situ* microbial mineralization of PHC and bioremediation of a petroleum hydrocarbon-contaminated aquifer. One method based on stable carbon isotope ratios (δ ^{13}C) was developed by Bolliger *et al.* (1999) who showed that 88% of the dissolved inorganic carbon (DIC) produced in the contaminated aquifer resulted when microbial PHC mineralization was linked to the consumption of oxidants such as O_2, NO^{3-}, and SO_4^{2-}. Other methods based on alkalinity, inorganic carbon in addition to measurements of stable isotope ratios were also proposed by combining data on oxidant consumption, production of reduced species, CH_4, alkalinity and DIC (Hunkeler *et al.* 1999).

SUMMARY

Petroleum contamination is a growing environmental concern that harms both terrestrial and aquatic ecosystems. Bioremediation is a potentially important option for dealing with oil spills and can be used as a cleanup method for this contamination by exploiting the activities of micro-organisms that occur naturally and can degrade these hydrocarbon contaminants. Biodegradation is the only process that leads to a considerable enzymatic reduction of the PHC or ideally results in complete mineralization of this contaminant. This degradation depends on several physical and chemical factors that need to be properly controlled to optimize the environmental conditions for the microorganisms and successfully remediate the contaminated sites. Among the developed and implemented technologies for cleaning up petroleum contamination those which may be conducted both *in situ* and on site. The *in situ* treatments of contaminated sites include bioventing, biosparging, extraction, phytoremediation and *in situ* bioremediation. On site treatment means that soil is excavated and treated above ground. The method involves land farming, biopiles, composting and bioreactors. Approaches to bioreme-diation of contaminated aquatic and terrestrial environments include two techniques. One method, biostimulation, uses the indigenous bacteria which are stimulated to grow by introducing nutrients into the soil or water environment, thereby enhancing the biodegradation process. The other method, bioaugmentation, involves culturing the bacteria independently and adding them to the site. Cometabolism is another course of action. Bioremediation of marine oil spills is usually approached by physical efforts followed by the application of fertilizers to stimulate the biodegradation of the remaining oil by the indigenous microbial populations. Bioremediation of oil-polluted soils and oily sludge that is generated by the petroleum industry involves environmental modification (fertilizer application plus tilling) in addition to seeding with microbial cultures. Phytoremediation and composting are alternative ways to clean contaminated soil. Bioremediation of oil contaminated groundwater and aquifers by naturally-occurring microorganisms (intrinsic bioremediation) is considered to be a cost-effective and environmentally sound remediation method. Effective petroleum biodegradation in such anaerobic contaminated sites requires a supply of oxygen and nutrients to stimulate the biodegradation of the leaked petroleum.

ACKNOWLEDGMENT

The authors would like to thank Jordan University of Science and Technology for the administrative support. A special thanks is extended to Prof. Khalid Hameed for reading the manuscript.

REFERENCES

Al-Awadhi, N., R. Al-Daher, M. Balba, H. Chino, and H. Tsuji. 1998a. Bioremediation of oil-contaminated desert soil: the Kuwait experience. *Environ. Int.* 24: 163-173.

Al-Awadhi, N., R. Al-Daher, and A. El-Nawawy. 1998b. Bioremediation of damaged desert environment using the windrow soil pile system in Kuwait. *Environ. Int.* 24: 175-180.

Al-Awadhi, N., M. Balba, M., A. El-Nawawy, and A. Yateem. 1998c. White rot fungi and their role in remediating oil-contaminated soil. *Environ. Int.* 24: 181-187.

Albrechtsen, H.J. 1994. Bacterial degradation under iron-reducing conditions. Pages 418-423 in *Hydrocarbon Bioremediation*, R.E. Hinchee, B.C. Alleman, R.E. Hoeppel and R.N. Miller, eds., CRC Press, Boca Raton, Florida.

Al-Gounaim, M.Y., and A. Diab. 1998. Ecological distribution and biodegradation activities of oil-degrading marine bacteria in the Arabian Gulf water at Kuwait. *Arab. Gulf J. Sci. Res.* 16: 359-377.

Al-Hasan, R., N. Sorkhoh, D. Al-Bader, and S. Radwan, 1994. Utilization of hydrocarbons by cyanobacteria from microbial mats on oily coasts of the Gulf. *Appl. Microbiol. Biotechnol.* 41: 615-619.

Al-Hasan, R., N. Sorkhoh, and S. Radwan. 1992. Self-cleaning the Gulf. *Nature* 359: 109.

Al-Lihaibi, S.S., and L. Al-Omran. 1995. Petroleum hydrocarbons in offshore sediments from the Gulf. *Mar. Poll. Bull.* 32: 65-69.

Al-Muzaini, S., and P.G. Jacob. 1996. An assessment of toxic metals content in the marine sediments of the Shuiba industrial area, Kuwait, after the oil spill during the Gulf War. *Water Sci. Technol.* 34: 203-210.

Al-Zarban, S., and C. Obuekwe. 1998. Bioremediation of crude oil pollution in the Kuwait Desert: the role of adherent microorganisms. *Environ. Int.* 24: 823-834.

Anon. 1989. Mishaps cause three oil spills off US. *Oil Gas J.* 87: 22.

API. 1980. Landfarming: an effective and safe way to treat/dispose of oily refinery wastes. American Petroleum Institute, Solid Wastes Management Committee, Washington, D.C.

Arcangeli, J.P., and E. Arvin. 1994. Biodegradation of BTEX compounds in a biofilm system under nitrate-reducing conditions. Pages 374-382 in *Hydrocarbon Bioremediation*, R.E. Hinchee, B.C. Alleman, R.E. Hoeppel and R.N. Miller, eds., CRC Press, Boca Raton, Florida.

Atlas, R.M. 1977. Stimulated petroleum biodegradation. *Crit. Rev. Microbiol.* 5: 371-386.

Atlas, R.M. 1981. Microbial degradation of petroleum hydrocarbons: an environmental perspective. *Microbiol. Rev.* 45: 180-209.

Atlas, R.M. 1984. *Petroleum Microbiology*. Macmillan Publishing Co., New York.

Atlas, R.M. 1991a. Microbial hydrocarbon degradation-bioremediation of oil spills. *J. Chem. Technol. Biotechnol.* 52: 149-156.

Atlas, R.M. 1991b. Bioremediation: using Nature's helpers-Microbes and enzymes-to remedy mankind's pollutants. Pages 255-264 in *Biotechnology in the Feed Industry, Proceedings of Alltech's Thirteenth Annual Symposium*, Lyons T.P. and K.A Jacpues, eds., Alltech Technical Publication, Nicholasville, Kentucky.

Atlas, R.M., and R. Bartha. 1972. Degradation and mineralization of petroleum in sea water: limitation by nitrogen and phosphorus. *Biotech. Bioeng.* 14: 319-330.

Atlas, R.M., and R. Bartha. 1992. Hydrocarbon biodegradation and oil spill bioremediation. *Adv. Microb. Ecol.* 12: 287-338.

Atlas, R.M., and R. Bartha. 1993. *Microbial Ecology, Fundamentals and Applications*. The Benjamin/Cummings Publishing Company, Inc., San Francisco, California.

Atlas, R.M., M.A. Horowitz, and M. Busdosh. 1978. Prudhoe crude oil in Arctic marine ice, water, and sediment ecosystems: degradation and interactions with microbial and benthic communities. *J. Fish. Res. Board Can.* 35: 585-590.

Atlas, R.M., E.A. Schofield, F.A. Morelli, and R.E. Cameron. 1976. Interaction of microorganisms and petroleum in the Arctic. *Environ. Pollut.* 10: 35-44.

Bailey, N.J.L., A.M. Jobson, and M.A. Rogers. 1973. Bacterial degradation of crude oil: comparison of field and experimental data. *Chem. Geol.* 11: 203-221.

Bartha, R. 1986. Biotechnology of petroleum pollutant biodegradation. *Microb. Ecol.* 12: 155-172.

Bartha, R., and I. Bossert. 1984. The treatment and disposal of petroleum refinery wastes. Pages 1-61 in *Petroleum Microbiology*, R.M. Atlas, ed., Macmillan Publishing Company, New York.

Bartha, R., H.G. Song, and X. Wang. 1990. Bioremediation potential of terrestrial fuel spills. *Appl. Environ. Microbiol.* 56: 652-656.

Beam, H.W., and J.J. Perry. 1974. Microbial degradation of cycloparaffinic hydrocarbons via cometabolism and commensalism. *J. Gen. Microbiol.* 82: 163-169.

Berwanger, D.J., and J.F. Barker. 1988. Aerobic biodegradation of aromatic and chlorinated hydrocarbons commonly detected in landfill leachates. *Water Pollut. Res. J. Can.* 23: 460-475.

Boopathy, R. 2003. Use of anaerobic soil slurry reactors for the removal of petroleum hydrocarbons in soil. *Int. Biodet. Biodeg.* 52: 161-166.

Bossert, I.D., and R. Bartha. 1984. The fate of petroleum in soil ecosystems. Pages 435-474 in *Petroleum Microbiology*, R.M. Atlas, ed., Macmillan Publishing Co., New York.

Bossert, I.D., W.M. Kachel, and R. Bartha. 1984. Fate of hydrocarbons during oily sludge disposal in soil. *Appl. Environ. Microbiol.* 47: 763-767.

Bossert, I.D., and G.C. Compeau. 1995. Cleanup of petroleum hydrocarbon contaminating in soil. Pages 77-128 in *Microbial Transformation* and *Degradation of Toxic Organic Chemicals*, L.Y. Young and C.E. Cerniglia, eds., Wiley-Liss, New York.

Bolliger, C., P. Hohener, D. Hunkeler, K. Haberli, and J. Zeyer. 1999. Intrinsic bioremediation of a petroleum hydrocarbon-contaminated aquifer and assessment of mineralization based on stable carbon isotopes. *Biodegradation* 10: 201-207.

Broderick, L.S., and J.J. Cooney. 1982. Emulsification of hydrocarbons by bacteria from freshwater ecosystems. *Dev. Ind. Microbiol.* 23: 425-434.

Brown, R.A., and R.D. Norris. 1994. The evolution of a technology: hydrogen peroxide in *in situ* bioremediation. Pages 148-162 in *Hydrocarbon Bioremediation*, R.E. Hinchee, B.C. Alleman, R.E. Hoeppel, and R.N. Miller eds., CRC Press, Boca Raton, Florida.

Bouwer, E.J. 1992. Bioremediation of organic contaminants in the subsurface. Pages 287-318 in *Environmental Microbiology*, R. Mitchel, ed., Wiley, New York.

Burback, B.L., and J.J. Perry. 1993. Biodegradation and biotransformation of groundwater pollutant mixtures by *Mycobacterium vaccae*. *Appl. Environ. Microbiol.* 59: 1025-1029.

Cerniglla, C.E. 1984. Microbial transformation of aromatic hydrocarbons. Pages 99-128 in *Petroleum Microbiology*, R.M. Atlas, ed., Macmillan Publishing Co., New York.

Chakrabarty, A.M. 1974. Microorganisms having multiple compatible degradative energy-generating plasmids and preparation thereof. *Off. Gaz. US Patent Office* 922: 1224.

Chakrabarty, A.M., G. Chou, and I.C. Gunsalus. 1973. Genetic regulation of octane dissimilation plasmid in *Pseudomonas*. *Proc. Nat. Acad. Sci. USA*, 70: 1137-1140.

Chakrabarty, A.M. 1976. Plasmids in *Pseudomonas*. *Annu. Rev. Genet.* 10: 7-30.

Chaney, R.L., M. Malik, Y.M. Li, S.L. Brown, E.P. Brewer, J.S. Angel, and A.J.M. Baker. 1997. Phytoremediation of soil metals. *Curr. Opin. Biotechnol.* 8: 279-284.

Chhatre, S., H. Purohit, R. Shanker, and P. Khanna. 1996. Bacterial consortia for crude oil spill remediation. *Water Sci. Technol.* 34: 187-193.

Chianelli, R.R., T. Aczel, R.E. Bare, G.N. George, M.W. Genowitz, M.J. Grossman, C.E. Haith, F.J. Kaiser, R.R. Lessard, R. Liotta, R.L. Mastracchio, V. Minak-Bern-ero, R.C. Prince, W.K. Robbins, E.I. Stiefel, J.B. Wilkinson, S.M. Hinton, J.R. Bragg, S.J. McMillan, and R.M. Atlas. 1991. Bioremediation technology development and application to the Alaskan spill. Pages 549-558 in *Proceedings of the 1991 International Oil Spill Conference*. American Petroleum Institute, Washington, D.C.

Cohen, Y. 2002. Bioremediation of oil by marine microbial mats. *Int. Microbiol.* 5: 189-193.

Colwell, R.R., and J.D. Walker. 1977. Ecological aspects of microbial degradation of petroleum in the marine environment. *Crit. Rev. Microbiol.* 5: 423-445.

Colwell, R.R., A.L. Mills, J.D. Walker, P. Garcia-Tello, and V. Campos-P. 1978. Microbial ecology of the *Metula* spill in the Straits of Magellan. *J. Fish. Res. Board Can.* 35: 573-580.

Compeau, G.C., W.D. Mahaffey, and L. Patras. 1991. Full-scale bioremediation of a contaminated soil and water. Pages 91-110 in *Environmental Biotechnology for Waste Treatment*, G.S. Sayler, R. Fox and J.W. Blackburn, eds., Plenum Press, New York.

CONCAWE. 1980. Sludge farming: A technique for the disposal of oily refinery wastes. Report No. 3/80.

Cooney, J.J. 1984. The fate of petroleum pollutants in fresh water ecosystems. Pages 399-434 in *Petroleum Microbiology*, R.M. Atlas, ed., Macmillan Publishing Co., New York.

Cooney, J.J., S.A. Silver, and E.A. Beck. 1985. Factors influencing hydrocarbon degradation in three freshwater lakes. *Microb. Ecol.* 11: 127-137

Cork D.J., and J.P. Krueger. 1991. Microbial transformation of herbicides and pesticides. *Adv. Appl. Microbiol.* 36: 1-66.

Dagley, S. 1975. A biochemical approach to some problems of environmental pollution. *Essays Biochem.* 11: 81-138.

Davis, S.J., and C.F. Gibbs. 1975. The effect of weathering on crude oil residue exposed at sea. *Water Res.* 9: 275-285.

Dean-Ross, D. 1989. Bacterial abundance and activity in hazardous waste-contaminated soil. *Bull. Environ. Contam. Toxicol.* 43: 511-517.

Desai, J.D., and I.M. Banat. 1997. Microbial production of surfactants and their commercial potential. *Microbiol. Mol. Rev.* 61: 47-64.

Diaz, M.P., S.J.W. Grigson, C. Peppiatt, and J.G. Burgess. 2000. Isolation and characterization of novel hydrocarbon degrading euryhaline consortia from crude oil and mangrove sediments. *Mar. Biotechnol.* 2: 522-532.

Dibble, J.T., and R. Bartha. 1979. Effect of environmental parameters on the biodegradation of oil sludge. *Appl. Environ. Microbiol.* 37: 729-739.

EPRI-EEI. 1989. *Remedial Technologies for Leaching Underground Storage Tanks,* Lewis Publishers, Chelsea, Michigan.

Erb, R.W., C.A. Eichner, I. Wangler-Dobler, and K.N. Timmis. 1997. Bioprotection of Microbial communities from toxic phenol mixtures by genetically designed pseudomonad. *Nature Biotechnol.* 15: 378-382.

El-Nemr, I., S. Radwan, and N. Sorkhoh. 1995. Oil biodegradation around roots. *Nature* 376: 302.

Farrell, R.E., C.M. Frick, and J.J. Germida. 2000. *PhytoPet©*: A database of plants that play a role in the phytoremediation of petroleum hydrocarbons. Pages 29-40 in *Proceedings of the Second Phytoremediation Technical Seminar,* Environment Canada, Ottawa.

Fedorak, P.M., and D.W.S. Westlake. 1981. Microbial degradation of aromatics and saturates in Prudhoe Bay crude oil as determined by glass capillary gas chromatography. *Can. J. Microbiol.* 27: 432-443.

Flathman, P.E., J.H. Carson, Jr., S.J. Whitenhead, K.A. Khan, D.M. Barnes, and J.S. Evans. 1991. Laboratory evaluation of the utilization of hydrogen peroxide for enhanced biological treatment of petroleum hydrocarbon contaminants in soil. Pages 125-142 in *In Situ Bioreclamation: Applications and Investigations for Hydrocarbon* and *Contaminated Site Remediation,* R.E. Hinchee and R.F. Olfenbuttel, eds., Butterworth-Heinemann, Stoneham, Mass.

Floodgate, G. 1984. The fate of petroleum in marine ecosystems. Pages 355-398 in *Petroleum Microbiology,* R.M. Atlas, ed., Macmillan Publishing Co., New York.

Flyvbjerg, J., E. Arvin, B.K. Jensen, and S.K. Olsen. 1991. Biodegradation of oil-and creosote-related aromatic compounds under nitrate-reducing conditions. Pages 471-479 in *In-Situ Bioreclamation: Applications and Investigations* for *Hydrocarbon and Contaminated Site Remediation,* R.E. Hinchee, and R.F. Olfenbuttel, eds., Butterworth-Heinemann, Stoneham, Mass.

Fowler, S.W. 1988. Coastal baseline studies of pollutants in Bahrain, UAE and Oman. Pages 155-180 in *Proceedings of Symposium on Regional Marine Pollution Monitoring and Research Programms,* ROPME IGC-412. Regional Organization for the protection of the Marine Environment, Kuwait.

Fox, J.E. 1991. Confronting doubtful oil cleanup data. *Biotechnology* 9: 14.

Ghosh, S., and Syed, H. 2001. Influence of soil characteristics on bioremediation of petroleum-contaminated soil. *Geological Society of America Annual Meeting,* Nov. 5-8, Boston, Massachusetts, USA.

Gibbs, C.F., K.B. Pugh, and A.R. Andrews. 1975. Quantitative studies on marine biodegradation of oil. II. Effect of temperature. *Proc. R. Soc. London Ser. B*, 188: 83-94.

Glazer, A.N., and H. Nikaido. 1994. *Microbial Biotechnology*, Freeman, New York.

Halmø, G., E. Sendstad, P. Sveum, A. Danielsen, and T. Hoddø. 1985. Enhanced biodegradation through fertilization. *SINTEF Report STF21 F85019.* Trondheim, Norway.

Halmø, G., and P. Sveum. 1987. Biodegradation and photooxidation of crude oil in arctic conditions. *SINTEF Report STF21 F87007.* Trondheim, Norway.

Head, I.M., and R.P.J. Swannell. 1999. Bioremediation of petroleum hydrocarbon contaminants in marine habitats. *Curr. Opin. Biotechnol.* 10: 234-239.

Hinchee, R.E., and S.K. Ong. 1992. A rapid *in Situ* respiration test for measuring aerobic biodegradation rates of hydrocarbons in soil. *J. Air Waste Manage. Assoc.* 42: 1305-1312.

Huddleston, R.L., and L.W. Cresswell. 1976. Environmental and nutritional constraints of microbial hydrocarbon utilization in the soil. Pages 71-72 in *Proceedings of the 1975 Engineering Foundation Conference: The Role of Microorganisms in the Recovery of Oil*, National Science Foundation, Washington, D.C.

Huesemann, M.H., and K.O. Moore. 1994. The effects of soil type, crude oil type and loading, oxygen, and commercial bacteria on crude oil bioremediation kinetics as measured by soil respirometry. Pages 58-71 in *Hydrocarbon Bioremediation*, Hinchee, R.E., B.C. Alleman, R.E. Hoeppel, and R.N. Miller, eds., CRC Press, Boca Raton, Florida.

Hunkeler, D., P. Höhener, S. Bernasconi, and J. Zeyer. 1999. Engineered *in situ* bioremediation of a petroleum hydrocarbon contaminated aquifer: assessment of mineralization based on alkalinity, inorganic carbon and stable carbon isotope balances. *J. Contaminant Hydrol.* 37: 201-223.

Ijah, U.J.J., and S.P. Antai. 2003. Removal of Nigerian light crude oil in soil over a 12-month period. *Int. Biodet. Biodeg.* 2: 93-99.

International Tanker Owners Pollution Federation (ITOPF). 1990. Response to marine oil spills. Witherby, London.

Jamison, V.M., R.L. Raymond, and J.O. Hudson, Jr. 1975. Biodegradation of high-octane gasoline in groundwater. *Dev. Ind. Microbiol.* 16: 305-312.

Jamison, V.M., R.L. Raymond, and J.O. Hudson, Jr. 1976. Biodegradation of high-octane gasoline in groundwater. Pages 187-196 in *Proceedings of the Third International Biodegradation Symposium*, J.M. Sharpley, and A.M. Kaplan, eds., Applied Science Publishers, Ltd. London.

Jensen, V. 1975. Bacterial flora of soil after application of oily waste. *Oikos* 26: 152-158.

Jobson, A., F.D. Cook, and D.W.S. Westlake. 1972. Microbial utilization of crude oil. *Appl. Microbiol.* 23: 1082-1089.

Jobson, A., M. McLaughlin, F.D. Cook, and D.W.S. Westlake. 1974. Effect of amendments on the microbial utilization of oil applied to soil. *Appl. Microbiol.* 27: 166-171.

Jones, D.M., A.G. Douglas, R.J. Parkes, J. Taylor, W. Giger, and C. Schuffner. 1983. The recognition of biodegraded petroleum-derived aromatic hydrocarbons in recent marine sediments. *Mar. Pollut. Bull.* 14: 103-108.

Jussara, P., D. Acro, and F.P. De Franca. 1999. Bioremediation of crude oil in sandy sediment. *Int. Biodet. Biodeg.* 44: 27-92.

Kästner, M., M.B. Jammali, and B. Mahro. 1998. Impact of inoculation protocols, salinity, and pH on the degradation of polycyclic aromatic hydrocarbons (PAHs) and survival of PAH-degrading bacteria introduced into soil. *Appl. Environ. Microbiol.* 64: 359-362.

Kostal, J., M. Suchaneck, H. Klierova, K. Demnerova, B. Kralova, and D.L. McBeth. 1998. *Pseudomonas* C12B, an SDS degrading strain, harbours a plasmid coding for degradation of medium chain length n-alkanes. *Int. Biodet. Biodeg.* 42: 221-228.

Krumholz, L.R., M.E. Caldwell, and J.M. Suflita. 1996. Biodegradation of 'BTEX" hydrocarbons under anaerobic conditions. Pages 61-99 in *Bioremediation: Principles and Applications*, Crawford, R.L., and D.L. Crawford, eds., Cambridge Univ. Press, Cambridge.

Ladousse, A., and B. Tramier. 1991. Results of 12 years of research in spilled oil bioremediation: Inipol EAP 22. Pages 577-581 in *Proceedings of the 1991 International Oil Spill Conference*, American Petroleum Institute, Washington, D.C.

Leadbetter, E.R., and J.W. Foster. 1958. Studies of some methane utilizing bacteria. *Arch. Microbiol.* 30: 91-118.

Leahy, J.G., and R.R. Colwell. 1990. Microbial degradation of hydrocarbons in the environment. *Microbiol. Rev.* 54: 305-315.

Leavitt, M.E., and K.L. Brown. 1994. Bioremediation versus bioaugmentation-there case studies. Pages 72-79 in *Hydrocarbon Bioremediation*, Hinchee, R.E., B.C. Alleman, R.E. Hoeppel, and R.N. Miller, eds., CRC Press, Inc., Boca Raton, Florida.

Lee, C.M., and D.F. Gongaware. 1997. Optimization of SFE conditions for the removal of diesel fuel. *Environ. Technol.* 18: 1157-1161.

Lee, M.D., J.M. Thomas, R.C. Borden, P.B. Bedient, and C.H. Ward. 1988. Biorestoration of aquifers contaminated with organic compounds. *CRC Crit. Rev. Environ. Control* 18: 29-89.

Lehtomaki, M., and S. Niemeia. 1975. Improving microbial degradation of oil in soil. *Ambio* 4: 126-129.

Lestan, D., and R.T. Lamar. 1996. Development of fungal inocula for bioaugmentation of contaminated soils. *Appl. Environ. Microbiol.* 62: 2045-2052.

Lianos, C., and A. Kjoller. 1976. Changes in the flora of soil fungi following oil waste application. *Oikos* 27: 377-382.

Lu, C.J. 1994. Effects of hydrogen peroxide on the *in situ* biodegradation of organic chemicals in a simulated groundwater system. Pages 140-147 *in Hydrocarbon Bioremediation*, R.E. Hinchee, B.C. Alleman, R.E. Hoeppel, and R.N. Miller, eds., CRC Press, Boca Raton, Florida.

Lu, C.J., and M.C. Hwang. 1992. Effects of hydrogen peroxide on the *in situ* biodegradation of chlorinated phenols in groundwater. *Water Environ. Federation 65th Annual Conference*, Sept. 20-24. New Orleans, Louisiana.

Mangan, K.S. 1990. University of Texas microbiologists seeks to persuade skeptical colleagues that bacteria could be useful in cleaning up major oil spills. *Chron. Higher Education.* 37: A5-A9.

Margesin, R. 2000. Potential of cold-adapted microorganisms for bioremediation of oil polluted Alpine soils. *Int. Biodet. Biodeg.* 46: 3-10.

Margesin, R., and F. Schinnur. 1997. Efficiency of endogenous and inoculated cold-adapted soil microorganisms for biodegradation of diesel oil in Alpine Soils. *Appl. Environ. Microbiol.* 63: 2660-2664.

Mathewson, J.R., R.B. Grubbs, and B.A. Molnaa. 1988. Innovative techniques for the bioremediation of contaminated soils. *California Pollution Control Association*, Oakland, CA, 7-8 June.

Mesarch, M.B., and L. Nies. 1997. Modification of heterotrophic plate counts for assessing the bioremediation potential of petroleum-contaminated soils. *Environ. Technol.* 18: 639-646.

Mihelcic, J.R., D.R. Lueking, R.J. Mitzelland, and J.M. Stapleton. 1993. Bioavailability of sorbed- and separate- phase chemicals. *Biodegradation* 4: 141-153.

Mille, G., M. Almalah, M. Bianchi, F. van Wambeke, and J.C. Bertrand. 1991. Effect of salinity on petroleum biodegradation. *Fresenius J. Anal. Chem.* 339: 788-791.

Miller, R.N., C.C. Vogel, and R.E. Hinchee. 1991. A field-scale investigation of petroleum hydrocarbon biodegradation in the *Vadose Zone* enhanced by bioventing at Tyndall Air Force Base, Florida. Pages 283-302 in *In-Situ Bioreclamation: Applications and Investigations for Hydrocarbon and Contaminated Site remediation*, R.E. Hinchee, and R.F. Olfenbuttel, eds., Butterworth-Heinemann, Boston, Mass.

Miljoplan, A.H. 1987. Soil decontamination. Meeting of the Working Party on Environmental Protection, 26-27 Nov., Genova.

Mishra, S., J. Jyot, R.C. Kuhad, and B. Lal. 2001. *In situ* bioremediation potential of an oily sludge-degrading bacterial consortiium. *Current Microbiol.* 43: 328-335.

Mulkins-Phillips, G.J., and J.E. Stewart. 1974. Effects of four dispersants on the biodegradation and growth of bacteria on crude oil. *Appl. Microbiol.* 28: 547-552.

National Research Council. 1989. *Using Oil Spill Dispersants on the Sea*, National Academy of Sciences, Washington, D.C.

Nedunuri, K.V., R.S. Govundaraju, M.K. Banks, A.P. Schwab, and Z. Chen. 2000. Evaluation of phytoremediation for field scale degradation of total petroleum hydrocarbons. *J. Environ. Eng.* 126: 483-490.

O'Brien, P.Y., and P.S. Dixon. 1976. The effects of oil and oil components on algae; a review. *Br. Phycol. J.* 11: 115-142.

Pardieck, D.L., E.J. Bouwer, and A.T. Stone. 1992. Hydrogen peroxide use to increase oxidant capacity for *in situ* bioremediation of contaminated soils and aquifers: A review. *J. Contaminant Hydrol.* 9: 221-242.

Perry, J.J. 1984. Microbial metabolism of cyclic alkanes. Pages 61-98 in *Petroleum Microbiology*, R.M. Atlas, ed., Macmillan Publishing Co., New York.

Pinholt, Y., S. Struwe, and A. Kjoller. 1979. Microbial changes during oil decomposition in soil. *Holaret Ecol.* 2: 195-200.

Polak, J., and B.C. Lu. 1973. Mutual solubilities of hydrocarbons and waters at 0° and 25 °C. *Can. J. Chem.* 51: 4018-4023.

Prince, R.C., J.R. Clark, and J.E. Lindstrom. 1990. Bioremediation monitoring report in the U.S. Coast Guard. *Alaska Department of Environmental Conservation*, Anchorage, Alaska.

Pritchard, H.P. 1990. Bioremediation of oil contaminated beach material in Prince William Sound, Alaska. 199th National meeting of the American Chemical Society, Boston, Massachusetts, 22-27 April, Abstract Environment 154.

Pritchard, H.P., and C.F. Costa. 1991. EPA's Alaska oil spill report. Part 5. *Env. Sci. Technol.* 25: 372-379.

Pritchard, H.P., J.G. Mueller, A. Kushmaro, R. Taube, E. Alder, and E.Z. Ron. 1992. Oil spill bioremediation: experiences, lessons and results from the Exxon Valdez oil spill in Alaska. *Biodegradation* 3: 315-335.

Providenti, M.A., C.A. Flemming, H. Lee, and J.T. Trevors. 1995. Effect of addition of rhamnolipid biosurfactants or rhamnolipid producing *Pseudomonas aeruginosa* on phenanthrene mineralization in soil slurries. *FEMS Microbiol. Ecol.* 17: 15-26.

Purvis, A. 1999. Ten largest oil spills in history. Planet Watch. *Time International.* 153: 12.

Radwan, S.S., N.A., Sorkhoh, F. Fardoun, and H. Al-Hasan. 1995. Soil management enhancing hydrocarbon biodegradation of the polluted Kuwaiti desert. *Appl. Microbiol. Biotechnol.* 44: 265-270.

Radwan, S.S., D. Al-Mailem, I. El-Nemr, and S. Salamah. 2000. Enhanced remediation of hydrocarbon contaminated desert soil fertilized with organic carbons. *Int. Biodet. Biodeg.* 46: 129-132.

Radwan, S.S., H. Awadhu, and I.M. El-Nemr. 2000. Cropping as a phytoremediation practice for oily desert soil with reference to crop safety as food. *Int. J. Phytoremed.* 2: 383-396.

Raghavan, P.U.M., and M. Vivekanandan. 1999. Bioremediation of oil-spilled sites through seeding of naturally adapted *Pseudomonas putida*. *Int. Biodet. Biodeg.* 44: 29-32.

Rahman, K.S.M., T. Rahman, P. Lakshmanaperumalsamy, and I. Bana. 2002. Occurrence of crude oil degrading bacteria in gasoline and diesel station soils. *J. Basic Microbiol.* 42: 286-293.

Rashid, M.A. 1974. Degradation of bunker C oil under different coastal environments of Chedabucto Bay, Nova Scotia. *Estuarine Coastal Mar. Sci.* 2: 137-144.

Rike, A.G., K.B. Haugen, M. Børresen, B. Engene, and P. Kolstad. 2003. *In situ* biodegradation of petroleum hydrocarbons in frozen arctic soils. *Cold Regions Sci. Technol.* 37: 97-120.

Rifai, H.S., R.C. Borden, J.T. Wilson, and C.H. Ward. 1995. Intrinsic bioattenuation for subsurface restoration. Pages 1-31 in *Intrinsic Bioremediation*, R.E. Hinchee, J. Wilson, and D.C. Downey, eds., Vol. 1, Battelle Press, Columbus, Ohio.

Robichaux, T.J., and H.N. Myrick. 192. Chemical enhancement of the biodegradation of crude oil pollutants. *J. Petrol. Technol.* 24: 16-20.

Roy, K.A. 1992. Petroleum company heals itself-and others. *Hazmat World,* May: 75-80.

Ruberto, L., S.C. Vazquez, and W.P. Mac Cormack. 2003. Effectiveness of the natural bacterial flora, biostimulation and bioaugmentation on the bioremediation of a hydrocarbon contaminated Antractic soil. *Int. Biodet. Biodeg.* 52: 115-125.

Saadoun, I. 2002. Isolation and characterization of bacteria from crude petroleum oil contaminated soil and their potential to degrade diesel. *J. Basic Microbiol.* 42: 420-428.

Saadoun, I. 2004. Recovery of *Pseudomonas* spp. from chronocillay fuel oil-polluted soils in Jordan and the study of their capability to degrade short chain alkanes. *World J. Microbiol. Biotechnol.* 20: 43-46.

Saeed, T., S. Al-Muzaini, and A. Al-Bloushi. 1996. Post-Gulf War assessment of the levels of PAHs in the sediments from Shuiba industrial area, Kuwait. *Water Sci. Technol.* 34: 195-201.

Saeed, Talaat, S, Al-Muzaini, and A. Al-Bloushi. 1996. Post-Gulf War assessment of the levels of PAHs in the sediments from Shuiba industrial area, Kuwait. *Water Sci. Technol.* 34: 195-201.

Šašsek, V., T. Cajthaml and M. Bhatt. 2003. Use of fungal technology in soil remediation: a case study. *Water Air Pollut. : Focus* 3: 5-14.

Schlegel, H.G. 1977. Aeration without air: oxygen supply by hydrogen peroxide. *Biotechnol. Bioeng.* 19: 413.

Schwarz, R.J., J.D. Walker, and R.R. Colwell. 1974a. Deep-sea bacteria: growth and utilization of hydrocarbons at ambient and *in situ* pressure. *Appl. Microbiol.* 28: 982-986.

Schwarz, R.J., J.D. Walker, and R.R. Colwell. 1974b. Growth of deep-sea bacteria on hydrocarbons at ambient and *in situ* pressure. *Dev. Ind. Microbiol.* 15: 239-249.

Schwarz, R.J., J.D. Walker, and R.R. Colwell. 1975. Deep-sea bacteria: growth and utilization of *n*-hexadecane at *in situ* temperature and pressure. *Can. J. Microbiol.* 21: 682-687.

Scragg, A. 1999. *Environmental Biotechnology*, Pearson Education, Essex, England.

Semprini, L. 1997. Strategies for the aerobic co-metabolism of clorinated solvents. *Curr. Opin. Biotechnol.* 8: 296-308.

Sendstad, E. 1980. Accelerated oil biodegradation of crude on arctic shorelines. *Proc. 3rd Arctic and Marine Oil Spill Program Tech,* pp. 402-416, Edmonton, Alberta.

Sendstad, E.T. Hoddø, P. Sveum, K. Eimhjellen, K. Josefen, O. Nilsen, and T. Sommer. 1982. Enhanced oil biodegradation on an arctic shorelines. *Proc. 5th Arctic and Marine Oil Spill Program Tech.,* pp. 331-340, Edmonton, Alberta.

Sendstad, E.T. Hoddø, P. Sveum, K. Eimhjellen, K. Josefen, O. Nilsen, and T. Sommer. 1984. Enhanced oil biodegradation in cold regions. *SINTEF Report STF21 F84032.* Trondheim, Norway.

Shiaris, M.P. 1989. Seasonal biotransformation of naphthalene, phenanthrene, and benzo[a]pyrene in suficial estuarine sediments. *Appl. Environ. Microbiol.* 55: 1391-1399.

Siciliano, S.D., and C. Greer. 2000. Plant-bacterial combinations to phytoremediate soil contaminated with high concentrations of 2,4,6-Trinitrotolene. *J. Environ. Quality* 29: 311-316.

Singer, M. and W. Finnerty. 1984 Microbial metabolism of straight-chain and branched alkanes. Pages 1-61 in *Petroleum Microbiology*, R.M. Atlas, ed., Macmillan Publishing Company, New York.

Song, H-G., X. Wang, and R. Bartha. 1990. Bioremediation of terrestrial fuel spills. *Appl. Environ. Microbiol.* 56: 652-656.

Spain, J.C., P.H. Pritchard, and A.W. Bourquin. 1980. Effects of adaptation on biodegradation rates in sediment/water cores from estuarine and freshwater environments. *Appl. Environ. Microbiol.* 40: 726-734.

Spain, J.C., and P.A. van Veld. 1983. Adaptation of natural microbial communities to degradation of xenobiotic compounds: effects of concentration, exposure time, inoculum, and chemical structure. *Appl. Environ. Microbiol.* 45: 428-435.

Sveum, P. L.G. Faksness, and S. Ramstad. 1994. Bioremediation of oil-contaminated shorelines: the role of carbon in fertilizers. Pages 163-174 in *Hydrocarbon Bioremediation,* Hinchee, R.E., B.C. Alleman, R.E. Hoeppel, and R.N. Miller, eds., CRC Press, Boca Raton, Florida.

U.S. National Research Council. 1993. *In situ* bioremediation, when does it work? National Academy Press, Washington, D.C.

van Eyk, J. 1994. Venting and bioventing for the *in situ* removal of petroleum from soil. Pages 234-251 in *Hydrocarbon Bioremediation*, Hinchee, R.E., B.C. Alleman, R.E. Hoeppel, and R.N. Miller, eds., CRC Press, Boca Raton, Florida.

Venosa, A.D., and X. Zhu. 2003. Biodegradation of crude oil contaminating marine sholelines and freshwater wetlands. *Spill Sci. Tec. Bul.* 8: 163-178.

Vasudevan, N., and P. Rajaram. 2001. Bioremediation of oil sludge-contaminated soil. *Environ. Int.* 26: 409-411.

Verstraete, W., R. Vanloocke, R. DeBorger, and A. Verlinde. 1976. Modelling of the breakdown and the mobilization of hydrocarbons in unsaturated soil layers. Pages 99-112 in *Proceedings of the 3rd International Biodegradation Symposium*, J.M. Sharpley, and A.M. Kaplan, eds., Applied Science Publishers Ltd., London.

von Wedel, R.J., J.F. Mosquera, C.D. Goldsmith, G.R. Hater, A. Wong, T.A. Fox, W.T. Hunt, M.S. Paules, J.M. Quiros, and J.W. Wiegand. 1988. Bacterial biodegradation of petroleum hydrocarbons in groundwater: *in situ* augmented bioreclamation with enrichment isolates in California. *Water Sci. Technol.* 20: 501-503.

Walker, J.D., and R.R. Colwell. 1976. Enumeration of petroleum-degrading microorganisms. *App. Environ. Microbiol.* 31: 198-207.

Walter, M.V., E.C. Nelson, G. Firmstone, D.G. Martin, M.J. Clayton, S. Simpson, and S. Spaulding. 1997. Surfactant enhances biodegradation of hydrocarbons: Microcosm and field study. *J. Soil Contam.* 6: 61-77.

Wang, X., and R. Bartha. 1990. Effect of bioremediation on residues: activity and toxicity in soil contaminated by fuel spills. *Soil Biol. Biochem.* 22: 501-506.

Wang, X., X. Yu, and R. Bartha. 1990. Effect of bioremediation on polycyclic aromatic hydrocarbon residues in soil. *Environ. Sci. Technol.* 24: 1086-1089.

Ward, D.M., and T.D. Brock. 1978. Anaerobic metabolism of hexadecane in marine sediments. *Geomicrobiol. J.* 1: 1-9.

Ward, D.M., and T.D. Brock. 1978. Hydrocarbon biodegradation in hypersaline environments. *Appl. Environ. Microbiol.* 35: 353-359.

Ward, D., R.M. Atlas, P.D. Boehm, and J.A. Calder. 1980. Microbial biodegradation and the chemical evolution of *Amoco Cadiz* oil pollutants. *Ambio* 9: 277-283.

Watt, I. 1994a. Shorelines clean-up procedures. A discussion pertaining to the Gulf Sanctuary. Pages 20-37 in *Establishment of a Marine Habitat and Wild Life Sanctuary for the Gulf Region*. Final Report for Phase II, E. Feltamp and F. Krupp, eds., Jubail and Frankfurt, CEC/NCWCD.

Watt, I. 1994b. An outline for the development of a contingency plan to combat oil pollution in the Gulf Sanctuary. Pages 38-80 in *Establishment of a Marine Habitat and Wild Life Sanctuary for the Gulf Region*. Final Report for Phase II, E. Feltamp, and F. Krupp, eds., Jubail and Frankfurt, CEC/NCWCD.

Williams, C.M., and M.T. Lieberman. 1992. Bioremediation of chlorinated and aromatic organic solvent waste in the subsurface. *The National Environ. J.* Nov/Dec.: 40-44.

Wilson, B.H., G.B. Smith, and J.F. Rees. 1986. Biotransformations of selected alkylbenzenes and halogenated aliphatic hydrocarbons in methanogenic aquifer material: A microcosm study. *Environ. Sci. Technol.* 20: 997-1002.

Xu, R, J.P. Obbard, and E.T.C. Tay. 2003. Optimization of slow-release fertilizer dosage for bioremediation of oil-contaminated beach sediment in a tropical environment. *World J. Microb. Biotech.* 19: 719-725.

ZoBell, C.E. 1973. The microbial degradation of oil pollutants. In Publ. No. LSU-SG-73-01. Center for Wetland Resources, D.G. Ahearn, and S.P. Meyers, eds., Louisiana State University, Baton Rouge, Louisiana.

Zucchi, M., L. Angiolini, S. Borin, L. Brusetti, N. Dietrich, C. Gigliotti, P. Barbieri, C. Sorlini, and D. Daffonchio. 2003. Response of bacterial community during bioremediation of an oil-polluted soil. *J. Appl. Microbiol.* 94: 248-257.

Bioremediation of BTEX Hydrocarbons (Benzene, Toluene, Ethylbenzene, and Xylene)

Hanadi S. Rifai

Department of Civil and Environmental Engineering, University of Houston, 4800 Calhoun Road, Houston, Texas 77204-4003, USA

Introduction

BTEX (benzene, toluene, ethylbenzene, and xylene) hydrocarbons are known to biodegrade under aerobic and anaerobic conditions in the subsurface. Biodegradation refers to the complete conversion of a chemical by living organisms to mineralized end products (e.g., CO_2 and water). In ground water aquifers, indigenous microorganisms undertake this conversion process and transform BTEX into innocuous products. Thus, the metabolism of BTEX is an extremely important fate process since it is the only one in ground water that has the potential to yield nonhazardous products instead of transferring contaminants from one phase in the environment to another. Researchers and professionals in the ground water industry have recognized the importance of biodegradation of BTEX for remediating hydrocarbon contaminated sites and have thus extensively studied intrinsic and enhanced bioremediation of these compounds. Intrinsic bioremediation refers to the biological processes that occur without human intervention in ground water and cause a reduction in BTEX concentration and mass over time. Enhanced bioremediation refers to engineered technologies that stimulate the indigenous microorganisms and accelerate their biodegradative capabilities.

In the decades of the 1970's and 1980's, research in biodegradation and bioremediation was focused on laboratory studies of aerobic biodegradation and on microbial characterization of aquifers. Researchers came to understand that soils and shallow sediments contain a large variety of microorganisms, ranging from simple bacteria to algae, fungi, and protozoa (McNabb and Dunlap 1975, Ghiorse and Wilson 1988). Studies also confirmed the ability of these microorganisms to degrade various organic compounds, including BTEX. The research focus shifted in the decade of the 1990's to studies involving anaerobic biodegradation and the use of natural biological processes as a remedy for contaminated sites because of the failure of engineered remedies in reaching cleanup goals in a reasonable timeframe.

In a similar fashion, bioremediation has come full circle from feasibility and pilot-scale testing in the 1970's and1980's to full-scale implementation in the 90's only to recognize the delivery and economic challenges associated with the technology. The heterogeneous nature of the subsurface and the relatively high electron acceptor demand of fuel spills have limited the use of aerobic bioremediation systems that relied on air sparging, or injection of liquid oxygen, for example. Thus focus has shifted in recent years to less energy intensive technologies such as biobarriers and more economical delivery methods for electron acceptors such as *Oxygen Releasing Compounds*.

The last decade has seen a plethora of laboratory BTEX biodegradation studies and quite a few field studies detailing aerobic and anaerobic biodegradation processes for these compounds. It is now commonly accepted that BTEX compounds biodegrade readily at most sites using aerobic and anaerobic electron acceptors and that their degradation is complete. Recent advances in BTEX bioremediation include the development of field protocols for assessing the natural biodegradation potential at field sites. These protocols rely on geochemical characterization of the subsurface, analysis of historical data, and estimating biodegradation and attenuation rates to assess the intrinsic biodegradation properties of the aquifer. Additional advances include developing analytical and numerical models for simulating biodegradation and bioremediation of BTEX. A promising novel development is the use of carbon isotope fractionation to determine *in situ* biodegradation. Essentially as microbial degradation proceeds, the contaminant concentration decreases while the $^{13}C/^{12}C$ isotope ratio in the residual substrate fraction increases. Researchers have studied this as a potential method for assessing biodegradation in field studies (Griebler *et al.* 2004, Richnow *et al.* 2003, Ahad *et al.* 2000, Dempster *et al.* 1997, Ward *et al.* 2000, Morasch *et al.* 2001).

Many challenges, however, remain. For instance, the anaerobic bio-degradation of benzene is not well understood and the same can be said for petroleum additives such as MTBE (methyl-tert butyl ether). MTBE has emerged as a serious concern because of its presence in surface soils, surface water and ground water supply systems (see, for example, Squillace *et al.* 1995), and because of its potentially recalcitrant nature. To date there is increasing evidence that MTBE biodegrades aerobically and to a lesser extent anaerobically (Salanitro *et al.* 1998, 2000, Landmeyer *et al.* 1998, Park and Cowan 1997, Yeh and Novak 1994, Mormile *et al.* 1994, Kolhatkar *et al.*

2000). Given the higher solubility of MTBE and its presence in gasoline at higher percentages than the other BTEX compounds, it would be expected that MTBE plumes would outstretch BTEX plumes unless biodegradation processes are effective at controlling MTBE plume extent and concentrations. This is an area for much research and study at the present time.

Ethanol has been proposed as an alternative additive to replace MTBE in fuel. However, little is known about how ethanol may affect BTEX biodegradation and BTEX plume extent in the subsurface. Lovanh *et al.* (2002) found lower biodegradation rates for BTEX at sites with high ethanol concentrations (e.g., at gasohol contaminated sites). This led them to conclude that high ethanol concentrations can cause longer BTEX plumes. Other researchers reported increased solubilization and cosolvency effects (Corseuil *et al.* 2004, Adam *et al.* 2002, Deeb *et al.* 2002).

This chapter will focus on the state-of-knowledge of biodegradation and bioremediation of BTEX. First, a discussion of metabolic pathways will be presented followed by a detailed presentation of BTEX biodegradation rates in subsurface media. The chapter then presents intrinsic remediation protocols and findings from multiple-plume studies. Existing and emerging *in situ* bioremediation methods are discussed next as are models for intrinsic remediation. An analytical as well as a numerical model for biodegradation and bioremediation are presented in detail.

Metabolic Pathways of BTEX

Organotrophs, organisms that use organic compounds as their energy source, oxidize BTEX thereby causing them to lose electrons. This electron loss is typically coupled with the reduction of an electron acceptor such as oxygen (O_2), nitrate (NO_3^-), ferric iron (Fe^{3+}), sulfate (SO_4^{2-}), and carbon dioxide (CO_2). During these oxidation-reduction reactions, both the electron donors and the electron acceptors are considered primary growth substrates because they promote microbial growth. Under aerobic conditions, i.e., in the presence of oxygen, BTEX compounds are rapidly biodegraded as primary substrates (Alvarez and Vogel 1991). In the absence of microbial cell production, the aerobic mineralization of benzene to carbon dioxide can be written as follows:

$$C_6H_6 + 7.5O_2 \rightarrow 6CO_2 + 3H_2O \qquad (1)$$

In equation 1, 7.5 moles of oxygen are required to biodegrade 1 mole of benzene. This translates to a mass ratio of oxygen to benzene of 3.1:1. Ground water aquifers typically have limited dissolved oxygen (<12 mg/L depending on ground water temperature) that is quickly depleted when fuel hydrocarbons are introduced into the ground water. Anaerobic conditions are thus established within the contaminated zone, and the anaerobic biodegradation of BTEX proceeds with denitrification followed by sulfate reduction, iron reduction and methanogenesis as shown in Figure 1.

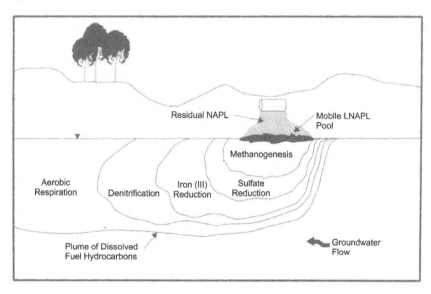

Figure 1 : Conceptualization of electron acceptor zones in the subsurface (Source: Wiedemeier *et al.* 1999).

Denitrification:
$$6NO_3^- + 6H^+ + C_6H_6 \rightarrow 6CO_2 + 6H_2O + 3N_2 \tag{2}$$
Sulfate Reduction:
$$7.5H^+ + 3.75SO_4^{2-} + C_6H_6 \rightarrow 6CO_2 + 3.75H_2S + 3H_2O \tag{3}$$
Iron Reduction:
$$60H^+ + 30Fe(OH)_3 + C_6H_6 \rightarrow 6CO_2 + 30Fe_2^+ + 78H_2O \tag{4}$$

During methanogenesis, BTEX compounds are fermented to compounds such as acetate and hydrogen (Wiedemeier *et al.* 1999). Organisms then use hydrogen and acetate as metabolic substrates and produce carbon dioxide and water. Methanogenic respiration is thought to be one of the most important anaerobic pathways in subsurface environments (Chapelle, 1993). The sequence of reactions for methanogenesis is given by:
$$C_6H_6 + 6H_2O \rightarrow 3CH_3COOH + 3H_2 \tag{5}$$

$$3CH_3COOH \rightarrow 3CH_4 + 3CO_2 \tag{6}$$

$$3H_2 + 0.75CO_2 \rightarrow 0.75CH_4 + 1.5H_2O \tag{7}$$

Table 1 presents the mass ratios for the above listed aerobic and anaerobic reactions for BTEX. Mass ratios for iron reduction and methanogenesis in Table 1 are presented in terms of ferrous iron and methane produced, respectively.

Table 1. Mass ratio of electron acceptors removed or metabolic by-products produced to total BTEX degraded, BTEX utilization factors, and number of electrons transferred for a given terminal electron-accepting process[a] (Source: Wiedemeier et al. 1999).

Terminal Electron Accepting Process	Average Mass Ratio of Electron Acceptor to Total BTEX	Average Mass Ratio of Metabolic By-product to Total BTEX	BTEX Utilization Factor. F (mg/mg)
Aerobic respiration	3.14:1	—	3.14
Denitrification	4.9:1	—	4.9
Fe(III) reduction	—	21.8:1	21.8
Sulfate reduction	4.7:1	—	4.7
Methanogenesis	—	0.78:1	0.78

[a]Simple average of all BTEX compounds based on individual compound stoichiometry.

It should be noted that the significance of anaerobic biodegradation for BTEX has not been fully appreciated and understood until recently. Only within the last decade have researchers begun to focus on studying the extent to which BTEX compounds can be degraded using anaerobic electron acceptors. One method for estimating the relative importance of the various biodegradation mechanisms has been presented by Wiedemeier et al. (1999). They define the biodegradation capacity as the amount of contamination that a given electron acceptor can degrade based on the electron-accepting capacity of the groundwater:

$$EBC_x = \frac{|C_B - C_P|}{F} \tag{8}$$

where EBC_x = expressed biodegradation capacity for given terminal electron accepting process (mg/L)

C_B = average background (upgradient) electron acceptor or metabolic by-product concentration (mg/L)

C_P = lowest measured (generally in NAPL source area) electron acceptor or metabolic by-product concentration (mg/L)

F   =   BTEX utilization factor (mg/mg)

Wiedemeier *et al.* (1999) presented biodegradation capacity calculations for 38 sites contaminated with BTEX and showed the estimated relative importance of the various mechanisms based on their calculations (Fig. 2). It can be seen from Figure 2 that methanogenesis and sulfate reduction are the most important of the biodegradation mechanisms although it should be noted that the iron reduction calculations may not be truly reflective of the true iron reduction capacity of the subsurface. This is because Wiedemeier *et al.* (1999) calculated the iron reduction capacity using the concentrations of ferrous iron in the ground water. Since ferrous iron reacts readily, it is preferable to estimate the iron reduction capacity by measuring the bioavailable iron in the solid matrix. Some methods for estimating bioavailable iron in soils have been developed (e.g., Hacherl *et al.* 2001) and research is on-going in this area.

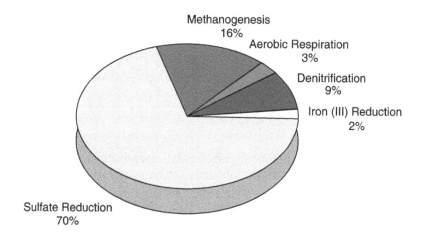

Figure 2. Relative importance of BTEX biodegradation mechanisms as determined from expressed biodegradation capacity (Source: Wiedemeier *et al.* 1999).

Overall, and based on the literature, it appears that anaerobic biodegradation rates are slower than their aerobic counterparts. Additionally, toluene is the most degradable of the BTEX compounds under anaerobic conditions, while benzene is the least degradable (Suarez and Rifai 1999). In fact, a number of laboratory studies did not observe the biodegradation of benzene under anaerobic conditions, whereas others have (e.g., Lovely *et al.* 1995, 1996). It has also been reported that benzene is relcalcitrant in the presence of nitrate (Kao and Borden 1997, Schreiber and Bahr 2002) and in landfill leachate (Thornton *et al.* 2000). Therefore, it appears that the factors affecting the rate and extent of anaerobic benzene biodegradation will continue to challenge researchers interested in biodegradation and bioremediation.

Notwithstanding the aforementioned lack of understanding of anaerobic biodegradation of benzene, most researchers agree that anaerobic rates of biodegradation are slower than their aerobic counterparts. However, it is also true that ground water aquifers have higher concentrations of anaerobic electron acceptors than oxygen making anaerobic biodegradation a more significant component of BTEX biodegradation at the field scale. Thus, much of the BTEX biodegradation research has been focused on estimating the rate of biodegradation using different electron acceptor regimes. Suarez and Rifai (1999) presented a valuable summary of aerobic and anaerobic BTEX biodegradation rates as did Aronson and Howard (1997).

Table 2 shows mean and recommended first-order rate coefficients from Aronson and Howard (1997) and Tables 3, 4, and 5 show biodegradation rates from Suarez and Rifai (1999). It is noted that Suarez and Rifai (1999) compiled Monod (or Michealis-Menton) kinetic variables, as well as zero-order rates and first-order rates for BTEX whereas Aronson and Howard

Table 2. Mean and recommended anaerobic first-order rate coefficients for selected petroleum hydrocarbons (Source: Wiedemeier *et al.* (1999), based on data from Aronson and Howard (1997).

| Compound | Mean of Field/In Situ Studies | | | Recommended First-Order Rate Constants | | | |
| | | | | Low End | | High End | |
	First-Order Rate Constant (day^{-1})	Half-Life (day^{-1})	Number of Studies Used for Mean	First-Order Rate Constant (day^{-1})	Half-Life (day^{-1})	First Order Rate Constant (day^{-1})	Half-Life (day^{-1})
Benzene	0.0036	193	41	0	No degradation	0.0036	193
Toluene	0.059	12	46	0.00099	700	0.059	12
Ethylbenzene	0.015	46	37	0.0006	1155	0.015	46
m-Xylene	0.025	28	33	0.0012	578	0.016	43
o-Xylene	0.039	18	34	0.00082	845	0.021	33
p-Xylene	0.014	49.5	26	0.00085	815	0.015	46

Table 3. Michaelis-Menten parameters for BTEX compounds, Source: Suarez and Rifai (1999). References cited in table are not included in reference list in this chapter.

Compound	Type of Study	Redox Environment	Culture	μ_{max} (day^{-1})	Doubling time (days)	Half-Saturation Ks (/mg/L)	Yield Y (mg/mg)	Max. Specific Degradation Rate μm/y (mg/mg-day)	Initial Concentration, S_0 (mg/L)	Reference
	Microcosm	Aerobic				12.20		8.30	<100	Alvarez et al. 1991
	Laboratory	Aerobic	Pseudomonas strain Bl	8.05	—	3.17	1.04	7.74	10.00	Chang et al. 1993
	Microcosm	Aerobic		8.30	—	12.20	0.50	16.60		Chen et al. 1992
	Microcosm	Aerobic		3.84	—	20.31			10-10	Goldsmith and Balderson 1998
	Respirometry	Aerobic		1.83	—	10.80	0.39	4.70		Grady et al. 1989
	Flow-through column	Aerobic		0.89-5.26	—	5.47				Kelly et al. 1996
Benzene	Microcosm	Aerobic		1.18	—	0.31	1.05	0.78	6.20	Kelly et al. 1996
	Microcosm	Aerobic	Pseudomonas strain PPOl	10.56		3.36	0.65	16.25		Oh et al. 1994
	Microcosm	Aerobic	Consortium	16.32		12.22	0.71	22.99		Oh et al. 1994
	Microcosm	Aerobic	Aerobic mixed	12.96		6.00	0.60	7.78		Park and Cowan 1997
	Laboratory, sludge from wastewater treament plant	Aerobic		6.77		6.57	0.27	25.07	20.100	Pabak et al. 1991

Compound	System	Condition	Organism						Reference
	Microcosm	Aerobic		0.65-6.0	<0.1	0.5-1.5	1.3-4.0	24-27	Allen-King et al. 1994
	Microcosm	Aerobic		0.00	17.40	0.01	9.90	<100	Alvarez et al. 1991
	Microcosm	Aerobic	Pseudomonas sp. T2 (marine strain). Uninduced cells		0.33		0.00	1-30	Button 1985
	Microcosm	Aerobic	Pseudomonas sp. T2 (marine strain) Induced cells	3.08	0.43	0.28	11.00	1-30	Button 1985
	Laboratory	Aerobic	Pseudomonas strain B2	13.03	1.22	1.96	6.65	10.00	Chang et al. 1993
	Laboratory	Aerobic	Pseudomonas strain XI	10.84	1.88	0.99	10.95	10.00	Chang et al. 1993
	Microcosm	Aerobic		9.90	17.40	0.50	19.80		Chen et al. 1992
	Microcosm	Sulfate-reducing		22.00		0.10		8-12	Edwards et al. 1992 Goldsmith and Balderson 1988
	Microcosm	Aerobic		6.24	20.27	0.29	16.80		Edwards et al. 1992
	Microcosm	Nitrate-reducing		4.80	0.15				Jorgensen et al. 1991
	Flow-through column	Aerobic		2.09-12.48	1.57				Kelly et al. 1996
Toluene	Microcosm	Aerobic		1.10	0.28	1.70	0.65	9.20	Kelly et al. 1996
	Column	Aerobic		0.21	0.65	0.43	0.49	3.00	MacQuarrie et al. 1990
	Microcosm	Aerobic	Pseudomonas strain PPO2	37.44	15.07	0.64	58.50		Oh et al. 1994
	Microcosm	Aerobic	Consortium	36.00	11.03	0.71	50.70		Oh et al. 1994
	Microcosm	Aerobic	Aerobic mixed	18.96	10.00	0.60	11.38		Park and Cowan 1997

Table 3 Contd.

Table 3 Contd.

Compound	System	Condition	Organism						Reference
	Microcosm	Aerobic	*Pseudomonas* sp. T2 (marine strain). Uninduced cells	0.00	0.03	0.10	0.01		Robertson and Button 1987
	Microcosm	Aerobic	*Pseudomonas* sp. T2 (marine strain). Induced cell	0.03	0.04	0.10	0.33		Robertson and Button 1987
	Microcosm	Aerobic		12.55	0.04mg/g.		0.00	1.96	Swindoll *et al.* 1988
	Laboratory, sludge from wastewater treament plant	Aerobic			7.75	0.36	34.87	20-100	Tabak *et al.* 1991
	Chemostat	Aerobic	*Pseudomonas*	4.22		1.42	2.97		Vecht *et al.* 1988
Ethyl-benzene	Laboratory, sludge from wastewater treament plant	Aerobic		5.18	10.07	0.34	15.25	20-100	Tabak *et al.* 1991
m-Xylene	Microcosm	Sulfate-reducing			20.00	0.14		8-12	Edwards *et al.* 1992
	Laboratory, sludge from wastewater treament plant	Aerobic		2.95	0.75	0.26	11.35	20-100	Tabak *et al.* 1991
o-Xylene	Microcosm	Aerobic		2.03	0.00	0.67	3.03	0.02	Corseuil and Weber 1994
p-Xylene	Microcosm	Aerobic		3.12	15.93				Goldsmith and Balderson 1988
	Laboratory, sludge from wastewater treament plant	Aerobic		3.36	2.47	0.36	9.33	20-100	Tabak *et al.* 1991

				All Studies	In situ & Laboratory Studies	Laboratory Studies	Aerobic/Anaerobic Field Studies	Field & Laboratory	Anaerobic Field/In Situ Studies	Reference
Xylenes	Laboratory	Aerobic	Pseudomonas strain X2	12.85		4.55		51.40	10.00	Chang et al. 1993
Xylenes	Flow-through column	Aerobic		1.49-9	0.85					Kelly et al. 1996
	Microcosm	Aerobic		9.19		13.27		7.07	6.40	Kelly et al. 1996
	Microcosm	Aerobic		6.67		5.52		5.13	6.40	Kelly et al. 1996

Table 4. Summary of zero-order decay rates for BTEX compounds (mg/L/day) (Source: Suarez and Rifai 1999).

	All Studies	In situ & Laboratory Studies	Aerobic In Situ Studies[a]	Laboratory Studies	Aerobic/Anaerobic Field Studies	Field & Laboratory	Anaerobic Field/In Situ Studies	Laboratory Studies
BENZENE								
Number of rates[b]	26	11	1	10	1	14	5	9
minimum	0.000	0.003		0.003		0.000	0.000	0.000
25th percentile	0.000	0.024		0.149		0.000	0.000	0.000
median	0.002	0.520		3.760			0.000	0.000
75th percentile	0.383	30.500		33.250		0.001	0.000	0.001
90th percentile	30.500	45.000		45.700		0.002	0.000	0.002
maximum	52.000	52.000		52.000		0.004	0.001	0.004
mean	6.389	15.099		16.607		0.001	0.000	0.001
standard deviation	15.018	20.474		20.928		0.001	0.000	0.001
geometric mean[e]	0.000	0.790	0.000	1.170	0.000	0.000	0.000	0.000
TOLUENE								
Number of rates[b]	28	9		9	2	18	5	13
minimum	0.000	0.004		0.004		0.000	0.007	0.000
25th percentile	0.042	0.400		0.400		0.059	0.007	0.090
median	0.285	5.000		5.000		0.154	0.090	0.230
75th percentile	0.480	20.000		20.000		0.375	0.108	0.380

Table 4 Contd.

Table 4 Contd.

90th percentile	20.900	28.000	28.000	0.484	0.367	0.454
maximum	239.000	48.000	48.000	239.000	0.540	239.000
mean	12.581	12.203	12.203	13.467	0.150	18.589
standard deviation	45.590	16.152	16.152	56.286	0.223	66.225
geometric mean[c]	0.147	1.309	19.040	0.055	0.047	0.058
ETHYLBENZENE						
Number of rates[b]	11	2	2	9	5	4
minimum	0.000			0.000	0.003	
25th percentile	0.004			0.003	0.005	
median	0.067			0.050	0.050	
75th percentile	0.240			0.130	0.067	
90th percentile	0.300			0.230	0.213	
maximum	0.310			0.310	0.310	
mean	0.122	0.285	0.285	0.086	0.087	0.085
standard deviation	0.127	0.021	0.021	0.110	0.128	0.103
geometric mean[c]	0.002	0.285	0.285	0.001	0.028	0.000
m-XYLENE						
Number of rates[b]	11	3	3	7	5	2
minimum	0.005			0.005	0.005	
25th percentile	0.027			0.053	0.006	
median	0.108			0.108	0.100	
75th percentile	0.240			0.165	0.108	
90th percentile	0.300			0.494	0.613	
maximum	0.950			0.950	0.950	
mean	0.195	0.206	0.206	0.214	0.234	0.165
standard deviation	0.272	0.154	0.154	0.331	0.403	0.035
geometric mean[c]	0.077	0.134	0.134	0.070	0.050	0.163
o-XYLENE						
Number of rates[b]	18	5	5	12	5	7
minimum	0.000	0.002	0.002	0.000	0.000	0.038

25th percentile	0.106	0.002	0.007	0.002	0.007	0.002
median	0.130	0.007	0.106	0.030	0.065	0.030
75th percentile	0.530	0.007	0.525	0.300	0.323	0.300
90th percentile	0.632	0.375	0.612	0.318	0.564	0.318
maximum	0.770	0.620	0.770	0.330	0.770	0.330
mean	0.316	0.127	0.237	0.133	0.196	0.133
standard deviation	0.288	0.275	0.287	0.167	0.253	0.167
geometric mean[c]	0.198	0.000	0.015	0.024	0.017	0.024
p-XYLENE						
Number of rates[b]	5	1	5	5	11	5
minimum	0.056		0.056	0.000	0.000	0.000
25th percentile	0.240		0.240	0.000	0.043	0.000
median	0.560		0.560	2.000	0.560	2.000
75th percentile	0.630		0.630	6.000	1.790	6.000
90th percentile	1.200		1.200	9.600	6.000	9.600
maximum	1.580		1.580	12.000	12.000	12.000
mean	0.613		0.613	4.000	2.100	4.000
standard deviation	0.589		0.589	5.099	3.725	5.099
geometric mean[c]	0.376		0.376	0.000	0.011	0.000

[a] In situ studies include in situ microcosms and in situ columns
[b] All the zero-order rates provided were calculated by the authors of the respective studies
[c] To calculate the geometric mean, values equal to zero were included as 10^{-10}.

Table 5. Summary of first-order decay rates for BTEX compounds (day^{-1}) Source : Suarez and Rifai (1999).

	All Studies	Field & Laboratory	Aerobic In Situ Studies[a]	Laboratory Studies	Aerobic/Anaer Field Studies	Field & Laboratory	Anaerobic Field/In Situ[a] Studies	Laboratory Studies
BENZENE								
Number of rates	150	26	3	23	20	104	45	59
Number of reported rates	80	14	3	11	15	51	32	19
Number of calculated rates[b]	70	12	0	12	5	53	13	40
mean	0.065	0.335	0.333	0.335	0.010	0.008	0.003	0.012
standard deviation	0.275	0.599		0.637	0.020	0.016	0.006	0.020
90th percentile	0.141	0.445	0.311	0.389	0.013	0.024	0.009	0.045
geometric mean[c]	0.000	0.025		0.018	0.001	0.000	0.000	0.000
range reported rates	0-2.5	0-2.5	0.2-0.5	0-2.5	0-0.087	0-0.089	0-0.023	0-0.089
TOLUENE								
Number of rates	135	16	3	13	13	106	43	63
Number of reported rates	65	12	3	9	8	45	27	18
Number of calculated rates[b]	70	4	0	4	5	61	16	45
mean	0.250	0.262	0.233	0.268	0.383	0.232	0.237	0.228
standard deviation	0.705	0.384		0.424	1.328	0.640	0.733	0.573
90th percentile	0.438	0.390	0.200	0.372	0.091	0.445	0.266	0.522
geometric mean[c]	0.009	0.142		0.132	0.002	0.007	0.013	0.005
range reported rates	0-4.8	0.016-1.63	0.1-0.4	0.016-1.63	0-4.8	0-4.32	0-4.32	0-3.28
ETHYLBENZENE								
Number of rates	82				13	69	33	36
Number of reported rates	41				9	32	21	11
Number of calculated rates[b]	41				4	37	12	25
mean	0.126				0.010	0.148	0.218	0.083
standard deviation	0.676				0.021	0.735	1.057	0.140
90th percentile	0.208				0.020	0.229	0.034	0.283
geometric mean[c]	0.000				0.001	0.000	0.000	0.000
range reported rates	0-6.048				0-0.078	0-6.048	0-6.048	0-0.48

m-XYLENE

Number of rates	90	4	13	73	30	43
Number of reported rates	38	0	8	30	18	12
Number of calculated rates[b]	52	4	5	43	12	31
mean	0.058	0.163	0.004	0.062	0.031	0.084
standard deviation	0.107		0.007	0.107	0.061	0.125
90th percentile	0.210		0.006	0.210	0.066	0.252
geometric mean[c]	0.001	0.066	0.001	0.001	0.001	0.000
range reported rates	0-0.49	0.008-0.43	0-0.025	0-0.49	0-0.32	0-0.49

o-XYLENE

Number of rates	92	10	3	7	12	70	27	43
Number of reported rates	45	6	3	3	7	32	21	11
Number of calculated rates[b]	47	4	0	4	5	38	6	32
mean	0.021	0.086	0.060	0.097	0.005	0.015	0.019	0.012
standard deviation	0.051	0.116		0.139	0.008	0.031	0.044	0.018
90th percentile	0.040	0.205	0.054	0.263	0.019	0.037	0.042	0.035
geometric mean[c]	0.000	0.046		0.043	0.001	0.000	0.000	0.000
range reported rates	0-0.38	0.008-0.38	0.04-0.1	0.008-0.38	0-0.023	0-0.214	0-0.214	0-0.075

p-XYLENE

Number of rates	65	3	15	47	25	22
Number of reported rates	42	0	10	32	21	11
Number of calculated rates[b]	23	3	5	15	4	11
mean	0.038	0.207	0.007	0.037	0.013	0.064
standard deviation	0.094		0.009	0.090	0.020	0.126
90th percentile	0.075		0.018	0.072	0.035	0.204
geometric mean[c]	0.000	0.086	0.002	0.000	0.001	0.000
range reported rates	0-0.44	0.008-0.43	0.0001-0.031	0-0.44	0-0.081	0-0.44

[a] In situ studies include in situ microcosms and in situ columns
[b] When enough information was provided by the authors of a study, the authors of this paper calculated the rate coefficient assuming first-order k_i
[c] To calculate the geometric mean, values equal to zero were included as 10^{-10}.

Table 6. Michaelis-Menten rate data for BTEX from recent studies.

Constituent	Study Type	Conditions	Culture	K_s	units	v_{max}	Error	Units	M_{max}	Units	Days	Reference
	microcosm	Aerobic		0.5±0.02	mg/L	1.2		g/g VSS day				Bielefeldt and Stensel 1999a
	chemostat			<0.19	mg/L							Rozkov et al. 1998
	chemostat	Aerobic-indirect	Pseudomonas putida F1	13	mg/L				0.35	1/hour		Lovanh et al. 2002
	laboratory	Aerobic-indirect	mixed	0.13±0.07	mg/L	0.6	0.16	g/g-day			20	Bielefeldt and Stensel 1999b
	laboratory	Aerobic-indirect	mixed	0.08±0.03	mg/L	0.88	0.24	g/g-day			5	Bielefeldt and Stensel 1999b
	laboratory	Aerobic-indirect	mixed	0.1±0.11	mg/L	0.63	0.22	g/g-day			5	Bielefeldt and Stensel 1999b
Benzene	laboratory	Aerobic-direct	mixed	0.13±0.07	mg/L	0.48	0.12	g/g-day			20	Bielefeldt and Stensel 1999b
	laboratory	Aerobic-direct	mixed	0.8±0.03	mg/L	0.78	0.12	g/g-day			5	Bielefeldt and Stensel 1999b
	laboratory	Aerobic-direct	mixed	0.1±0.11	mg/L	1.21	0.29	g/g-day			5	Bielefeldt and Stensel 1999b
	fibrous-bed bioreactor	Anaerobic	Pseudomonas putida/Pseudomonas fluorescens	600	mg/L							Shim and Yang 1999
Ethyl-benzene	microcosm	Aerobic		0.23±0.05	mg/L	0.77		g/g VSS day				Bielefeldt and Stensel 1999a
	laboratory	Aerobic-indirect	mixed	0.32±0.24	mg/L	0.75	0.22	g/g-day			20	Bielefeldt and Stensel 1999b
	laboratory	Aerobic-indirect	mixed	0.21±0.13	mg/L	1.05	0.31	g/g-day			5	Bielefeldt and Stensel 1999b
	laboratory	Aerobic-indirect	mixed	0.29±0.3	mg/L	0.49	0.21	g/g-day			5	Bielefeldt and Stensel 1999b

Compound	System	Process	Organism	Value	Unit			Unit	Rate	Unit	No.	Reference
Ethylbenzene	laboratory	Aerobic-direct	mixed	0.32±0.24	mg/L	0.52	0.11	g/g-day			20	Bielefeldt and Stensel 1999b
	laboratory	Aerobic-direct	mixed	0.21±0.13	mg/L	0.78	0.12	g/g-day			5	Bielefeldt and Stensel 1999b
	laboratory	Aerobic-direct	mixed	0.29±0.3	mg/L	0.75	0.29	g/g-day			5	Bielefeldt and Stensel 1999b
	fibrous-bed bioreactor	Anaerobic	*Pseudomonas putida/ Pseudomonas*	236	mg/L							Shim and Yang 1999
m-Xylene	alluvial sand column Aerobic			1.04±0.07	g/m^3	0.96	0.39	g/g cell day			23	Hohener et al. 2003
	microcosm	Aerobic		0.79	mg/L				4.13	1/day		Schirmer et al. 1999
m-Xylene/ Ethylbenzene	chemostat			<0.19	mg/L							Rozkov et al. 1998
	microcosm	Aerobic		0.16±0.08	mg/L	0.61		g/g VSS day				Bielefeldt and Stensel 1999a; Rozkov et al. 1998
	chemostat			<0.19	mg/L							
	laboratory	Aerobic-indirect	mixed	0.26±0.2	mg/L	0.71	0.21	g/g day			20	Bielefeldt and Stensel 1999b
	laboratory	Aerobic-indirect	mixed	0.18±0.18	mg/L	1.16	0.44	g/g-day			5	Bielefeldt and Stensel 1999b
	laboratory	Aerobic-indirect	mixed	0.49±0.15	mg/L	0.64	0.14	g/g-day			5	Bielefeldt and Stensel 1999b
o-Xylene	laboratory	Aerobic-indirect	mixed	0.26±0.2	mg/L	0.62	0.14	g/g-day			20	Bielefeldt and Stensel 1999b
	laboratory	Aerobic-indirect	mixed	0.18±0.18	mg/L	0.81	0.42	g/g-day			5	Bielefeldt and Stensel 1999b
	laboratory	Aerobic-direct	mixed	0.49±0.15	mg/L	0.82	0.09	g/g-day			5	Bielefeldt and Stensel 1999b
	fibrous-bed bioreactor	Anaerobic	*Pseudomonas putida/Pseudo- monas fluorescens*	80	mg/L							Shim and Yang 1999

Table 6 Contd.

Table 6 Contd.

Compound	System	Condition	Organism	Conc.	Unit	Rate	Yield	Rate unit	n	Reference
	microcosm	Aerobic		0.23±0.07	mg/L	0.73		g/g VSS day		Bielefeldt and Stensel 1999a
	chemostat			<0.19	mg/L					Rozkov et al. 1998
p-Xylene	laboratory	Aerobic-indirect	mixed	0.3±0.27	mg/L	0.32	0.09	g/g-day	20	Bielefeldt and Stensel 1999b
	laboratory	Aerobic-indirect	mixed	0.29	mg/L	0.39		g/g-day	5	Bielefeldt and Stensel 1999b
	laboratory	Aerobic-direct	mixed	0.3±0.27	mg/L	0.22	0.07	g/g-day	20	Bielefeldt and Stensel 1999b
	laboratory	Aerobic-direct	mixed	0.29	mg/L	0.18	0.02	g/g-day	5	Bielefeldt and Stensel 1999b
		alluvial sand column.Aerobic		<0.3	g/m^3	0.29	0.05	g/g cell day	23	Hohener et al. 2003
	microcosm	Aerobic		0.47±0.04	mg/L	2.09		g/g VSS day		Bielefeldt and Stensel 1999a
	chemostat			<0.19	mg/L					Rozkov et al. 1998
	laboratory	Aerobic-indirect	mixed	0.18±0.13	mg/L	0.77	0.2	g/g-day	20	Bielefeldt and Stensel 1999b
	laboratory	Aerobic-indirect	mixed	0.2±0.04	mg/L	1.32	0.43	g/g-day	5	Bielefeldt and Stensel 1999b
Toluene	laboratory	Aerobic-indirect	mixed	0.22±0.16	mg/L	0.99	0.41	g/g-day	5	Bielefeldt and Stensel 1999b
	laboratory	Aerobic-direct	mixed	0.18±0.13	mg/L	0.58	0.15	g/g-day	20	Bielefeldt and Stensel 1999b
	laboratory	Aerobic-direct	mixed	0.2±0.04	mg/L	0.91	0.23	g/g-day	5	Bielefeldt and Stensel 1999b
	laboratory	Aerobic-direct	mixed	0.22±0.16	mg/L	1.31	0.39	g/g-day	5	Bielefeldt and Stensel 1999b
	fibrous-bed bioreactor	Anaerobic	*Pseudomonas putida/ Pseudomonas fluorescens*	462	mg/L					Shim and Yang 1999

Table 7. Zero order rates for BTEX from recent studies.

Constituent	Rate	Units	Study Type	Reference
Benzene	14-20.5	mg/g day	field	O'Leary et al. 1995
Ethylbenzene	1.3-1.9	mg/g day	field	O'Leary et al. 1995
m-Xylene	3.4-5	mg/g day	field	O'Leary et al. 1995
o-Xylene	2-3.3	mg/g day	field	O'Leary et al. 1995
p-Xylene	1.3-1.9	mg/g day	field	O'Leary et al. 1995
	13[a]±3	mg/L day	field	Schreiber and Bahr 2002
Toluene	9.1-13.7	mg/g day	field	O'Leary et al. 1995

[a]-nitrate-reducing conditions

Table 8. First order rates for BTEX from recent studies.

Constituent	Rate	Units	Condition	Study type	Reference
	0.13	1/day	Anaerobic	field	Bockelmann et al. 2001
	0.0768	1/day	Anaerobic ax = 3m	field	Zamfirescu and Grathwohl 2001
	0.068	1/day	Anaerobic ax = 0	field	Zamfirescu and Grathwohl 2001
Benzene	0.0886	%1/day	Anaerobic, iron-reducing	field	Kao and Wang 2000
	0.0781	%1/day	Anaerobic, iron-reducing and nitrate-reducing	field	Kao and Wang 2000
	0.0663	%1/day	Anaerobic, nitrate-reducing	field	Kao and Wang 2000
	0.0123	%1/day	Anaerobic, iron-reducing	field	Kao and Wang 2000
Total BTEX	0.105	%1/day	Anaerobic, iron-reducing and nitrate-reducing	field	Kao and Wang 2000
	0.085	%1/day	Anaerobic, nitrate-reducing	field	Kao and Wang 2000
	0.014	1/week		field	Kampbell et al. 1996
	0.079±0.026	1/day	Anaerobic, nitrate-reducing	field	Schreiber and Bahr 2002
	0.051	1/day	Anaerobic	field	Bockelmann et al. 2001
Ethylbenzene	0.0536	1/day	Anaerobic ax = 3m	field	Zamfirescu and Grathwohl 2001
	0.0493	1/day	Anaerobic ax = 0	field	Zamfirescu and Grathwohl 2001
	0.0867	%1/day	Anaerobic, iron-reducing	field	Kao and Wang 2000
	0.0845	%1/day	Anaerobic, iron-reducing and nitrate-reducing	field	Kao and Wang 2000
	0.082	%1/day	Anaerobic, nitrate-reducing	field	Kao and Wang 2000
	0.041±0.012	1/day	Anaerobic, nitrate-reducing	field	Schreiber and Bahr 2002
m,p-Xylenes	0.119	%1/day	Anaerobic, iron-reducing	field	Kao and Wang 2000
	0.11	%1/day	Anaerobic, iron-reducing and nitrate-reducing	field	Kao and Wang 2000
	0.1	%1/day	Anaerobic, nitrate-reducing	field	Kao and Wang 2000
	3.28	1/day	Anaerobic	column	Hohener et al. 2003
m-Xylene	4.42±0.94	1/day	Aerobic-live	microcosm	Hohener et al. 2003
	0.74±0.77	1/day	Aerobic-abiotic	microcosm	Hohener et al. 2003
	0.014	1/day	Anaerobic	field	Hohener et al. 2003
o-Xylene	0.141	%1/day	Anaerobic, iron-reducing	field	Kao and Wang 2000
	0.108	%1/day	Anaerobic, iron-reducing and nitrate-reducing	field	Kao and Wang 2000
	0.071	%1/day	Anaerobic, nitrate-reducing	field	Kao and Wang 2000
	0.038	1/day	Anaerobic	field	Bockelmann et al. 2001

Compound	Value	Units	Conditions	Type	Reference
p-Xylene	0.0134	1/day	Anaerobic ax = 3m	field	Zamfirescu and Grathwohl 2001
	0.0131	1/day	Anaerobic ax = 0	field	Zamfirescu and Grathwohl 2001
	1.31	1/day	Aerobic	column	Hohener et al. 2003
	2.86±0.08	1/day	Aerobic – live	microcosm	Hohener et al. 2003
	0.91±0.52	1/day	Aerobic-abiotic	microcosm	Hohener et al. 2003
	6.4	1/day	Aerobic-methanogenic	column	Thornton et al. 2000
	0.073±0.021	1/day	Anaerobic-nitrate-reducing	field	Schreiber and Bahr 2002
Toluene	0.031	1/day	Aerobic	field	Bockelmann et al. 2001
	0.0138	1/day	Anaerobic ax = 3m	field	Zamfirescu and Grathwohl 2001
	0.0135	1/day	Anaerobic ax = 0	field	Zamfirescu and Grathwohl 2001
	0.155	%1/day	Anaerobic, iron-reducing	field	Kao and Wang 2000
	0.144	%1/day	Anaerobic, iron-reducing and nitrate-reducing	field	Kao and Wang 2000
	0.131	%1/day	Anaerobic, nitrate-reducing	field	Kao and Wang 2000

focused on first-order anaerobic rates only. These data compilation efforts are very useful for predicting future concentrations at field sites and for use in ground water fate and transport models. Tables 6, 7, and 8 provide a summary of additional reported biodegradation rate data in the general literature since the publication of the aforementioned studies. The next section will discuss what a biodegradation rate is and how it might be obtained so that the rate data presented earlier can be better understood and used.

Biodegradation Rates

Biodegradation reactions involving BTEX occur at specific rates. These rates are a function of a number of environmental factors such as temperature, pH, and the availability of electron donors and acceptors. Quantifying this biodegradation rate is important because biodegradation is a key mechanism affecting BTEX distribution in the subsurface. Typically, and assuming that organisms are in the stationary phase of growth (i.e., bacterial numbers are constant), the rate of limiting substrate utilization can be predicted by the Monod expression or saturation kinetics (Fig. 3) given by:

$$-\frac{dS}{dt} = \frac{kSX}{k_s + S} \tag{9}$$

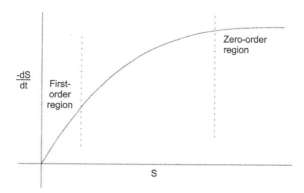

Figure 3. Observed rate of limiting substrate utilization (dS/dt) in a stationary phase bacterial culture. At low concentrations of S, -dS/dt is directly proportional to ΔS and the reaction rate is "saturated" (Source: Bedient *et al.* 1999).

In this equation, S is the limiting substrate concentration (mg/L), X is the biomass concentration (mass per volume or number per volume), k is the maximum substrate utilization rate ($S * (X * \text{time})^{-1}$), and Ks is the half-saturation coefficient (mg/L). At low concentrations, the rate of substrate utilization is linearly proportional to S (1st order) whereas at high concentrations, it is independent of S (zero-order). In many cases, BTEX concentrations in ground water are relatively low thereby allowing the use of a first-order expression to represent biodegradation:

$$-\frac{dS}{dt} = \left(\frac{kX}{Ks}\right)S = k'S \tag{10}$$

The first-order rate constant k' in Equation 10 is often expressed in terms of a half-life for the chemical:

$$t_{1/2} = \frac{0.693}{k'} \tag{11}$$

Biodegradation rates can be determined in a variety of ways, including laboratory columns or microcosms. Microcosm experiments involve obtaining soil and ground water samples from a contaminated area of interest and transferring these media into bottles that can be sealed and incubated. Samples can then be taken periodically to evaluate the fate of the chemical in the microcosm and calculate the biodegradation rate for the chemical. Column experiments, on the other hand, are not static and have the advantage of accounting for flow through the porous medium, even though it is one-dimensional flow. Data from column experiments, however, are slightly more complicated to analyze and will usually involve using a model to simulate the column experimental data and estimate the various fate and transport variables including the biodegradation rate.

Biodegradation rates can also be estimated from field studies and using models. Wiedemeier *et al.* (1996) detail two methods for extracting biodegradation rates from field data. The first method normalizes changes in concentration of BTEX to those of a non-reactive tracer (1,3,5-trimethylbenzene). The second method assumes that the plume has evolved to a steady-state equilibrium and uses a one-dimensional analytical solution (Buscheck and Alcantar 1995) to extract the biodegradation rate.

It should be noted that field-reported rate constants generally represent an overall estimate for aerobic and anaerobic reactions together and will incorporate the specific environmental conditions prevalent at the field site. Laboratory data, on the other hand, are derived under controlled environmental conditions and using individual electron acceptors. Thus laboratory data may not be directly transferable to the field.

Intrinsic Bioremediation Protocols

As mentioned in the Introduction section of this chapter, several technical protocols have been developed to demonstrate the natural biodegradation of BTEX. Of those, three are noteworthy as they are widely used: EPA's Monitored Natural Attenuation guidance (EPA 1997), ASTM's Remediation by Natural Attenuation or RNA standard (ASTM 1998), and the Air Force Protocol (Wiedemeier *et al.* 1995). All protocols involve developing an understanding of the geochemistry of ground water at BTEX sites and evaluating the correlations, if any, between the concentrations of the electron acceptors and by-products of the biodegradation reactions with BTEX concentrations. So for example, and in the case of aerobic biodegra-dation, a pattern of depleted oxygen within the plume and an oxygen-rich ground water in pristine areas indicates oxygen utilization. Similarly, depleted nitrate and sulfate concentrations within the contaminated zone indicate denitrification and sulfate reduction. Additionally, the production of ferrous iron and methane are considered to be indirect evidence of iron reduction and methanogenesis. Figures 4 and 5 illustrate these patterns at the Keesler Air Force Base in Mississippi. The main benefit of these developed protocols is the emphasis placed on characterizing the biodegradation potential at field sites in addition to the standard transport characterization that is traditionally undertaken. It should be noted, however, that most protocols place less emphasis on microbial characterization at the field scale since it does not provide direct information that can be used to study and understand the fate and transport of BTEX.

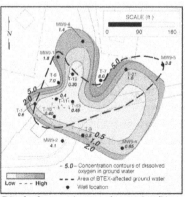

Dissolved oxygen in ground water (mg/L)

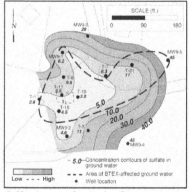

Sulfate in ground water (mg/L)

Figure 4. Distribution of electron acceptors in groundwater at the Keesler Air Force Base, April 1995 (Source: Wiedemeier *et al.* 1999).

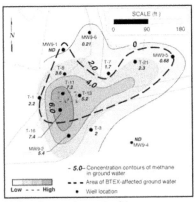 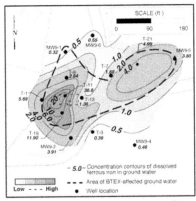

Methane in ground water (mg/L)   *Dissolved ferrous iron in ground water (mg/L)*

Figure 5. Metabolic by-products in groundwater Keesler Air Force Base, April 1995 (Source: Wiedemeier *et al.* 1999).

Multiple Plume Studies

A novel type of studies has emerged in the past decade that shed light on the intrinsic bioremediation potential at BTEX sites. These studies are referred to as *plume-athon* or *multiple-plume* studies. Essentially the studies involve gathering and statistically analyzing data from numerous sites in an attempt to draw global conclusions about the behavior of BTEX in the subsurface. Multiple-plume studies have had a tremendous impact on the management of BTEX sites, particularly those with soil and ground water contamination resulting from underground storage tank spills (Rice *et al.* 1995, Mace *et al.* 1997, and GSI 1997). Rice *et al.* (1995), for instance, coined the terms *expanding, stable, shrinking* and *exhausted* plumes in describing a BTEX plume life cycle. A BTEX plume expands as a result of a source load that overwhelms the bioremediation capacity of the aquifer, while a plume stabilizes when the rate of attenuation equals the rate of loading from the source. Once the BTEX source is depleted, a plume begins to shrink until the contamination is exhausted. Rice *et al.* (1995) and Mace *et al.* (1997) both reported that more than 70% of the studied sites in California and Texas had stable or shrinking BTEX plumes (Fig. 6).

Another significant finding from the aforementioned studies has to do with the lack of correlation between plume extent and commonly characterized hydrogeologic parameters and the lack of discernible reductions in concentration and mass at sites with active remediation systems relative to those without. Wiedemeier *et al.* (1999) provide a succinct summary of key findings from the three studies as well as results

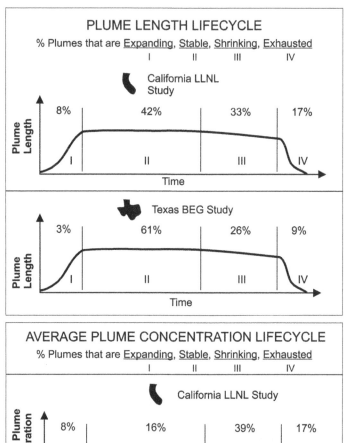

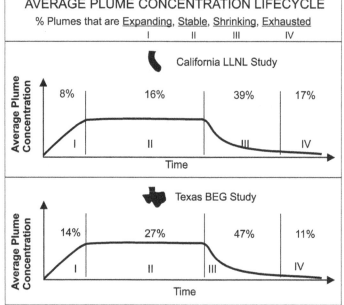

Figure 6. Summary of trends for plume length and plume concentration from California and Texas (Source: Rice *et al.* 1995 and Mace *et al.* 1997).

from an earlier database effort by Newell *et al.* (1990) and from 38 Air Force petroleum sites. Newell and Connor (1998) combined the results for dissolved hydrocarbon plume lengths from four of the studies as shown in Figure 7. The data in Figure 7 indicate that a typical BTEX plume would extend no more than 300 ft and that very few BTEX plumes (<2%) exceed 900 ft long. The most significant impact of these findings has been the regulatory shift to risk-based remediation at BTEX sites and increased acceptance of using intrinsic processes as a remedy for the contamination.

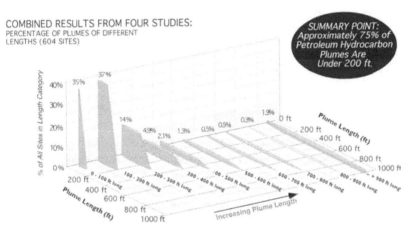

Figure 7. Dissolved hydrocarbon plume lengths from four studies (Source: Newell and Conner 1998).

Enhanced Bioremediation of BTEX

Over the last 30 years, ground water scientists and engineers have devised a number of remediation technologies to contain or remediate soil and ground water contamination. Ground water remediation, however, has changed direction since 1993 due to a number of complicating issues that were discovered at numerous waste sites. Pump-and-treat systems that were aimed at removing dissolved contamination, for instance, failed to clean up ground water to acceptable water quality levels (EPA 1989,1992). A National Research Council publication (NRC 1994) indicated that many of the existing remediation technologies were largely ineffective, and emerging methods may be required. This was attributed to the presence of NAPLs (Non-Aqueous Phase Liquids), continued leaching from source areas, high sorption potential, and hydrogeologic factors such as heterogeneities, low permeability units, and fractures. Design factors such as pumping rates, recovery well locations and screened intervals were also issues impeding successful remediation. The reports from EPA and NRC provided more justification for using bioremediation technologies for BTEX since bioremediation mineralizes these compounds and can be used to treat residual and NAPL sources and dissolved plumes at the same time. Bioremediation as a remedial technology, however, itself underwent a transformation as well due to limitations inherent in its use. These issues are discussed below.

Enhanced bioremediation can be accomplished using a variety of electron acceptors and delivery methods to accelerate the cleanup process. Several excellent reviews have been presented in the general literature on bioremediation (NRC 1993, Norris *et al.* 1994 among others). The literature is also replete with bioremediation field experiments (see for example the classic articles by Raymond *et al.* (1976) and Raymond (1978)). Figure 8 illustrates an in-situ or in-place oxygen injection system for aerobic bioremediation. Contaminated ground water is pumped, treated and mixed with oxygen and nutrients prior to re-injection. There are two key considerations before such a system can be used at a field site. First, the subsurface hydrogeology must be sufficiently transmissive (i.e., relatively high hydraulic conductivity) to allow the transport of the electron acceptors and the nutrients. Second, microorganisms must be present in sufficient numbers and types to degrade the contaminants of interest. In addition, placement of pumping and injection wells must be designed in such a manner to allow hydraulic control of the contaminated zone. The injection and pumping rates for the system must allow adequate distribution of the nutrients and electron acceptors within the subsurface while maintaining the required residence time for biodegradation to occur. A bioremediation system such as the one shown in Figure 8 has the added complication of dealing with regulatory requirements regarding re-injection of treated ground water. While it is possible to design a bioremediation system that does not involve re-injection or uses other sources of water for injection, disposal of the pumped ground water still presents a problematic consideration. The key limitations to the successful use of the bioremediation system in Figure 8, however, are some of the same limitations expressed by EPA and NRC for pump-and-treat: heterogeneities in the subsurface, and high costs.

Aerobic bioremediation has been preferred over other electron acceptors because of the short aerobic half-lives for BTEX. The key challenge in an aerobic system is the delivery of the required amounts of oxygen in a reasonable time frame. Air or liquid oxygen provide relatively lower concentrations of oxygen than hydrogen peroxide. These various sources of oxygen can be delivered using several techniques. Air sparging

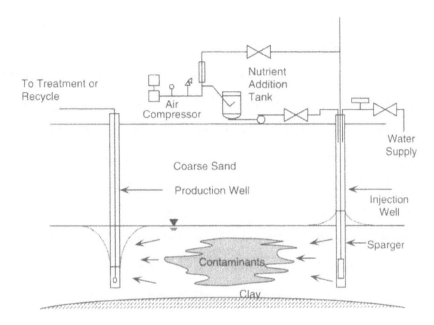

Figure 8. Injection system for oxygen (Source: Bedient *et al.* 1999).

(Ahlfeld *et al.* 1993, Goodman *et al.* 1993, Johnson *et al.* 1994) has the advantage of simplicity but is limited because of the preferential migration of air in channels, and the relatively high-energy requirements. Hydrogen peroxide, on the other hand, provides relatively high concentrations of oxygen but may be harmful to native microorganisms in the injection zone because of its oxidizing characteristics. Additionally, much of the oxygen present in peroxide maybe lost due to its relatively high rate of decomposition when compared to the rate of oxygen diffusion in the subsurface.

A variation on bioremediation using air sparging as an oxygen source emerged when soil vapor extraction systems were being deployed to treat residual and NAPL sources. Soil vapor extraction (SVE) (Batchelder *et al.* 1986, Bennedsen *et al.* 1985, Baehr *et al.* 1989) involves decontaminating the soil by pulling air through it (Fig. 9). As air is drawn through the pores, it will carry away the existing vapors. This technology was adapted to bioremediate BTEX in ground water aquifers. In a *bioventing* application, the SVE system is operated to deliver oxygen at a slow flow rate to the indigenous microbes, thereby promoting degradation of the organics in the pore space.

Another oxygen delivery method involves the use of a permeable

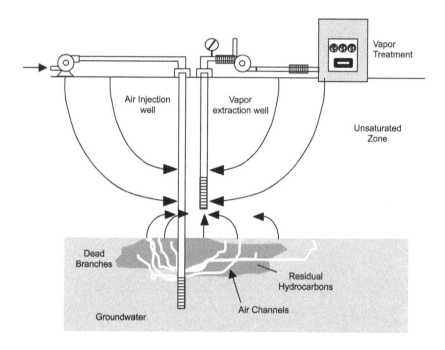

Figure 9. Schematic of in-situ air sparging-SVE (Soil-Vapor Extraction) system.

reactive barrier or PRB (Mackay *et al.* 2001, Wilson *et al.* 2002). PRBs consist of permeable walls that are installed across the flow path of a contaminant plume. The PRBs contain a zone of reactive material that is designed to act as a passive in-situ treatment zone for specific contaminants as ground water flows through it. While PRBs have been shown to create an aerobic zone in the subsurface, their use may be limited because of the economical aspects (continuous oxygen supply and replacing native media with non-native permeable media). Gibson *et al.* (1998) also proposed oxygen delivery via diffusion from silicone tubing, and demonstrated increased oxygen levels at distances up to 7.5 ft downgradient.

Mainly because of the many challenges encountered in designing and implementing in-situ bioremediation systems but also because of their relatively high costs, there has been continued interest in other delivery technologies for electron acceptors. Oxygen releasing compounds (ORC) serve as an alternative aerobic bioremediation technology that has the advantage of not involving pumping and disposal of contaminated water or requiring an energy source or high maintenance. ORC is a patented magnesium peroxide formulation (or calcium peroxide formulation) that releases oxygen upon hydration. The compound is in powder form and can be poured or injected directly into the subsurface as a slurry or installed in "socks" or casings in monitoring wells (Fig. 10). Additionally, ORC has been introduced into aquifers via trenches and by mixing it into concrete to form concrete chunks or *briquettes*. ORC has been studied both in the laboratory (Waite *et al.* 1999, Schmidtke *et al.* 1999) and at the field scale (Bianchi-Mosquera *et al.* 1994, Borden *et al.* 1997, Chapman *et al.* 1997, Barcelona *et al.* 1999, Landemeyer *et al.* 2001). ORC studies have confirmed the slow release of oxygen from ORC for up to a period of two years but they have also pointed to limiting considerations. Bridging or "lockup" of

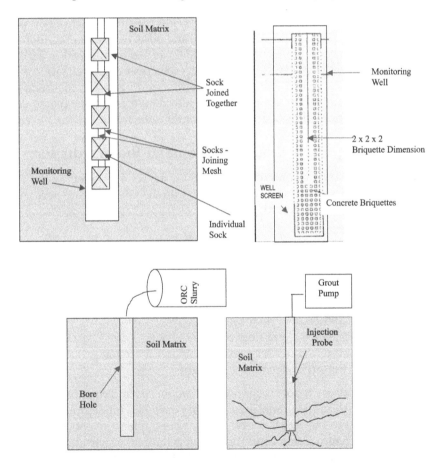

Figure 10.   A.   ORC Socks Placed in a Well
          B.   ORC-Concrete Briquettes
          C.   ORC Slurry Backfill
          D.   ORC Slurry Injection

oxygen within the ORC has been observed due to the formation of magnesium hydroxide that prevents all the oxygen from being released from ORC. Other observations include an increased pH in the vicinity of the ORC zone and clogging due to iron precipitation.

While not studied as extensively as aerobic technologies, enhanced bioremediation using anaerobic electron acceptors has been undertaken at field sites. Cunningham *et al.* (2000), for instance augmented a site in California with sulfate and with sulfate and nitrate. They found that certain fuel hydrocarbons were removed preferentially over others, but the order of preference was dependent upon the geochemical conditions. They also found that the electron acceptors were quickly consumed. Schreiber and Bahr (2002) added nitrate to the subsurface and observed biodegradation of TEX but not benzene. Stempvoort *et al.* (2002) added humic acid in a pilot scale test involving diesel fuel. They observed increased solubilization and biodegradation of BTEX. Thus, it appears that adding anaerobic electron acceptors to the subsurface is a promising alternative, albeit one that is also dependant upon the flow characteristics of the subsurface.

Modeling BTEX Biodegradation and Bioremediation

As discussed earlier, a common approach for modeling biodegradation is to introduce a first-order decay expression into the fate and transport one-dimensional transport equation:

$$\frac{\partial C}{\partial t} = \frac{D_x}{R}\frac{\partial^2 C}{\partial x^2} - \frac{v_x}{R}\frac{\partial C}{\partial x} - \lambda C \tag{12}$$

where C   = solute concentration
       t   = time
       D_x = hydrodynamic dispersion along flow path
       R   = coefficient of retardation
       x   = distance along flow path
       v_x = groundwater seepage velocity in x direction
       λ = first-order decay rate constant

The first-order decay model is simple mathematically and requires only one input parameter, λ. The model assumes that the BTEX biodegradation rate is proportional to the BTEX concentration. The higher the concentration, the higher the degradation rate. The model is limited, however, because it does not convey the electron acceptor limitations that exist in subsurface environments. Additionally, the first-order decay expression is only applicable when BTEX concentrations are much smaller than the half-saturation constant for the chemical. Another difficulty associated with using first-order decay to model biodegradation has to do with estimating the rate of biodegradation to use in the model. Data from laboratory studies are not directly transferable to the field since microcosm and column studies are typically undertaken in controlled and idealized settings and do not reflect the heterogeneities exhibited in subsurface environments. Field data can be used to extract overall bulk attenuation rates but not a biodegradation rate (unless the plume is stable or a non-reactive tracer is present at the site). Modelers typically use the first-order decay coefficient as a calibration parameter, and adjust this variable until the modeled data match the field observations for BTEX. This approach, obviously, introduces further uncertainty in model results and predictions.

Borden and Bedient (1986) proposed an alternate approach for modeling biodegradation in lieu of using a first-order decay expression. They argued that biodegradation kinetics are fast relative to the rate of transport at field sites and suggested that the kinetics of the reaction can be neglected. Borden and Bedient (1986) used an instantaneous reaction to simulate the biodegradation process and considered only the stoichiometry of the reaction with oxygen to estimate biodegradation:

$$\Delta C_R = -\frac{O}{F} \tag{13}$$

where ΔC_R = change in contaminant concentration due to biodegradation

O = the concentration of oxygen

F = the utilization factor, or the ratio of oxygen to contaminant consumed

Their study was the basis for the development of the BIOPLUME (I and II) numerical model (Borden et al. 1986, Rifai et al. 1988) for BTEX biodegradation. Since Equation (13) does not involve a rate, the transport of BTEX and oxygen can be modeled independently of each other. The BTEX plume and the oxygen plume can then be allowed to "react" at specified points in time using superposition. Wiedemeier et al. (1999) analyzed BTEX and anaerobic electron acceptor data from numerous sites and concluded that the instantaneous approach can also be applied to anaerobic electron acceptors. They support this conclusion by observing the pattern of anaerobic electron acceptors and metabolic by-products across the plume (Figs. 11 and 12). This finding was incorporated into the Bioplume III model (Rifai et al. 1997).

In addition to first-order, and instantaneous, Monod-kinetics have been integrated into ground water models (Rifai et al. 1997, Essaid et al. 2003, Clements et al. 1998). The one-dimensional transport equation is modified to include the Monod expression for substrate utilization (Equation 9) as follows:

If microbial kinetics were limiting the rate of biodegradation:	If microbial kinetics were relatively fast (instantaneous):
• Anaerobic electron acceptors (nitrate and sulfate) would be constantly **decreasing** in concentration as one moved downgradient from the source zone, and	• Anaerobic electron acceptors (nitrate and sulfate) would be mostly or totally **consumed in the source zone,** and
• Anaerobic by-products (ferrous iron and methane) would be constantly **increasing** in concentration as one moved downgradient from the source zone.	• Anaerobic by-products (ferrous iron and methane) would be **found in the highest concentrations in the source zone.**

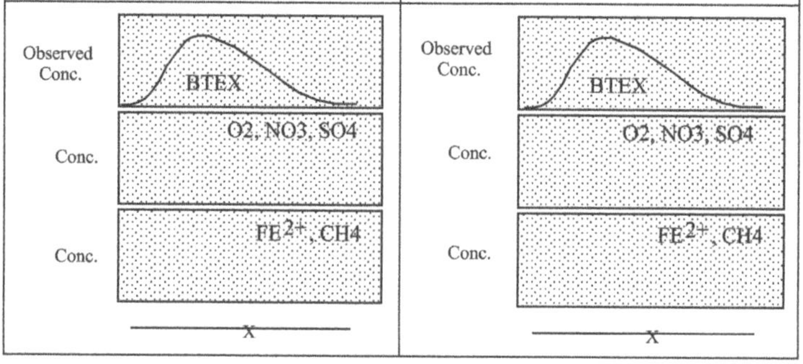

Figure 11. Conceptual models for the relationship between BTEX, electron acceptors, and metabolic by-products versus distance along centerline of plume (Source: Wiedemeier *et al.* 1999).

$$\frac{\partial C}{\partial t} = D_x \frac{\partial^2 C}{\partial x^2} - v \frac{\partial C}{\partial x} - M_t \mu_{max} \left(\frac{C}{K_c + C} \right) \tag{14}$$

where D_x =   dispersion coefficient
   C =   contaminant concentration
   M_t =   total microbial concentration
   μ_{max} =   maximum contaminant utilization rate per unit mass microorganisms
   K_c =   contaminant half saturation constant
   v =   seepage velocity

   It can be seen from Equation 14 that this model requires an estimate of the maximum utilization rate, the concentration of the microbial population and the half-saturation constant for the chemical. Additionally, and because the microbial population varies spatially and temporally, it is necessary to simulate its fate and transport in the subsurface along with BTEX and the electron acceptors. Based on the studies in the literature (see

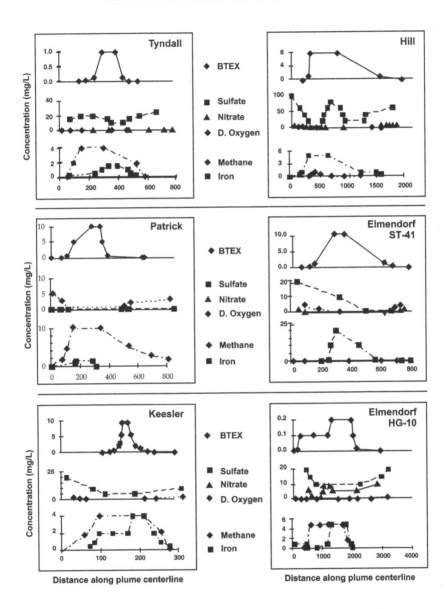

Figure 12. Distribution of BTEX, electron acceptors, and metabolic by-products versus distance along centerline of plume (Source: Wiedemeier *et al.* 1999). [*Sampling date and source of data:* Tyndall 3/95, Keesler 4/95 (Groundwater Services, Inc.), Patrick 3/94 (note: one NO$_3$ outlier removed, sulfate not plotted), Hill 7/93, Elmendorf site ST41 6/94, Elmendorf site HG10 6/94 (Parsons Engineering Science)].

Suarez and Rifai 1999) and the biodegradation rate tables in this chapter), Monod kinetic variables have yet to be determined for electron acceptors and all the BTEX components. Additionally, characterizing the microbial population spatially and temporally at a field site is rarely undertaken, thus using this model is not justified.

The remainder of this section will describe the BIOSCREEN analytical model and the BIOPLUME numerical model for intrinsic remediation. Other models for biodegradation and bioremediation have been developed and the reader is referred to Molz *et al.* (1986), Srinivasan and Mercer (1988), Widdowson *et al.* (1988), Celia *et al.* (1989), Essaid *et al.* (2003), and Clement *et al.* (1998).

BIOSCREEN Analytical Model. The BIOSCREEN model (Newell *et al.* 1996) is a screening tool for simulating the natural attenuation of petroleum hydrocarbons in groundwater. The model, which uses an Excel spreadsheet interface, is based on the Domenico (1987) analytical solution that includes first-order decay during solute transport. The Domenico solution simulates groundwater flow using a fully penetrating vertical plane perpendicular to groundwater flow, a linear isotherm sorption, and three-dimensional dispersion. Newell *et al.* (1996) modified the original Domenico analytical solution to include a decaying source and an electron-acceptor limited instantaneous reaction for biodegradation (Fig. 13). The authors added an instantaneous reaction expression because some of their modeling efforts indicated that the instantaneous reaction assumption may be more appropriate for natural attenuation simulations (Connor *et al.* 1994).

Because BIOSCREEN is an analytical model, it assumes simple groundwater flow conditions and approximates more complicated processes that occur in the field. As such, few input variables are required to run the model, and users can learn and implement the model with relative ease. BIOSCREEN has two intended applications. First, it can be used as a screening model to determine if natural attenuation is a feasible remedial alternative at a site. By using BIOSCREEN early in a remedial investigation, it can help direct a field program and aid the development of long-term monitoring plans. Second, BIOSCREEN can be used as the primary natural attenuation groundwater model at smaller, less complicated, and lower risk sites. The model cannot be used to simulate sites with pumping systems that create a complicated flow field, nor can it be applied to sites where contaminant transport is affected by vertical flow.

BIOSCREEN has a user-friendly interface for data entry and visualization of model results as mentioned earlier. The model provides centerline graphs of the extent of the BTEX plume (Fig. 14), as well as 3D plots of plume concentrations and mass balance data.

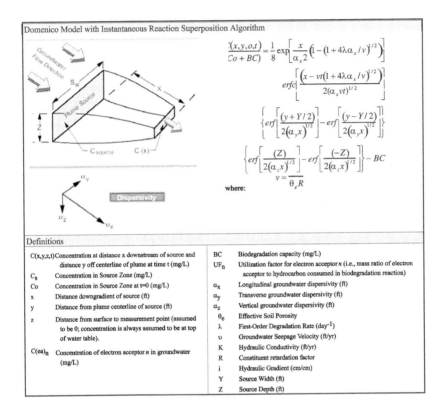

Figure 13. Domenico Model used in BIOSCREEN.

BIOPLUME Numerical Model. The BIOPLUME I model was developed by Borden and Bedient (1986) based on their work at the United Creosoting Company Superfund Site in Conroe, Texas. BIOPLUME I relies on the premise that the availability of dissolved oxygen in groundwater often limits the biodegradation of dissolved hydrocarbons. Borden and Bedient combined Monod kinetics with the advection-dispersion equation to simulate aerobic biodegradation. Later, they modified the model by replacing the Monod kinetics with an instantaneous reaction between the hydrocarbon and oxygen based on the stoichiometry of the reaction.

Rifai *et al.* (1987) incorporated the concepts developed by Borden and Bedient (1986) into the two-dimensional USGS solute transport model (Konikow and Bredehoeft 1978), known as the method of characteristics (MOC) model. The resulting model, BIOPLUME II, traces both the oxygen and hydrocarbon plumes. These plumes are superimposed at each time step to determine the resulting oxygen and hydrocarbon concentrations

DISSOLVED HYDROCARBON CONCENTRATION ALONG PLUME CENTERLINE (mg/L at Z = 0)

TYPE OF MODEL	Distance from Source (ft)										
	0	32	64	96	128	160	192	224	256	288	320
No. Degradation	13.544	6.575	5.280	4.581	4.107	3.754	3.474	3.241	3.040	2.861	2.697
Ist Order Decay	13.544	3.117	1.186	0.488	0.208	0.090	0.040	0.018	0.008	0.004	0.002
Inst. Reaction	12.021	5.463	4.248	3.500	2.860	2.257	1.678	1.114	0.559	0.004	0.000
Field Data from Site	12.000	5.000	1.000				0.500			0.001	

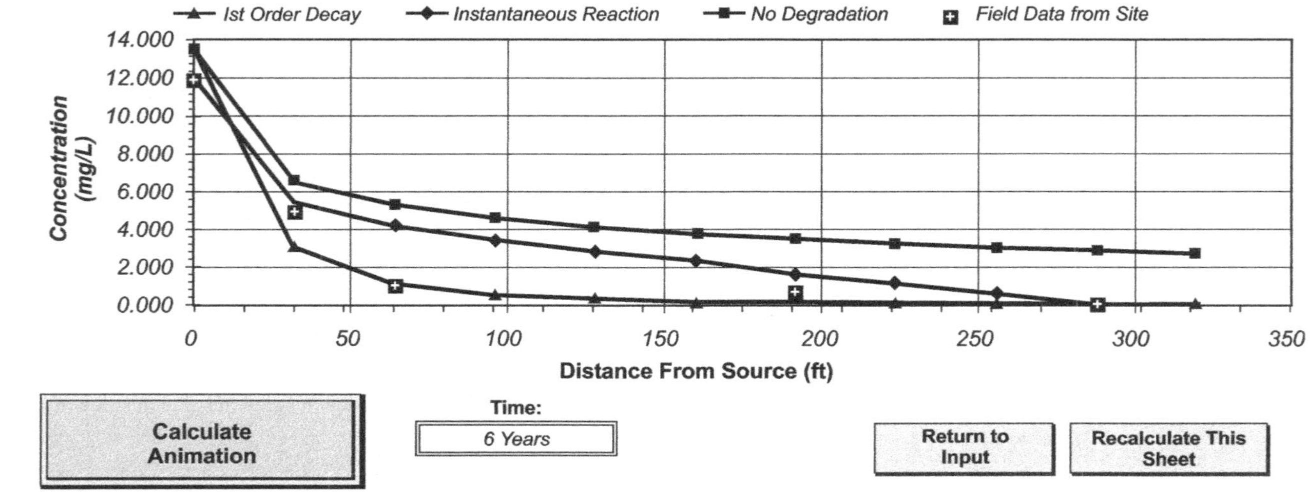

Figure 14. BIOSCREEN centerline output.

assuming that an instantaneous reaction occurs. Anaerobic biodegradation is modeled in BIOPLUME II as a first-order decay in hydrocarbon concentrations.

With the same approach used to develop the BIOPLUME II model, Rifai *et al.* (1997) modified the 1989 version of the MOC (Konikow and Bredehoeft 1989) into BIOPLUME III. BIOPLUME III is a two-dimensional, finite difference model for simulating aerobic and anaerobic biodegradation of hydrocarbons in groundwater in addition to advection, dispersion, sorption, and ion exchange. In BIOPLUME III, three different kinetic expressions can be used to simulate biodegradation. These are first-order decay, instantaneous reaction, and Monod kinetics.

The model simulates hydrocarbon biodegradation using five different electron acceptors: oxygen, nitrate, iron (III), sulfate, and carbon dioxide. BIOPLUME III solves the transport equation six times to determine the fate and transport of the hydrocarbons and the electron acceptors/byproducts. When iron (III) is used as an electron acceptor, the model simulates the production and transport of iron (II), which is the soluble species. The hydrocarbon and electron acceptor plumes are combined using the principle of superposition. Biodegradation occurs sequentially in the following order:

$$oxygen \rightarrow nitrate \rightarrow iron\ (III) \rightarrow sulfate \rightarrow carbon\ dioxide \qquad (15)$$

The BIOPLUME III model was developed primarily to model the natural attenuation of BTEX in groundwater due to advection, dispersion, sorption, and biodegradation. The model can also simulate the bioremediation of dissolved hydrocarbons by the injection of electron acceptors (with the exception of ferrous iron), remediation using air sparging at low injection rates, and pump-and-treat systems.

Development of BIOPLUME IV is currently underway by the authors of BIOPLUME III. Conceptually, BIOPLUME IV will differ from its predecessors for modeling BTEX fate and transport. Monod kinetics are not used since the Monod variables have not been measured for all electron acceptors and donors and environmental conditions. The model, instead, includes a zero-order expression for oxygen, nitrate, sulfate and iron reduction. Methanogenesis is simulated strictly as a first-order reaction with no limitations and is allowed to occur in conjunction with iron reduction. Thus the sequence of reactions in BIOPLUME IV is:

$$oxygen \rightarrow nitrate \rightarrow sulfate \rightarrow iron\ and\ methanogenesis \qquad (16)$$

The BIOPLUME model has been widely used and applied to numerous field sites (e.g., Rifai *et al.* 1988, 2000). The model is preferred over an analytical model for use at complex sites with non-uniform hydrogeologic properties. Additionally, a numerical model such as BIOPLUME, can simulate a varying source over time as well as remediation systems. Both analytical and numerical models, however, are important tools for understanding the biodegradation and bioremediation of BTEX.

SUMMARY

The past three or four decades have seen a tremendous growth in the knowledge base for biodegradation and bioremediation for BTEX. Researchers determined that the subsurface readily supports active microbial populations that can biodegrade these soluble components in fuels. They also determined that the microbial biodegradation of BTEX is limited by the electron acceptor supply. Early bioremediation efforts involved pumping-and-injection circulation systems aimed at adding oxygen and nutrients to contaminated soils and ground water. Such systems proved costly and limited by the inherent heterogeneities in subsurface media. More recent efforts involve adding anaerobic electron acceptors, injecting air and introducing solids that release oxygen into the subsurface and using biobarriers in lieu of circulation systems. The past decade has also seen a marked acceptance of relying on intrinsic bioremediation for controlling and treating BTEX plumes. Numerous protocols and models have been developed to aid in assessing the biodegradation potential at field sites and predicting future plume status with intrinsic bioremediation. The biodegradation and bioremediation of BTEX still pose many challenges including LNAPL contamination, fate of BTEX in the unsaturated zone, and determining field based biodegradation rates.

REFERENCES

Adam, G., K. Gamoh, D.G. Morris, and H. Duncan. 2002. Effect of alcohol addition on the movement of petroleum hydrocarbon fuels in soil. *Sci. Total Environ.* 286: 15-25.

Ahad, J., B. Sherwood Lollar, E.A. Edwards, G.F. Slater, and B. Sleep. 2000. Carbon isotope fractionation during anaerobic biodegradation of toluene: Implications for intrinsic bioremediation. *Environ. Sci. Technol.* 34: 892-896.

Ahlfeld, D., A. Dahami, E. Hill, J. Lin, and J. Wei. 1993. Laboratory study on air sparging: Air flow visualization. *Ground Water Monitoring Remediation* 4: 115-126.

Allen-King, R.M., K.E. O'Leary, R.W. Gillham, and J.F. Barker, 1994. Limitations on the biodegradation rate of dissolved BTEX in a natural unsaturated, sand soil: Evidence from field and laboratory experiments. Pages 175-191 in *Hydrocarbon Bioremediation*. R.E. Hinchee, B.C. Alleman, R.E. Hoeppel, and R.N. Miller, eds., Lewis Publishers, Boca Raton, Florida.

Alvarez, P.J., and T.M. Vogel, 1991. Substrate interaction of benzene, toluene, and paraxylene during microbial degradation by pure cultures and mixed culture aquifer slurries. *Appl. Environ. Microbiol.* 57: 2981-2985.

Alvarez, P.J., and T.M. Vogel. 1991. Kinetics of aerobic benzene, toluene in sandy aquifer material. *Biodegradation* 2: 43-51.

American Society for Testing and Materials (ASTM). 1998. ASTM Guide for Remediation by Natural Attenuation at Petroleum Release Sites, ASTM, Philadelphia.

Aronson, D., and P.H. Howard. 1997. Anaerobic Biodegradation of Organic Chemicals in Groundwater: A Summary of Field and Laboratory Studies, Draft Final Report, American Petroleum Institute, Washington, DC.

Baehr, A.L., G.E. Hoag, and M.C. Marley. 1989. Removing volatile contaminants from the unsaturated zone by inducing advective air-phase transport. *Contaminant Hydrol.* 4: 1-26.

Barcelona, M., J. Jaglowski, and R. David. 1999. Subsurface Fate and Transport of MTBE in a Controlled Reactive Tracer Experiment. Pages 123-137 in proceedings of 1999, Petroleum Hydrocarbons & Organic Chemicals in Groundwater, Houston Texas.

Batchelder, G.V., W.A. Panzeri, and H.T. Phillips. 1986. Soil Ventilation for the Removal of Adsorbed Liquid Hydrocarbons in the Subsurface. Pages 672-688 in Petroleum Hydrocarbons and Organic Chemicals in Ground Water, NWWA, Dublin, Ohio.

Bedient, P.B., H.S. Rifai, and C.J. Newell. 1999. *Ground Water Contamination: Transport and Remediation*, Prentice Hall, Upper Saddle River, New Jerey.

Bennedsen, M.B., J.P. Scott, and J.D. Hartley. 1985. Use of Vapor Extraction Systems for In Situ Removal of Volatile Organic Compounds from Soil. Pages 92-95, National Conference on Hazardous Wastes and Hazardous Materials, HMCRI.

Bianchi-Mosquera, G.C., R.M Allen-King, and D.M. Mackay. 1994. Enhanced degradation of dissolved benzene and toluene using a solid oxygen-releasing compound. *Groundwater Monitoring Remediation* 14: 120-128.

Bielefeldt, A.R., and H.D. Stensel. 1999a. Biodegradation of aromatic compounds and TCE by a filamentous bacteria-dominated consortium. *Biodegradation* 10: 1-13.

Bielefeldt, A.R., and H.D. Stensel. 1999b. Evaluation of Biodegradation Kinetic Testing Methods and Longterm Variability in Biokinetics for BTEX Metabolism. *Water Res.* 33: 733-740.

Bockelmann, A., T. Ptak, and G. Teutsch. 2001. An analytical quantification of mass fluxes and natural attenuation rate constants at a former gasworks site. *Contaminant Hydrol.* 53: 429-453.

Borden, R.C., and P.B. Bedient. 1986. Transport of dissolved hydrocarbons influenced oxygen limited biodegradation: 1. Theoretical development. *Water Resources Res.* 22: 1973-1982.

Borden, R.C., P.B. Bedient, M.D. Lee, C.H. Ward, and J.T. Wilson. 1986. Transport of dissolved hydrocarbons influenced by oxygen limited Biodegradation: 2. Field Development" *Water Resources Res.* 22: 1983-1990.

Borden, R.C., R.T. Goin, and C.M. Kao. 1997. Control of BTEX migration using a biologically enhanced permeable barrier. *Ground Water Monitoring Remediation* 17: 70-80.

Buscheck, T.E., and C.M. Alcantar. 1995. *Regression Techniques and Analytical Solutions to Demonstrate Intrinsic Bioremediation. Intrinsic Bioremediation*, R. E. Hinchee, J.T. Wilson, and D.C. Downey, eds., Battelle Press, Columbus, Ohio.

Celia, M.A., J.S. Kindred, and I. Herrara. 1989. Contaminant transport and biodegradation: 1. A numerical model for reactive transport in porous media, *Water Resources Res.* 25: 1141-1148.

Chang, M.K., T.C. Voice, and C.S. Criddle, 1993. Kinetics of competitive inhibition and cometabolism in the biodegradation of benzene, toluene, and p-xylene by two *Pseudomonas* isolates. *Biotechnol. Bioeng.* 41: 1057-1065.

Chapelle, F.H. 1993. *Groundwater Microbiology and Geochemistry*, Wiley, New York.

Chapman, S.W., B.T. Byerely, D.J.A. Symth, and D.M. Mackay. 1997. A pilot test of passive oxygen release for enhancement of in situ bioremediation of BTEX contaminated ground water. *Groundwater Monitoring Remediation* 17: 93-105.

Chen, Y.M., L.M. Abriola, P.J.J. Alvarez, P.J. Anid, and T.M. Vogel, 1992. Modeling transport and biodegradation of benzene and toluene in sandy aquifer material: Comparisons with experimental measurements. *Water Resour. Res.* 28: 1833-1847.

Clement, T.P., Y. Sun, B.S. Hooker, and J.N. Petersen. 1998. Modeling multispecies reactive transport in ground water. *Ground Water Monitoring Remediation.* 18: 79-92.

Connor, J.A., C.J. Newell, J.P. Nevin, and H.S. Rifai. 1994. Guidelines for Use of Groundwater Spreadsheet Models in Risk-Based Corrective Action Design. National Ground Water Association, Proceedings of the Petroleum Hydrocarbons and Organic Chemicals in Ground Water Conference, Houston, Texas, November 1994, pp. 43-55.

Corseuil, H.X., and W.J. Weber, 1994. Potential biomass limitations on rates of degradation of monoaromatic hydrocarbons by indigenous microbes in surface soils, *Water Resources*, 28: 1415-1423.

Corseuil, H.X., B.I.A. Kaipper, and M. Fernandes. 2004. Cosolvency effect in subsurface systems contaminated with petroleum hydrocarbons and ethanol. *Water Res.* 38: 1449-1456.

Cunningham, J.A., G.D. Hopkins, C.A. Lebron, and M. Reinhard. 2000. Enhanced anaerobic bioremediation of groundwater contaminated by fuel hydrocarbons at Seal Beach, California. *Biodegradation* 11: 159-170.

Deeb, R.A., J.O. Sharp, A. Stocking, S. McDonald, K.A. West, M. Laugier, P.J.J. Alvarez, Pedro, M.C. Kavanaugh, and L. Alvarez-Cohen. 2002. Impact of ethanol on benzene plume lengths: Microbial and modeling studies. *J. Environ. Engineer.* 128: 868-875.

Dempster, H.S., B. Sherwood Lollar, and S. Feenstra. 1997. Tracing organic contaminants in groundwater: A new methodology using compound-specific isotopic analysis. *Environ. Sci. Technol.* 31: 3193-3197.

Domenico, P.A. 1987. An analytical model for multidimensional transport of a decaying Contaminant Species. *J. Hydrol.* 91: 49-58.

Edwards, E.A., L.E. Willis, M. Reinhard, and D. Gribić-Galić, 1992. Anaerobic degradation of toluene and xylene by aquifer microorganisms under sulfate-reducing conditions. *Appl. Environ. Microbiol.* 58: 794-800.

Essaid, H.I., I.M. Cozzarelli, R.P. Eganhouse, W.N. Herkelrath, B.A. Bekins, and G.N. Delin. 2003. Inverse modeling of BTEX dissolution and biodegradation at the Bemidji, MN, crude-oil spill site. *Contaminant Hydrol.* 67: 269-299.

Ghiorse, W.C., and J.T. Wilson. 1988. Microbial ecology of the terrestrial subsurface. *Advanced Applied Microbiol.* 26: 2213-2218.

Goldsmith, C.D., and R.K. Balderson, 1988. Biodegradation and growth kinetics of enrichment isolates on benzene, toluene, and xylene. *Water Sci. Technol.* 20: 505-507.

Goodman, I., R.E. Hinchee, R.L. Johnson, P.C. Johnson, and D.B. McWhorter. 1993. An overview of in situ air sparging. *Ground Water Monitoring Remediation* 4: 127-135.

Grady, C.P.L., Jr., G. Aichinger, S.F. Cooper, and M. Naziruddin, 1989. Biodegradation kinetics for selected toxic/hazardous compounds. In *1989 A&WMA/EPA International Symposium on Hazardous Waste Treatment: Biosystems for Pollution Control.* Pp. 141-153, Cincinnati, OH: February.

Griebler, C., M. Safinowski, A. Vieth, H.H. Richnow, and R.U. Meckenstock. 2004. Combined Application of stable carbon isotope analysis and specific metabolite determination for assessing in situ degradation of aromatic hydrocarbons in a tar oil-contaminated aquifer. *Environ. Sci. Technol.* 38: 617-631.

Groundwater Services, Inc. (GSI). 1997. Florida RBCA Planning Study, Impact of RBCA Policy Options on LUST Site Remediation Costs, report prepared for Florida Partners in RBCA (PIRI), 24 pp.

Hacherl E.L., D.S. Kosson, L.Y. Young, and R.M. Cowan, 2001. Measurement of iron(III) bioavailability in pure iron oxide minerals and soils using anthraquinone-2,6-disulfonate oxidation. *Environ. Sci. Technol.* 35: 4886-4893.

Hohener, P., C. Duwig, G. Pateris, K. Kaufmann, N. Dakhel, and H. Harms. 2003. Biodegradation of petroleum hydrocarbon vapors: Laboratory studies on rates and kinetics in unsaturated alluvial sand. *Contaminant Hydrol.* 66: 93-115.

Johnson, P.C., A. Baehr, R.A. Brown, R. Hinchee, and G. Hoag 1994. Innovative Site Remediation Technology: Vacuum Vapor Extraction, American Academy of Environmental Engineers.

Jorgensen, C., E. Mortensen, B.K. Jensen, and E. Arvin, 1991. Biodegradation of toluene by a dentrifying enrichment culture. Pages 480-487 in *In Situ Bioreclamation: Applications and Investigations for Hydrocarbon and Contaminated Site Remediation.* Butterworth-Heinemann, Stoneham, Massachusetts.

Kampbell, D.H., T.H. Wiedemeier, and J.E. Hansen. 1996. Intrinsic bioremediation of fuel contamination in ground water at a field site. *J. Hazardous Material* 49: 197-204.

Kao, C.M., and C.C. Wang 2000. Control of BTEX migration by intrinsic bioremediation at a gasoline spill site. *Water Res.* 34: 3413-3423.

Kao, C.M., and R.C. Borden. 1997. Site specific variability in BTEX biodegradation under denitrifying conditions. *Ground Water* 35: 305-311.

Kelly, W.R., G.M. Hornberger, J.S. Herman, and A.L. Mills, 1996. Kinetics of BTX biodegradation and mineralization in batch and column studies. *J. Contam. Hydrol.* 23: 113-132.

Konikow, L.F., and J.D. Bredehoeft. 1978. Computer Model of Two-Dimensional Solute Transport and Dispersion in Ground Water, Automated Data Processing and Computations, Techniques of Water Resources Investigations of the USGS, Washington, DC, 100 pp.

Konikow, L.F., and J.D. Bredehoeft. 1989. Computer Model of Two-Dimensional Solute Transport and Dispersion in Ground Water. In Techniques of Water Resources Investigation of the United States Geological Survey, Reston, Virginia, Book 7.

Landmeyer, J.E., F.H. Chapelle, H.H. Herlong, and P.M. Bradley. 2001. Methyl tert-butyl ether biodegradation by indigenous aquifer microorganisms under natural and artificial oxic conditions. *Environ. Sci. Technol.* 35: 1118-1126.

Landmeyer, J.E., F.H. Chapelle, P.M. Bradley, J.F. Pankow, C.D. Church, and P.G. Tratnyek. 1998. Fate of MTBE relative to benzene in a gasoline-contaminated aquifer (1993-98). *Ground Water Monitoring Rev.* 18: 93-102.

Lovanh, N., C.S. Hunt, and P.J.J. Alvarez. 2002. Effect of ethanol on BTEX biodegradation kinetics: Aerobic continuous culture experiments. *Water Res.* 36: 3739-3746.

Lovley, D.R, J.D. Coates, J.C. Woodward, and E.J.P. Phillips. 1995. Benzene oxidation coupled to sulfate reduction. *Appl. Environ. Microbiol.* 61: 953-958.

Lovley, D.R., J.C. Woodward, and F.H. Chappelle. 1996. Rapid anaerobic benzene oxidation with a variety of chelated Fe(III) forms, *Appl. Environ. Microbiol.* 62: 288-291.

Mace, R.E., R.S. Fisher, D.M. Welch, and S.P. Parra. 1997. Extent, Mass, and Duration of Hydrocarbon Plumes from Leaking Petroleum Storage Tank Sites in Texas, Bureau of Economic Geology Geologic Circular 97-1, Bureau of Economic Geology, Austin, TX, 52 pp.

Mackay, D., R.D. Wilson, K. Scow, M. Einarson, B. Flower, and I. Wood.(2001). "In Situ Remediation of MTBE at Vandenberg Air Force Base, California." Contaminated Soil Sediment and Water, 43-46.

Mackay, D.M., and R.M. Cohen. 1990. A review of immiscible fluids in the subsurface. *J. Contaminant Hydrol.* 6: 107-163.

MacQuarrie, K.T.B., E.A. Sudicky, and E.O. Frind, 1990. Simulation of biodegradable organic contaminants in ground water: 1. Numerical formulation in principal directions, *Water Res.* 26: 207-222.

McNabb, J.F., and W.J. Dunlap. 1975. Subsurface biological activity in relation to groundwater pollution. *Ground Water* 13: 33-44.

Molz, F.J., M.A. Widdowson, and L.D. Benefield. 1986. Simulation of microbial growth dynamics coupled to nutrient and oxygen transport in porous media. *Water Resources Res.* 22: 1207-1216.

Morasch, Barbara, H.H. Richnow, B. Schink, and R.U. Meckenstock. 2001. Stable hydrogen and carbon isotope fractionation during microbial toluene degradation: Mechanistic and environmental aspects. *App. Environ. Microbiol.* 67: 4842-4849.

Mormile, M.R., S. Liu, and J.M. Suflita. 1994. Anaerobic biodegradation or gasoline oxygenates: extrapolation of information to multiple sites and redox conditions. *Environ. Sci. Technol.* 28: 1727-1732.

National Research Council (NRC). 1993. *In Situ Bioremediation*. National Academy Press, Washington, DC.

National Research Council (NRC). 1994. *Alternatives for Ground Water Cleanup*, Washington, DC. National Research Council.

Newell, C.J., and J.A. Conner. 1998. Characteristics of Dissolved Petroleum Hydrocarbon Plumes: Results from Four Studies, API Tech Transfer Bulletin, 8 pp.

Newell, C.J., L.P. Hopkins, and P.B. Bedient. 1990. A hydrogeologic Database for ground-water modeling. *Ground Water* 28: 703-714.

Newell, C.J., R.K. McLeod, and J.R. Gonzales. 1997. BIOSCREEN, Natural Attenuation Decision Support System, Version 1.4 Revisions.

Newell, C.J., R.K., McLeod and J.R. Gonzales, 1996. BIOSCREEN Natural Attenuation Decision Support System User's Manual, Version 1.3, EPA/600/R-96/087, August, Robert S. Kerr Environmental Research Center, Ada, OK.

Norris, R.D., R.E. Hinchee, R. Brown, P. L. McCarty, L. Semprini, J. T. Wilson, D. H. Kampbell, M. Reinhard, E.J. Bouwer, R.C. Borden, T.M. Vogel, J.M. Thomas, and C.H. Ward, eds. 1994. *Handbook of Bioremediation,* Lewis Publishers, Boca Raton, Florida.

Oh, Y.S., Z. Shareefdeen, B.C. Baltzis, and R. Bartha, 1994. Interations between Benzene, Toluene, and p-Xylene (BTX) during their Biodegradation. *Biotechnol. Bioeng.* 44: 533-538.

O'Leary, K.E., J.F. Barker, and R.W. Gillham. 1995. Remediation of dissolved BTEX through surface application: A prototype field investigation. *Ground Water Monitoring Remediation.* 15: 99-109.

Park, K., and R.M. Cowan. 1997. Effects of Oxygen and Temperature on the Biodegradation of MTBE. in American Chemical Society Division of Environmental Chemistry preprints of papers, 213th, San Francisco, California: ACS, 37: 421-424.

Raymond, R.L. 1978. Environmental Bioreclamation. Presented at 1978 Mid-Continent Conference and Exhibition of Control of Chemicals and Oil Spills, Detroit, Mich., September 1978.

Raymond, R.L., V.W. Jamison, and J.O. Hudson. 1976. Beneficial stimulation of bacterial activity in groundwaters containing petroleum products. AIChE Symp. Ser., 73, 390 pp.

Rice, D.W., R.D. Grose, J.C. Michaelsen, B.P. Dooher, D.H. Macqueen, S.J.Cullen, W.E. Kastenberg, L.G. Everett, and M.A. Marino. 1995. California Leaking Underground Fuel Tank (LUFT) Historical Case Analyses, Environmental Protection Department: Lawrence Livermore National Laboratory, Livermore, California, UCRL-AR-122207.

Richnow, H.H., R.U. Meckenstock, L.A. Reitzel, A. Baun, A. Ledin, and T.H. Christensen. 2003. In situ biodegradation determined by carbon isotope fractionation of aromatic hydrocarbons in an anaerobic landfill leachate plume (Vejen, Denmark). *J. Contaminant Hydrol.* 64: 59-72.

Rifai, H.S. and P.B. Bedient, 1987, BIOPLUME II - Two Dimensional Modeling for Hydrocarbon Biodegradation and In-Situ Restoration, in Proceedings of the NWWA/API Conference on Petroleum Hydrocarbons Hydrocarbons and Organic Chemicals in Groundwater: Prevention, Detection, and Restoration, November 17-19, Houston, TX, pp. 431-450, National Ground Water Association, Westerville, OH.

Rifai, H.S., C.J. Newell, J.R. Gonzales, and J.T. Wilson. 2000. Modeling Natural Attenuation of Fuels with BIOPLUME III, ASCE. *J. Environ. Engineer.* 126: 428-438.

Rifai, H.S., C.J. Newell, J.R. Gonzales, S. Dendrou, L. Kennedy, and J. Wilson. 1997. BIOPLUME III Natural Attenuation Decision Support System Version 1.0 User's Manual, prepared for the U.S. Air Force Center for Environmental Excellence, San Antonio, TX, Brooks Air Force Base.

Rifai, H.S., P.B. Bedient, J.T. Wilson, K.M. Miller, and J.M. Armstrong. 1988. Biodegradation modeling at aviation fuel spill site. *Environ. Engineer.* 5: 1007-1029.

Robertson, B.R., and, D.K. Button, 1987. Toluene induction and uptake kinetics and their inclusion in the specific affinity relationship for describing rates of hydrocarbon metabolism. *Appl. Environ. Microbiol.*, 53: 2193-2205.

Rozkov, A., A. Kaard, and R. Vilu, 1998. Biodegradation of Dissolved Jet Fuel in Chemostat by a Mixed Bacterial Culture Isolated from a Heavily Polluted Site. *Biodegradation* 8: 363-369.

Salanitro, J.P., C.-S. Chou, H.L. Wisniewski, and T.E. Vipond. 1998. Perspectives on MTBE Biodegradation and the Potential for In Situ Aquifer Bioremediation. In Proceedings of the Southwest Focused Ground Water Conference: Discussing the Issue of MTBE and Perchlorate in Ground Water, Anaheim, California, June 3-4, 1998. Westerville, Ohio: National Ground Water Association, p. 40-54.

Salanitro, P.J., P.C. Johnson, G.E. Spinnler, P.M. Manner, H.L. Wisniewski, and C. Bruce. 2000. Field-Scale Demonstration of Enhanced MTBE Bioremediation through Aquifer Bioaugmentation and Oxygenation. *Environ. Sci. Technol.*, 34: 4152-4162.

Schirmer, M., B.J. Butler, J.W. Roy, E.O. Frind, and J.F. Barker, 1999. A relative-least-squares technique to determine unique monod kinetic parameters of BTEX compounds using batch experiments. *J. Contaminant Hydrol.* 37: 69-86.

Schmidtke, T., D. White, and C. Woolard. 1999. Oxygen release kinetics from solid phase oxygen in Arctic Alaska. *J. Hazardous Materials.* B 64: 157-165.

Schreiber, M.E., and J.M. Bahr. 2002. Nitrate-enhanced bioremediation of BTEX-contaminated groundwater: Parameter estimation from natural-gradient tracer experiments. *J. Contaminant Hydrol.* 55: 29-56.

Shim, H., and S.-T. Yang. 1999. Biodegradation of benzene, toluene, ethyl-benzene, and o-xylene by a coculture of *Pseudomonas putida* and *Pseudomonas fluorescens* immobilized in a fibrous-bed bioreactor. *J. Biotechnol.* 67: 99-112.

Squillace, P.J., D.A. Pope, and C.V. Price. 1995. Occurrence of the Gasoline Additive MTBE in Shallow Ground Water in Urban and Agricultural Areas, U.S. Geological Survey: FS-1449-9.

Srinivasan, P., and J.W. Mercer. 1988. Simulation of biodegradation and sorption processes in ground water. *Ground Water* 4:475-487.

Stempvoort, D.R.V., S. Lesage, K.S. Novakowski, K. Millar, S. Brown, and J.R. Lawrence. 2002. Humic acid enhanced remediation of an emplaced diesel source in groundwater: 1. Laboratory-based pilot scale test. *Contaminant Hydrol.* 54: 249-276.

Suarez, M.P., and H.S. Rifai. 1999. Biodegradation rates for fuel hydrocarbons and chlorinated solvents in groundwater. *J. Bioremediation* 3: 337-362.

Swindoll, C.M., C.M. Aelion, D.C. Dobbins, O. Jiang, S.C. Long, and F.K. Pfaender, 1988. Aerobic biodegradation of natural xenobiotic organic compounds by subsurface microbial communities. *Environ. Toxicol. Chem.* 7: 291-299.

Tabak, H.H., Desai, S. and Govind, R., 1991, Development and application of a multilevel respirometric protocol to determine biodegradability and biodegradation kinetics of toxic organic pollutant compounds, Pages 324-340. *On-Site Bioreclamation: Processes for Xenobiotic and Hydrocarbon Treatment*, R.E. Hinchee and R.F. Olfenbuttel, eds., Butterworth-Heinemann, Stoneham, Massachusetts.

Thornton, S.F., M.I. Bright, D.N. Lerner, and J.H. Tellam. 2000. Attenuation of landfill leachate by UK triassic sandstone aquifer materials 2. Sorption and degradation of organic pollutants in laboratory columns. *J. Contaminant Hydrol.* 43: 355-383.

U.S. Environmental Protection Agency (EPA). 1989. *Evaluation of Ground-Water Extraction Remedies, Volume 1 Summary Report*, Office of Emergency and Remedial Response, EPA/540/2-89/054, Washington, DC.

U.S. Environmental Protection Agency (EPA). 1992. *Evaluation of Ground-Water Extraction Remedies: Phase II, Volume 1 Summary Report*, Office of Emergency and Remedial Response, 9355.4-05, Washington, DC.

U.S. Environmental Protection Agency (EPA). 1997. *Monitored Natural Attenuation at Superfund, RCRA Corrective Action, and Underground Storage Tank Site, Draft Interim Final Policy*, Office of Solid Waste and Emergency Response (OSWER), Washington, DC. 9200-4-17.

Vecht, S.E., M.W. Platt, Z. Er-El, and I. Goldberg, 1988. The growth of *Pseudomonas putida* on m-toluic acid and toluene in batch chemostat cultures. *Appl. Microbiol. Biotechnol.* 27: 587-592.

Waite, A.J., J.S. Bonner, and R. Autenrieth. 1999. Kinetics and stoichiometry of oxygen release from solid peroxides. *Environ. Engineer. Sci.* 16: 187-199.

Ward, Julie A.M., J.M.E. Ahad, G. Lacrampe-Couloume, G.F. Slater, E.A. Edwards, and B. Sherwood Lollar. 2000. Hydrogen isotope fractionation during methanogenic degradation of toluene: potential for direct verification of bioremediation. *Environ. Sci. Technol.* 34: 4577-4581.

Widdowson, M.A., F.J. Molz, and L.D. Benefield. 1988. A numerical transport model for oxygen and nitrate-based respiration linked to substrate and nutrient availability in porous media. *Water Resources Res.* 9: 1553-1565.

Wiedemeier, T.H., H.S. Rifai, C.J. Newell, and J.T. Wilson. 1999. *Natural Attenuation of Fuels and Chlorinated Solvents in the Subsurface*, John Wiley & Sons, Inc., New York.

Wiedemeier, T.H., M.A. Swanson, J.T. Wilson, D.H. Kampbell, and R.N. Miller. 1996. Approximation of biodegradation rate constants for monoaromatic hydrocarbons (BTEX) in ground water. *Groundwater Monitoring Remediation* 16: 186-194.

Wiedemeier, T.H., J.T. Wilson, D.H. Kampbell, R.N. Miller, and J.E. Hansen. 1995. *Technical Protocol for Implementing Intrinsic remediation with Long-Term Monitoring for Natural Attenuation of Fuel Contamination Dissolved in Groundwater*, U.S. Air Force Center for Environmental Excellence, San Antonio, TX.

Wilson, R.D., D.M. Mackay, and K.M. Scow. 2002. In Situ MTBE biodegradation supported by diffusive oxygen release. *Environ. Sci. Technol.* 36: 190-199.

Yeh, C.K., and J.T. Novak. 1994. Anaerobic biodegradation of gasoline oxygenates in soils. *Water Environ. Res.* 66: 744-752.

Zamfirescu, D., and P. Grathwohl. 2001. Occurrence and attunutation of specific organic compounds in the groundwater plume at a former gasworks site. *J. Contaminant Hydrol.* 53: 407-427.

Remediating RDX and HMX Contaminated Soil and Water

Steve Comfort

School of Natural Resources, University of Nebraska,
256 Keim Hall, Lincoln, Nebraska 68583-0915, U.S.A.

Introduction

RDX (hexahydro-1,3,5-trinitro-1,3,5-triazine) and HMX (octahydro-1,3,5,7-tetranitro-1,3,5,7-tetrazocine) are two types of heterocyclic nitramine compounds that have been manufactured and used worldwide as military explosives. RDX was first synthesized in 1899 for medicinal purposes but later recognized for its value as an explosive in 1920 (Akhavan 1998). In the 1940s, the Bachmann synthesis was developed and used for large scale production of RDX during World War II (Bachmann and Sheehan 1949). At the time of its development, the Bachmann synthesis was considered an efficient reaction with high yields but the products of the reaction also contained an impurity (i.e., HMX) that was later recognized and utilized as an explosive (Akhavan 1998). By varying the temperature and reagent concentrations, the Bachmann synthesis could produce large yields of HMX (Urbanski 1984). This led to the introduction of octols in 1952, (mixtures of HMX and TNT), which increased HMX use by the military. Depending on where RDX was manufactured and used, it has also been known as Research Development eXplosive (U.S.), Research Department eXplosive (Britain), or Royal Demolition eXplosive (Canada). Other common names for RDX include cyclonite, hexogen, and cyclotrimethyl-enetrinitramine. HMX has a higher melting point than RDX and is thus known as High Melting explosive, octogen, or cyclotetramethyl-enetetranitramine (Urbanski 1984, Akhavan 1998).

RDX and HMX are classified as secondary explosives (also known as high explosives), which means they cannot be detonated readily by heat or shock but require a primary explosive for initiation. RDX and HMX have been used by the military for a variety of purposes. RDX is commonly used in press-loaded projectiles, cast loadings with TNT, plastic explosives, or base charges in blasting caps and detonators. As an explosive, HMX is superior to RDX because its ignition temperature is higher and has greater chemical stability. HMX is commonly used as a booster charge in mixtures or as an oxidizer in solid rocket and gun propellants (Island Pyrochemical Industries 2004). HMX is also used to implode fissionable material to achieve critical mass in nuclear devices (Yinon 1990). Like other energetic compounds, RDX and HMX possess multiple nitro groups but are structurally unique in that the nitro moieties of nitramines are bonded to the central ring via single nitrogen-nitrogen bonds (Fig. 1). This distinction is important because unlike aliphatic and aromatic nitro compounds that can be produced biosynthetically by some plant and microbial species (Turner 1971), nitramines appear to be true xenobiotics and occur in nature solely from human activities (Coleman et al. 1998).

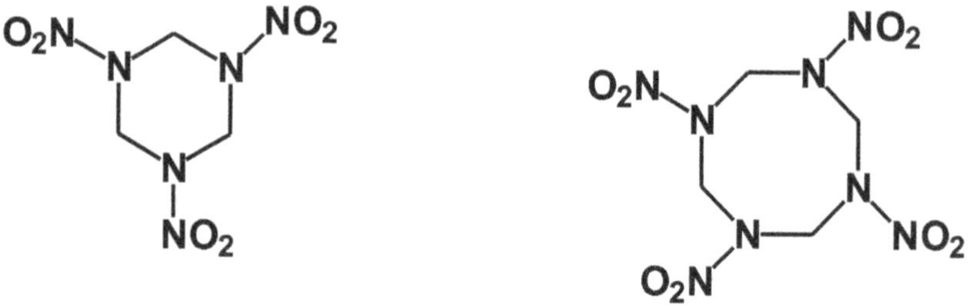

RDX   **HMX**

Figure 1. Chemical structures of RDX and HMX.

Examples of military activities that have contaminated many former and current defense sites include the improper disposal of wastewaters generated during the manufacturing and assembling of munitions or the repeated discharge of live ammunition at military training grounds. Training ranges have the additional problem of unexploded ordnances (UXOs), which are caused when munitions fail to explode or have incomplete detonations. These so-called "dud rates" can be as high as 10% (Defense Science Board 1998) and are a major concern because UXOs can potentially leak and cause high concentrations directly around the projectile, whereas incomplete detonation disperses undetonated materials around the impact site. The types of UXOs vary widely and include small arms ammunition, bombs, artillery rounds, mortars, air-craft cannon, tank-fired projectiles, rockets, guided missiles, grenades, torpedoes, mines, chemical munitions, bulk explosives, and pyrotechnics (MacDonald 2001).

Getting a precise figure on the number of sites contaminated with munitions is difficult because not all federal lands have been extensively sampled but it is known that at least geographically, the Department of Defense (DoD) has the largest cleanup program. Through fiscal year 1996, the DoD has spent $9.4 billion on environmental cleanup and identified 12,000 contaminated sites (not all munitions-contaminated) at 770 active or recently closed installations, plus 3,523 contaminated sites at 2,641 former facilities (Siegel 1998). Specific estimates on the number of sites with UXOs also vary. Data compiled by the EPA in 1999 indicated >7500 sites already transferred or slated for transfer from military control could contain UXOs (MacDonald 2001). The Defense Science Board (1998) on UXOs estimated 1500 sites existed but acknowledged this number is uncertain because of the absence of surveys. These sites ranged from small parcels of land to vast tracts covering thousand of acres. Consequently, the Department of Defense's cleanup challenges could cover somewhere between 10 to 20 million acres in the U.S. (Siegel 1998). The Defense Science Board projected that if only 5% of the acreage suspected of containing UXOs required remediation, cleanup costs could exceed $15 billion.

At sites where munitions were manufactured or assembled, soil contamination has typically resulted from the once common practice of releasing explosive-tainted wastewater to drainage ditches, sumps, settling ponds or impoundments. TNT manufacturing for example, required large volumes of water for purification. The aqueous waste produced from this process, known as red water, has been found to contain up to 30 additional compounds besides TNT (Urbanski 1984). Similar practices occurred at loading, packing and assembling plants, where wastewater (also known as pink water) generated during plant operations was routinely discarded outside into sumps and drainage ditches. Left untreated, surface soils laden with wastewater constituents eventually became point sources of ground water contamination. One study showed that of the numerous sites sampled, >95% contained TNT and 87% exceeded permissible ground water concentrations (Walsh et al. 1993).

Environmental risk assessments of military facilities have determined that contaminated soils often contain mixtures of energetic compounds rather than a single explosive. In addition to nitramines (RDX, HMX), other classes of contaminants commonly observed include nitroaromatics (TNT, dintro- and nitrotoluenes) and nitrate esters (nitroglycerin, nitrocellulose). Of these, TNT, RDX and HMX have been the most frequently detected largely because of their prevalence in manufacturing specific compositions (e.g., octol contains ~ 75% HMX and 25% TNT; Akhavan 1998, Composition C-4, 91% RDX, Smith-Simon and Goldhaber 1995) and synthesis impurities. The Bachmann synthesis of RDX (Bachmann and

Sheehan 1949) commonly results in HMX impurities of 8 to 12% (Fedoroff and Sheffield 1966) whereas the two grades of HMX used for military purposes contain between 2 and 7% (w/w) RDX (Island Pyrochemical Industries 2004). Consequently, to effectively remediate munitions-contaminated soil and water, treatment technologies must be robust enough to treat multiple energetic compounds rather than a single explosive.

Environmental Fate and Toxicity

Structural differences between energetic classes (i.e., nitroaromatics vs nitramines) are manifested in their recalcitrance and environmental fate. For example, when soils contain low to modest concentrations of TNT (e.g., <500 mg kg^{-1}), several biotic and abiotic transformations can promote natural attenuation and detoxification (Comfort et al. 1995, Hundal et al. 1997a, Peterson et al. 1996, 1998, Kreslavski et al. 1999). Microbial degradation of TNT has been demonstrated by aerobic, anaerobic, or combined pathways with practically every study observing amino degradation products (reduction of one or more nitro moieties; e.g., McCormick et al. 1976, Isbister et al. 1980, Schackmann and Müller 1991, Walker and Kaplan 1992, Funk et al. 1993, Marvin-Sikkema and de Bont 1994, Bradley and Chapelle 1995, Gilcrease and Murphy 1995, Bruns-Nagle et al. 1996, Pasti-Grigsby et al. 1996). The aromatic amines derived from TNT (e.g., 2,4-diaminonitrotoluene) can partition to soil organic matter and eventually irreversibly bind to soil humic matter through imine linkages resulting from condensation with carbonyl groups (Bartha and Hsu 1974, Hsu and Bartha 1976, Bollag et al. 1983). Also, the electron donating character of NH_2-substituents makes these products more prone to attack by dioxygenases and subject to further degradation (Dickel et al. 1993). Hundal et al. (1997a) studied the long-term sorption of TNT in soils and observed that after 168 d of equilibration, 32 to 40% of the sorbed ^{14}C-TNT was irreversibly bound (unextractable). In a detailed study using size-exclusion chromatography, Achtnich et al. (1999) showed that reduced derivatives of TNT formed during an anaerobic/aerobic soil treatment were irreversibly bound to a wide range of molecular size humic acids (>5,000 daltons).

By contrast, the reduction of the nitramines (i.e., RDX, HMX) typically produces nitroso rather than amino derivatives. Because RDX and its nitroso derivatives are stable under aerobic conditions, reports of irreversibly bound RDX have been less prevalent. Price et al. (2001) systematically studied the short-term fate of RDX and concluded that RDX was fundamentally different from TNT by being relatively more recalcitrant in aerobic soils slurries. Only under highly reducing conditions was RDX subject to extensive mineralization. Price *et al.* (2001) also concluded that under aerobic conditions, most RDX was associated with the solution phase and did not bind in unextractable forms. Sheremata *et al.* (2001), however, studied the long-term fate of RDX and reported that although RDX was not extensively sorbed by surface soils (Kd = 0.83 L kg^{-1}), what was sorbed was nearly irreversible with no appreciable difference between sterile and nonsterile soils. By contrast, Singh *et al.* (1998a) observed conditions under which unextractable RDX residues did and did not form and suggested soil concentration might be a determining factor. In a long-term soil slurry study, Singh *et al.* (1998a) observed that 34% of the added RDX (32 mg L^{-1}) was sorbed within 30 min, with sorption increasing to only 37% after 168 d. Approximately 84% of the sorbed RDX was readily extractable and only 8% of the initial ^{14}C was unextractable. Interestingly, no bound residue formed when the soils were highly contaminated and contained solid-phase RDX - a condition that is readily observed in surface soils surrounding loading and packing facilities. The presence of solid-phase RDX in a soil matrix would keep the soil solution saturated and severely limit microbial activity (i.e., biotic transformations). Oh *et al.* (2001) also observed bound-residue formation while treating RDX with zerovalent iron and hypothesized that upon ring fission, the amine-containing products (e.g., hydroxymethylnitramine) could bind to the carbonyl functional groups of humics to form unextractable residues. Therefore, it appears that transformation (biotic or abiotic) beyond the nitroso derivatives is needed before bound or unextractable residues of RDX will be observed but in many cases, RDX will remain recalcitrant in the solution phase and readily available for transport.

Although TNT, RDX, and HMX have been detected in ground water, RDX appears to pose the greater environmental concern for aquifers because of its prevalent use and sorption characteristics. Despite having a lower aqueous solubility than TNT (RDX = 34.4 mg L^{-1} at 25°C, TNT = 128.5 mg L^{-1}; Park *et al.* 2004; Table 1), RDX is more mobile in soils than TNT. This characteristic has resulted in the observance of larger RDX plumes than TNT beneath sites that have been heavily contaminated with both compounds (Spalding and Fulton 1988). In comparing RDX with HMX, both are structurally similar by consisting of multiples of the $CH_2=N-NO_2$ monomeric unit but these polynitramines differ with HMX being less water soluble than RDX (Table 1) and chemically more stable and resistant to attack by strong base (Akhavan 1998). Recent biodegradation studies have also confirmed that HMX is more resistant to microbial attack than RDX (Shen *et al.* 2000).

Brannon and Pennington (2002) compiled solubility and sorption

Table 1. Compiled solubility and adsorption coefficients for RDX and HMX.

Compound	Statistic	Solubility and Temperature (mg L⁻¹)	Solubility Reference[1]	Partition Coefficient (L kg⁻¹)	Total Organic Carbon (%)	Organic Carbon Coefficient Koc (L kg⁻¹)	Partition Reference[1]
RDX		28.9 (10°C)	Sikka et al. 1980	0.43	0.6	73	Tucker et al. 1985
		42.3 (20°C)	Sikka et al. 1980	0.16	0.5	32	Tucker et al. 1985
		38.4 (20°C)	Spanggord et al. 1983	0.93	0.5	186	Tucker et al. 1985
		59.9 (25°C)	Banerjee et al. 1980	1.21	1.4	85	Tucker et al. 1985
		59.9 (26.5°C)	Sikka et al. 1980	1.65	2.0	81	Tucker et al. 1985
		75.7 (30°C)	Sikka et al. 1980	2.39	6.0	40	Tucker et al. 1985
		27.2 (20°C)	Bier et al. 1999	0.81	0.5	150	Tucker et al. 1985
		34.5 (25°C)	Bier et al. 1999	2.42	6.3	38	Tucker et al. 1985
		43.8 (30°C)	Bier et al. 1999	7.30	3.1	235	Tucker et al. 1985
		89.7 (45°C)	Bier et al. 1999	0.74	1.2	61	Tucker et al. 1985
		113.9 (50°C)	Bier et al. 1999	0.57	1.0	58	Tucker et al. 1985
				0.87	0.7	132	Tucker et al. 1985
				1.20	1.0	125	Brannon et al. 1992
				3.50	2.4	146	Brannon et al. 1992
				0.95	2.4	40	Myers et al. 1998
				0.77	0.6	135	Myers et al. 1998
				0.97	1.7	57	Singh et al. 1998a
				6.38	4.9	130	Singh et al. 1998a
				1.40	3.3	42	Spanggord et al. 1980
				4.20	3.3	127	Spanggord et al. 1980
	Mean					99	
	Median					83	
	Std Dev.					58	
HMX		1.21 (10°C)	Spanggord et al. 1982	12.10	2.4	504	Brannon et al. 1999
		2.60 (20°C)	Spanggord et al. 1982	4.25	0.6	670	Brannon et al. 1999
		5.00 (22-25°C)	Glover and Hoffsommer 1973	1.60	2.4	67	Myers et al. 1998
		5.70 (30°C)	Spanggord et al. 1982	1.17	0.6	205	Myers et al. 1998
		6.60 (20°C)	McLellan et al. 1988b	8.70	1.3	669	McGrath 1995
		2.00 (20°C)	Park et al. 2004				
		8.00 (45°C)	Park et al. 2004				
		22.0 (55°C)	Park et al. 2004				
	Mean					423	
	Median					504	
	Std Dev.					275	

[1] Many references initially compiled and cited in Brannon and Pennington (2002).

coefficients of energetic compounds from several references and reported linear adsorption coefficients (Kd) between 0 and 8.4 L kg^{-1} for RDX and <1 to 18 L kg^{-1} for HMX. While previous studies have revealed that sorption of RDX and HMX appear to be governed more by clay content than organic matter (Sheremata *et al.* 2001, Monteil-Rivera *et al.* 2003), differences in organic carbon partition coefficients (Koc) among the nitramines can be ascertained. Using only Kd values where the organic carbon content of the sorbent was ≥0.5% and more representative of surface soils, the calculated average Koc value was 99 L kg^{-1} for RDX and 423 L kg^{-1} for HMX (Table 1). By comparison, the recently developed polycyclic nitramine CL-20 (2,4,6,8,10,12-hexanitro-2,4,6,8,10,12-hexaazaisowurtzitane), which is being considered by the military for large scale production, had an average Koc of 745 L kg^{-1} (n=5) (Balakrishnan *et al.* 2004). Szecsody *et al.* (2004) compared the sorption characteristics of RDX and CL-20 on six different soils and observed mixed results in a 24-h batch experiment that monitored changes in solution phase concentrations. On three of the soils, RDX sorption was similar to two-fold greater than CL-20 but on the other three soils, CL-20 sorption was 3 to 9 fold greater than RDX. Calculated Koc values for CL-20 on soils with organic carbon contents ≥0.5% averaged 367 L kg^{-1} (range: 84 - 680 L kg^{-1}, n=4). The lower Koc for RDX indicates that it would be the first to migrate through the surface soil layers and underscores why it has been the most frequently observed nitramine in ground water.

The driving force behind all remedial efforts is a concern for the environment and human health. In this regard, RDX has been shown to adversely affect the central nervous system, gastro-intestinal tract and kidneys (Etnier 1989). Common symptoms of RDX intoxication include nausea, vomiting, hyperirritability, headaches, and unconsiousness (Kaplan *et al.* 1965, Etnier 1989, Etnier and Hartley 1990). Liver tumors have been reported in mice fed RDX for 3 months (ATSDR 1996) and the EPA has classified RDX as a possible human carcinogen. The EPA established a lifetime health advisory guidance level of 2 ug L^{-1} for RDX in drinking water for adults. Information on the adverse effects of HMX on humans is limited but the EPA has recommended that drinking water be less than 400 ug L^{-1} (ATSDR 1997).

In addition to human health concerns, the dissemination of nitramines into waterways and soil pose ecological concerns. Consequently, risk assessments and cleanup activities of munitions-contaminated sites require extensive exposure and effects data so that accurate and realistic decisions can be made (Steevens *et al.* 2002). Considerable research on this topic has been published in the last five years (see Steevens *et al.* 2002, Gong *et al.* 2001a, b, and Robidoux *et al.* 2002 and references cited within) and includes work by Robidoux *et al.* (2002), who measured the sublethal and chronic toxicities of RDX and HMX on earthworms. They found that reproduction parameters such as number of juveniles and biomass for the earthworm species *Eisenia andrei* were significantly decreased by RDX (soil concentration $\geq 46.7 \pm 2.6$ mg/kg) and HMX ($\geq 15.6 \pm 4.6$ mg/kg). Gong *et al.* (2001a, b) examined the ecotoxicological effects of RDX and HMX on indigenous soil microbial processes and concluded that extractable soil concentrations as high as 12500 mg HMX kg^{-1} did not significantly influence soil microorganisms whereas RDX showed significant inhibition on several microbial activities (e.g., potential nitrification, basal respiration). Because soil microorganisms will likely be more affected by what is in the soil solution rather than the total soil concentration (typically determined by acetonitrile extractions), differences in aqueous solubilities between RDX and HMX may explain their results. Using HMX at its solubility limit (<6.5 mg L^{-1}), Sunahara *et al.* (1998) observed no toxic effects to *Vibrio fisheri* (Microtox) and a green alga (*Selenastrum capricornutum*) whereas the cell density of the green alga was reduced by 40% at RDX concentrations near its solubility limit (40 mg L^{-1}). While nitroaromatic compounds (e.g., TNT) can adversely affect aquatic organisms, Lotufo *et al.* (2001) found that HMX and RDX had no significant effect in survival or growth of benthic invertebrates. Earlier research by Bentley *et al.* (1977a, b) also determined that RDX was more toxic than HMX to bluegills (*Lepomis macrochirus*), fathead minnows (*Pimephales promelas*) and aquatic algae (*S. capricornutum, A. flos-aquae*).

The recalcitrance of nitramines in contaminated soils combined with their capacity to leach and impart toxicity concerns underscore the importance of designing remedial treatments that rapidly transform RDX and HMX and render these compounds harmless. To this end, this chapter presents some site-specific examples of RDX and HMX contamination and reviews some laboratory, pilot, and field-scale remediation studies specifically aimed at mitigating soil and ground water contamination.

Examples of RDX and HMX Contamination at Military Sites

Soil and water contaminated with munitions have resulted from a variety of military operations. Examples included: (i) explosive manufacturing, (ii) load, assemble and packing facilities, (iii) munitions maintenance and demilitarization, and (iv) training ranges were firing of live ammunition from small arms, artillery mortar fire, and explosive detonation occurred over multiple acreages. To illustrate the environmental and financial ramifications of these past activities, three defense sites are highlighted with a brief overview of the contamination, cleanup costs, and the formidable challenges that lie ahead in remediating the contaminated soil and water.

Nebraska Ordnance Plant

The former Nebraska Ordnance Plant (NOP, Mead, NE) was a military loading, assembling, and packing facility that produced bombs, boosters, and shells during World War II and the Korean War. Ordnances were loaded with TNT, amatol (TNT and NH_4NO_3), tritional (TNT and Al), and Composition B (~60% RDX and 40% TNT) (Comfort *et al.* 1995). During ordnance production, process wastewater was routinely discharged into sumps and drainage ditches. These ditches became grossly contaminated with TNT and RDX with soil concentrations exceeding 5000 mg kg^{-1} near the soil surface (Hundal *et al.* 1997b). When rainfall exceeded infiltration rates, ponded water that formed in the drainage ditches literally became saturated with munitions residues (i.e., reached HE solubility limits) before percolating through the profile (Fig. 2). Considering this process proceeded unabated for more than 40 years, it is no surprise that the ground water beneath the NOP eventually became contaminated. Further complicating ground water concerns were the extensive use of trichloroethylene (TCE) to degrease and clean pipelines by the U.S. Air Force in the early 1960s. As a

Figure 2. Photograph of drainage ditch following heavy precipitation. Drainage ditch adjoined munitions load line building at former Nebraska Ordnance Plant (Mead, NE).

result, the RDX/TCE contaminant plume under the NOP facilities is estimated in the billions of gallons and covering several square miles (Fig. 3).

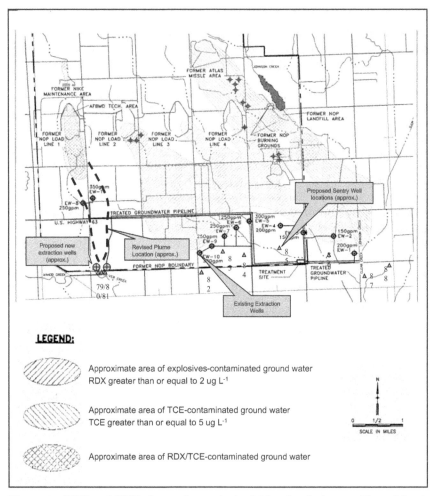

Figure 3 : RDX and TCE plumes beneath the Nebraska Ordnance Plant (Mead, NE).

To prevent the contaminated plume from migrating offsite and in the direction of municipal well fields, an elaborate series of eleven extraction wells and piping networks were constructed to hydraulically contain the leading edge of the RDX/TCE plume (Fig. 3). Currently this $33 million dollar facility treats approximately 4 million gallons of ground water per day with granular activated carbon (GAC). Annual operating costs are approximately $800,000/year. Hydraulic containment plus additional remediation efforts may take 125 years to remediate the ground water plumes. Future costs will also involve the installation of additional wells to contain a larger than originally anticipated plume under one of the load lines (Fig. 3). The ground water treatment costs are in addition to costs of incinerating the soils that were laden with TNT, RDX, and HMX. An incineration system consisting of rotary kiln followed by a secondary combustion chamber incinerated approximately 16,449 tons of contaminated soil at a technology cost of $6.5 million ($394/ton) and a total cost of $10.7 million ($650/ton). Additional costs were also incurred to remove the contaminated load line buildings and site restoration.

Department of Energy Pantex Plant

The U.S. Department of Energy's (USDOE) Pantex plant near Amarillo, Texas, was constructed during World War II by the U.S. Army for the production of conventional ordnances. In the 1950s, portions of the original plant were renovated and new facilities constructed so that HEs could be manufactured and used for assembly of nuclear weapons. Pre-1980 industrial operations included on-site disposal of high explosives and wastewater into unlined ditches. Surface runoff from these ditches into an aquifer-recharging playa (i.e., closed drainage basins that is periodically wet and dry during the year) has contaminated the perched aquifer beneath the Pantex Plant (Fig. 4). The perched aquifer is contaminated with RDX, HMX, TNT, TCE, 2,4-DNT (2,4-dinitrotoluene), 1,2-DCA (1,2-dichloroethane), PCE (tetrachloroethene) and chromium. Of these, considerable attention has focused on the high explosive RDX because it is the most widespread. The plume is estimated at 1.5 billion gallons and covers approximately 5 to 6 square miles (Fig. 4). While a ground water pump and treat system is currently in place to capture contaminants of potential concern in a section of the perched aquifer, hydrological characteristics of the site make implementing additional remedial technologies extremely formidable. Foremost is that the perched aquifer is ~90 m (300 ft) below the surface and 30 m (100 ft) above the High Plains aquifer, one of the largest aquifers in the world. Second, the saturated thickness of the perched aquifer is less than 4.5 m (15 ft) in many locations, making pump and treat systems ineffective. Migration of the contaminated plume beyond the bound of the Pantex site and into privately owned lands has further exacerbated the problem.

The USDOE Innovative Treatment and Remediation Demonstration (ITRD) program was initiated to evaluate emerging technologies that may potentially replace inefficient or ineffective technologies. In 1998, the ITRD

A. Ground water flow

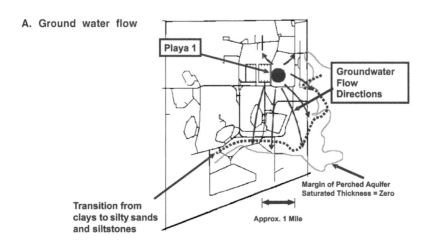

B. RDX plume

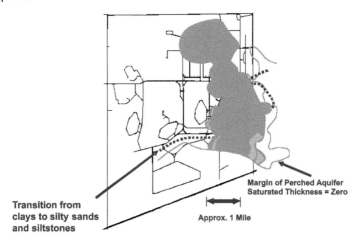

Figure 4. Ground water flow (A) and RDX plume (B) beneath the Pantex Plant (Amarillo, TX). Figure courtesy of Aquifer Solutions, Inc. (Evergreen, CO).

process for the Pantex Plant recommended three in situ technologies for further testing: (i) oxidation by $KMnO_4$; (ii) anaerobic biodegradation; and (iii) chemical reduction by dithionite-treated (reduced) aquifer material. Bench-scale feasibility studies of all three technologies have been conducted and deployment scenarios developed. Well construction costs and spacing are principal driving variables in cost estimates with a number of data gaps still present (Aquifer Solutions Inc. 2002).

Massachusetts Military Reservation

Military testing and training grounds provide vital lands for preparing military troops for combat and maintaining readiness. While an important resource for military exercises, site commanders must delicately balance these lands so that training operations proceed without the environmental consequences associated with repeated release of energetic compounds. Decades of continuous discharge of live ammunition from small arms, artillery mortar fire, and explosive detonations have contaminated surface soils and impacted ground water at several locations across the U.S. This type of contamination is exemplified at the Massachusetts Military Reservation (MMR) where more than 40 years of military and law enforcement training has contaminated Cape Cod's sole aquifer with RDX. Additional investigations have uncovered propellants, metals, pesticides, volatile and semi-volatile organic compounds, and unexploded ordnances. By impacting 200,000 year-round residents and >500,000 summertime residents who rely on Cape Cod's aquifer for drinking water, the financial costs of cleaning up MMR, coupled with public and political outcry, have forced the Department of Defense to seek proactive remedial technologies that prevent situations like MMR from reoccurring, yet still allow the training grounds to be used for preparing U.S. troops.

Remediating military training ranges present unique challenges because they typically encompass thousands of acres that are under constant barrage from training exercises. For example, the training ranges at MMR cover approximately 144,000 acres, with multiple target areas. Recent records indicate that before the EPA halted military activities at MMR, 1,770,000 small arms and more than 3000 rounds of artillery and mortar were fired annually. Extensive investigations at military training grounds have determined that soil contamination is extremely variable with concentrations ranging from "no detect" to isolated "hot spots." In a recent study characterizing contamination at military firing ranges, Jenkins *et al.* (2001) found examples of contamination by showing that surface soils concentrations ranged between 458 and 175,000 ug 2,4-dintrotoluene kg^{-1} in front of a single 105-mm howitzer that had fired about 600 rounds within 30 d. This same study observed that soil samples collected below and adjacent to a 155-mm howitzer shell that had undergone low-order detonation were heavily contaminated with TNT and its biological degradation products (Jenkins *et al.* 2001). These reports as well as others (Thiboutot *et al.* 1998), confirm that military testing and training ranges, although vital to preparing Armed Forces, must be characterized for environmental contaminants and in many instances, remedial efforts taken to control the leaching of explosives and prevent ground water contamination.

The Nebraska Ordnance Plant, Pantex, and Massachusetts facilities exemplify just three of the potentially thousands of munitions-contaminated sites currently in need of remediation. Clearly, the enormous environmental and financial ramifications associated with these facilities provide ample justification for the development of cost-effective and environmentally sound treatment technologies. To date, the most demonstrated remediation technology for explosive-containing soils is incineration. Although incineration is effective, it is expensive, produces an unusable ash byproduct, and has poor public acceptance due to safety concerns regarding air emissions (Lechner 1993). Incineration would also be impractical for impact ranges that may require treatment of thousands of acres. Likewise, the use of granular activated carbon in pump and treat systems for ground water is effective in removing a wide variety of HEs but at a considerable annual cost and, as demonstrated at the Nebraska Ordnance Plant, could take more than a century to complete. Furthermore, the spent activated carbon can usually not be reused and requires incineration (Sisk 1993). The hydrological characteristics of some sites (i.e., Pantex) also make GAC unfeasible and in-situ technologies more desirable. While many researchers have taken a biological approach to solving HE contamination (bioremediation, phytoremediation, composting, etc.), there is also a large contingency that have tried an abiotic or chemical approach. Chemical approaches are usually performed by adding one or more chemicals reagents (reductant, oxidant) or altering the physiochemical properties of the soil-water environment. Chemical methods offer several advantages to biological methods because they are often faster, can treat highly contaminated environments, and are less sensitive to ambient conditions. The goal of a chemical approach is to either transform the xenobioitc into carbon dioxide, H_2O and mineral elements or structurally transform the parent compound into a product that is more biodegradable (i.e., abiotic-biotic approach). Because excellent reviews of biological approaches to remediating munitions contaminated soil and water are currently available (Gorontzy et al. 1994, Hawari 2000, Hawari et al. 2000, Rosser et al. 2001, Spain 1995, Spain et al. 2000, Van Aken and Agathos 2001), this chapter focuses on reviewing some abiotic approaches that have been used to remediate RDX and HMX contaminated soil and water.

Abiotic Remediation Treatments for RDX/HMX-Contaminated Soil and Water

Chemical Reduction Using Zerovalent Iron

From a historical perspective, metals have been used to transform and synthesize organic chemicals since the late 1800s. The use of zerovalent metals in environmental research, however, did not surface until fifteen to twenty years ago when Sweeny (1979, 1981), followed by Senzaki and Kumagai (1988, 1989), reported that metallic iron could be used to degrade organic contaminants such as chlorinated solvents in water. The more recent idea that iron metal could be used for *in situ* remediation of subsurface contaminants grew primarily from work carried out at the University of Waterloo. In a project involving sorption of organic compounds to well casings, it was noted that the concentration of the halogenated compound, bromoform, declined when in contact with steel and aluminum casing materials. This 1984 observation was attributed to a dehalogenation reaction, but the environmental significance of this work was not realized until a few years later when the results were re-evaluated and published (Reynolds *et al.* 1990). Today, the use of zerovalent metals has become an alternative to the common pump-and-treat and air-sparging technologies, and the emergence of the so-called permeable reactive barriers (PRBs), consisting of scrap Fe^0 cuttings, has proven to be a highly cost effective treatment for contaminated ground water (Wilson 1995). Since these initial reports by the University of Waterloo, a flurry of research activity on the use of zerovalent metals in environmental research has ensued with more than 500 publications currently available (http://cgr.ese.ogi.edu/ironrefs/). Furthermore, >80 PRBs have been installed in the U.S. and 100 world-wide (EnviroMetal Technologies, Inc. 2004), with the majority of applications targeting chlorinated solvents. This heightened interest and field-scale deployment has helped to make Fe^0 the most widely studied chemical reductant for environmental applications (Tratnyek *et al.* 2003).

Iron is an effective remediation tool because when placed in water, metallic iron (Fe^0) becomes an avid electron donor and its oxidation ($E^0_h = -0.409$ V) can drive the reaction of many redox-sensitive contaminants. Researchers have used Fe^0 as the bulk reductant for the reduction of nitroaromatic compounds to anilines (Agrawal and Tratnyek 1996), which can be further degraded biologically (Dickel *et al.* 1993) or incorporated into natural organic matter via enzyme-catalyzed coupling reactions (Bollag 1992, Hundal *et al.* 1997a). Through studies on the dehalogenation of chlorinated methanes, Matheson and Tratnyek (1994) proposed three possible reduction mechanisms, with direct electron transfer at the iron interface as the most probable reaction pathway. Later, Scherer *et al.* (1999) expanded this theory by proposing that mineral precipitates formed during iron corrosion may influence the efficacy of iron to transform contaminants by acting as a physical barrier, semiconductor, or reactive surface. Others have shown that iron can work in conjunction with naturally occurring electron transfer mediators (quinone moieties in humic and fulvic acids) to facilitate contaminant destruction in contaminated soils and sediments (Weber 1996).

Several recent studies have also indicated the importance of surface-bound Fe(II)-species as electron donors in redox transformations of organic compounds (Klausen *et al.* 1995, Heijman *et al.* 1993, Heijman *et al.* 1995, Amonette *et al.* 2000, Satapanajaru *et al.* 2003a, b). The catalytic activity of Fe(II) in the presence of oxides is believed to be the result of complexation of Fe(II) with surface hydroxyl groups and the formation of inner-sphere bonds, which increases the electron density of the adsorbed Fe(II). Klausen *et al.* (1995) demonstrated that surface bound Fe(II) on iron (hydr)oxide surfaces or surface coatings plays an important role in the reductive transformation of nitroaromatic compounds. Gregory *et al.* (2004) observed complete transformation of RDX by adsorbed Fe(II) on magnetite with reaction rates increasing as a function of adsorbed Fe(II). While Klausen *et al.* (1995) observed that unbound Fe(II) species were not reactive for nitroaromatics, the lack of reactivity to unbound Fe(II) appears to be compound specific because Eary and Rai (1988) report chromate reduction by ferrous iron and Gregory *et al.* (2004) reported that 72 uM of RDX was transformed by 1.5 mM Fe(II) ($FeCl_2$) at pH 8.0. Our laboratory has also observed RDX degradation in aqueous solutions by Fe(II) alone (100 mg L^{-1} $FeSO_4$ • $7H_2O$) at pH 8.5 (unpublished data). A confounding factor in these experiments however, is that magnetite or green rusts can also form at alkaline pH and thus a homogenous solution of Fe(II) can become a mixed phase system (Fe(II) + magnetite or other precipitates).

In aerobic environments, oxygen is the normal electron acceptor during iron corrosion while under anaerobic conditions, such as those encountered in ground water or waterlogged soils, electron transfer during iron corrosion can be coupled to redox sensitive organic contaminants. For this reason, use of zerovalent iron is generally implemented under fully anoxic conditions because the presence of oxygen is expected to lower the efficiency of the process by competing with the target contaminants (Joo *et al.* 2004), accelerating iron aging (passivation), and cause loss of reactivity (Gaber *et al.* 2002). Ironically, examples exist where destruction kinetics of certain contaminants by Fe^0 have been accelerated by exposure to air. Tratnyek *et al.* (1995) observed a higher rate of CCl_4 degradation by Fe^0 in an air-purged system ($t_{1/2}$ = 48 min) than in a nitrogen-purged ($t_{1/2}$ = 3.5 h) or oxygen-purged environment ($t_{1/2}$ = 111 h). Satapanajaru *et al.* (2003a) found that Fe^0-mediated destruction of metolachlor [2-chloro-N-(2-ethyl-6-methylphenyl)-N-(2-methoxy-1-methyl ethyl) acetamide] was faster in batch reactors shaken under aerobic than anaerobic conditions and contributed this increase to the formation and facilitating effects of green rusts, mixed Fe(II)-Fe(III) hydroxides with interlayer anions that impart a greenish-blue color. Joo *et al.* (2004) also observed that the herbicide molinate (S-ethyl hexahydro-1H-azepine-1-carothioate) was much more readily transformed by Fe^0 when shaken in the presence of air than when purged with N_2. Interestingly, they showed that the transformation occurring was actually an oxidation caused by a two-electron reduction of oxygen to form hydrogen peroxide, which subsequently caused the formation of strongly oxidizing substrates (i.e., hydroxyl radicals). All these observations lend credence to using zerovalent iron in microaerophillic environments, such as those that might be encountered in treating soils.

Earlier work with zerovalent zinc demonstrated the utility of metals to treat soils contaminated with DDT (Staiff *et al.* 1977), methyl parathion (Butler *et al.* 1981) and polychlorinated biphenyls (Cuttshall *et al.* 1993). More recent research indicates the tremendous potential of Fe^0 to degrade high explosives (Hundal *et al.* 1997b, Singh *et al.* 1998b, 1999, Wildman and Alvarez 2001, Oh *et al.* 2001, Oh and Alvarez 2002, Comfort *et al.* 2003, Park *et al.* 2004) and a variety of pesticides (atrazine, Singh *et al.* 1998c; dicamba, Gibb *et al.* 2004; metolachlor, Comfort *et al.* 2001; Satapanajaru *et al.* 2003a, b).

RDX/HMX-Contaminated Soil. One of the biggest obstacles to treating contaminated soils at former loading, packing and manufacturing facilities (e.g., Nebraska Ordnance Plant) is the sheer magnitude of contamination present in the impacted surface soils. It is not uncommon for surface soils to contain energetic compounds in percentage concentrations and approach detonation potential (Crockett *et al.* 1996, Talmage *et al.* 1999, Comfort *et al.* 2003, Schrader and Hess 2004). Because of the equilibrium relationship between the soil solution and solid phase explosive, remediating soils containing solid-phase HEs will not only require treatments that demonstrate rapid destruction in solution but also those that continue to transform RDX/HMX as dissolution and desorption occurs from the soil matrix. To evaluate Fe^0 as a remedial treatment for RDX-contaminated soil, Singh *et al.* (1998b) began by initially determining the effectiveness of zerovalent iron to remove or transform RDX in a near-saturated solution. Treating a 32 mg L^{-1} RDX solution (144 uM) with 10 g Fe^0 L^{-1} resulted in complete RDX removal from solution within 72 h (Fig. 5A). Simultaneous tracking of ^{14}C in solution provided a carbon mass balance for the RDX. At Fe^0 concentrations ≤ 2 g Fe^0 L^{-1}, solution ^{14}C activity remained unchanged (Fig. 5B), indicating that RDX transformation products produced from the Fe^0 treatment (measured as ^{14}C activity) were watersoluble and not strongly sorbed by the Fe^0. At 100 g Fe^0 L^{-1}, 80% of initial ^{14}C activity was lost from solution. More than 95% of the ^{14}C lost, however, was recovered from the Fe^0 surface through a series of extraction and oxidation procedures. Oh *et al.* (2002) treated RDX with scrap iron and high-purity iron under anaerobic conditions. They observed that RDX was readily transformed by both iron

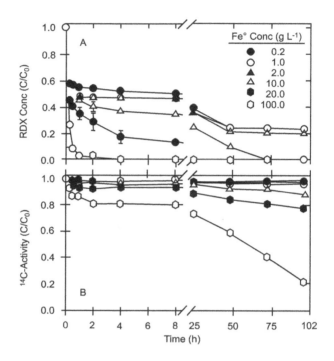

Figure 5. Changes in aqueous RDX (A) and ^{14}C (B) concentrations after treating an aqueous solution containing 32 mg RDX L-1 with various Fe^0 concentrations. Originally printed in *Journal of Environmental Quality* 27:1240-1245.

sources with no appreciable buildup of identifiable degradates. By measuring changes in TOC they also confirmed that RDX transformation products were not sorbed to the iron surface.

Major biological reduction products of RDX include mono-, di, and tri-nitroso degradates of RDX, specifically MNX (1,3-dinitro-5-nitroso-1,3,5-triazacyclohexane), DNX (1,3-dinitroso-5-nitro-1,3,5-triazacyclohexane), and TNX (1,3,5-trinitroso-1,3,5-triazacyclohexane) (McCormick *et al.* 1981). Singh *et al.* (1998b) monitored these compounds, as well as inorganic N species (NH_4^+, NO_2^-, and NO_3^-), in an experiment that treated 20 mg L^{-1} of RDX with 10 g Fe^0 L^{-1}. Results indicated a rapid initial decrease in RDX concentration that slowed by 24 h (Fig. 6A). While a buildup of MNX, DNX, and TNX was observed and could be considered a potential concern given the general toxicity of N-nitroso compounds (Mirvish *et al.* 1976, George *et al.* 2001), the nitroso products of RDX were eventually degraded (Fig. 6B). After 24 h, a slight increase in rate of RDX loss was observed and corresponded with the time when the pH had increased >8.8 (Fig. 6A).

Other researchers have observed a somewhat similar situation during Fe⁰
treatments where destruction rates were initially slow for 12 to 24 h (i.e., lag

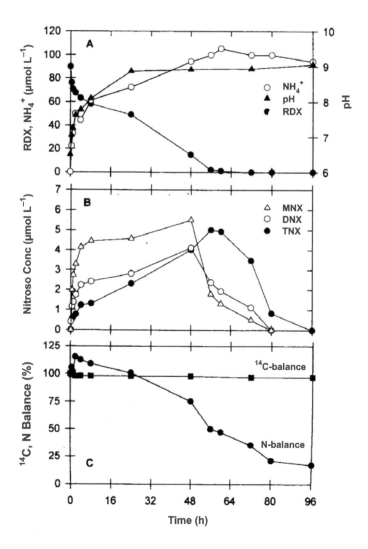

Figure 6. Changes in pH, RDX, and NH_4^+ (A) and MNX, DNX, and TNX (B)
concentrations following treatment of 20 mg RDX L^{-1} (90 uM) with 10 g Fe⁰ L^{-1}.
Carbon-14 balance of added ¹⁴C-RDX was determined by ¹⁴C-activity remaining
in solution (C); N-balance of added RDX-N was determined by summing RDX-N,
NH_4^+-N, and N associated with nitroso degradation products (C). Originally
printed in *Journal of Environmental Quality* 27:1240-1245.

time) before an increase in destruction was observed. These observations typically occur when the pH is alkaline (>8.0) and been attributed to the formation of reactive Fe(II) precipitates such as green rust, magnetite, and secondary reductants [Fe(II) or Fe(II)-containing oxides and hydroxides] coordinated on the oxides of the Fe^0 surface (Satapanajaru et al. 2003, Alowitz and Scherer 2002, Gregory et al. 2004).

Most research studies performed with Fe^0 have been aimed at potential applications to in situ permeable reactive barriers (PRB). As a result, the majority of laboratory experiments have been conducted inside an anaerobic chamber. While anaerobic conditions can be easily obtained inside a PRB, it will be more difficult to exclude oxygen when treating soils ex situ, even if treatments involve soil slurries. To optimize conditions under which zerovalent iron could be used, a few researchers have investigated Eh/pH conditions under which Fe^0 was most effective in transforming RDX (Price et al. 2001, Singh et al. 1999). By using an Eh/pH-stat, Singh et al. (1999) showed that RDX destruction kinetics by Fe^0 increased as the Eh and pH decreased with no appreciable increase in destruction rates when Eh <0 mV (Fig. 7).

Although numerous reports now confirm that $Fe°$ can effectively transform RDX in solution and soil slurries (Hundal et al. 1997b, Singh et al. 1998b, 1999, Wildman and Alvarez 2001, Oh et al. 2001, Oh and Alvarez 2002, Comfort et al. 2003, Park et al. 2004), working with soil slurries is problematic for several reasons. The equipment required for continuous agitation is expensive and limits the volume of soil that can be treated at any given time. Dewatering of treated soil is also required. A desirable alternative to slurry treatment in situ applications or on-site treatment in soil windrows. Using soil windrows allows much greater volumes of soil to be treated and is constrained by only the size of the windrows and acreages available (Comfort et al. 2003). However, for Fe^0 to be effective in static soil windrows, contaminant destruction must occur in the soil solution before the intermixed iron in the soil matrix becomes passivated by exposure to air. As stated earlier, because strictly anoxic conditions are not required for Fe^0 to transform contaminants and examples exist where destruction kinetics were faster in microaerophillic than anaerobic conditions (Tratnyek et al.1995, Satapanajaru et al. 2003a, Joo et al. 2004), it is reasonable to assume that Fe^0 can be used as a soil treatment for remediation purposes. Initial laboratory work with RDX-contaminated soil from the Nebraska Ordnance Plant showed that Fe^0 intermixed with moist soil (0.30-0.40 kg H_2O kg^{-1} soil) could transform RDX under static unsaturated conditions (Singh et al. 1998b). Results showed that a single addition of 5% Fe^0 (w/w) transformed 57% of the initial RDX (3600 mg kg^{-1}) following a 12 month incubation.

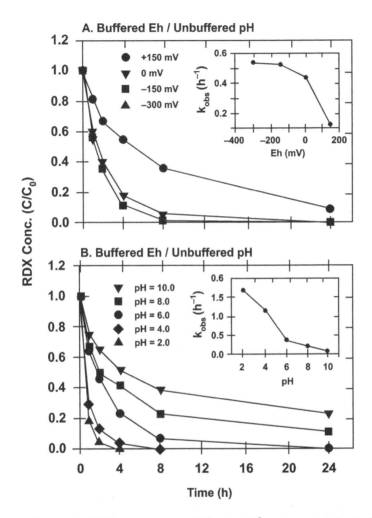

Figure 7. Changes in RDX concentration following Fe^0 treatment under buffered (A) Eh (+150, 0, -150, and -300 mV vs. Ag/AgCl reference electrode) and (B) pH (10, 8, 6, 4, 2). Originally printed in *Environmental Science and Technology* 33:1488-1494.

The effectiveness of Fe^0 to transform RDX in unsaturated soil opened the door for field-scale applications. But using zerovalent iron at the field scale requires the machinery that can thoroughly mix iron throughout the soil matrix. The importance of good mixing cannot be understated because unlike slurries were continual agitation would allow constant movement and contact with Fe^0, the radius of influence for Fe^0 in a static windrow is relatively stationary. The Microenfractionator® (H&H Eco Systems, North

Bonneville, WA) is the trade name of a high-speed mixer that has been specifically augmented to mix windrows of soil (Fig. 8). In 1999, we successfully utilized a pull-behind-tractor version of this mixer to remediate 1000 yd^3 of pesticide-contaminated soil with Fe^0 at the field scale (Comfort *et al.* 2001). In 2000, we attempted to evaluate the use of Fe^0 for treating HE-contaminated soil by conducting pilot-scale experiments (70 kg soil) with a bench-top replica of the field-scale unit (Fig. 8). Contaminated soil containing RDX, TNT, and HMX from an outwash pond that had previously been used for munitions wastewater disposal (Los Alamos National Laboratory, NM) was treated with Fe^0 and some acidifying amendments. Zerovalent iron effectively removed 98% of the RDX and TNT within 120 d under static unsaturated conditions (Comfort *et al.* 2003). Because HMX is considered less toxic than RDX (Bentley *et al.* 1977a, b, McLellan *et al.* 1988a, b), Los Alamos personnel did not initially considered it a contaminant of concern. Further soil analysis, however, revealed that HMX was present at very high concentrations (>30 000 mg kg^{-1}) and that this energetic compound was not effectively destroyed by the Fe^0 treatment.

To determine if low solubility was responsible for the inability Fe^0 to transform HMX, Park *et al.* (2004) attempted to increase HMX solubility with higher temperatures and surfactants. While higher temperatures increased the aqueous solubility of HMX (2 mg L^{-1} at 20°C; 8 mg L^{-1} at 45°C, 22 mg L^{-1} at 55°C), increasing temperature did not increase HMX destruction by Fe^0 when RDX and TNT were also present in the soil slurry

Figure 8. Photograph of field-scale soil mixer. Photo courtesy of H&H Eco Systems, Inc. (North Bonneville, WA).

matrix. Furthermore, by conducting batch experiments with single and binary mixtures of RDX and HMX, Park *et al.* (2004) showed that when RDX and HMX were present at equal molar concentrations, RDX was a preferential electron acceptor over HMX; consequently, iron-based remedial treatments of RDX/HMX-contaminated soils may need to focus on removing RDX first. The rationale for using surfactants is typically to get more of the contaminant in solution so that it can be degraded. Park *et al.* (2004) found that the cationic surfactants didecyl (didecyldimethyl ammonium bromide) and HDTMA (hexadecyltrimethyl ammonium bromide) could increase HMX solubility ($\sim$200 mg L^{-1}) and that both RDX and HMX were effectively transformed by Fe^0 in the surfactant matrix (Fig. 9). Preliminary laboratory studies also showed that didecyl plus Fe^0 could be used to treat HMX-contaminated soil under unsaturated conditions (unpublished data).

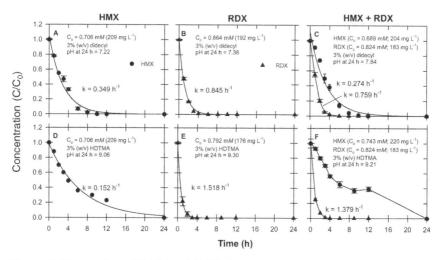

Figure 9. Destruction of HMX and RDX alone or in combination by unannealed Fe^0 in a 3% didecyl or HDTMA matrix. Originally printed in *Journal of Environmental Quality* 33:1305-1313.

Fe^0 Treatment of RDX/HMX-Contaminated Ground Water. Although more than 100 permeable reactive barriers have been installed worldwide (EnviroMetal, Inc. 2004) the majority of PRBs have been targeted for chlorinated compounds and only recently has research been aimed at using PRB for environmental contaminants with multiple nitro groups (e.g., TNT and RDX). Widman and Alvarez (2001) evaluated the potential benefits of an integrated microbial-Fe^0 system to intercept and treat RDX-contaminated ground water. They found that a combined Fe^0-based bioremediation system may offer significant advantages over either Fe^0 or biodegradation when used alone. Specifically, anaerobic Fe^0 corrosion by water produces cathodic hydrogen, which can then serve as an electron donor for the biotransformation of RDX. Oh and Alvarez (2002) used flow-through columns to evaluate the efficacy of permeable reactive barriers to treat RDX-contaminated ground water. They found that extensive RDX removal (>99%) occurred by Fe^0 columns for more than one year. Through a variety of treatments, they also showed that the Fe^0 could interact with indigenous aquifer microcosms and produce hydrogen gas and acetate, which subsequently facilitated RDX degradation. Column experiments with TNT have shown that permeable iron barriers can reduce TNT to triaminotoluene (Miehr et al. 2003), which would be more prone to biotic oxidations (i.e., more biodegradable) in aerobic environments.

Chemical Reduction Using *In Situ* Redox Barriers

In situ redox manipulation (ISRM) is a technology that injects a chemical reductant (sodium dithionite buffered at high pH) into an aquifer. Because dithionite is a strong reductant, particularly in alkaline solutions (reduction potential of -1.12 V), it chemically dissolves and abiotically reduces amorphous and some crystalline Fe(III) oxides (Rueda et al. 1992, Chilakapati et al. 2000, Szecsody et al. 2001, Szecsody et al. 2004, US Patent 5,783,088), leaving behind several possible Fe(II) species such as structural Fe(II), adsorbed Fe(II), $FeCO_3$ precipitates, and FeS. The simple reaction describing the reduction of iron by dithionite is:

$$S_2O_4^{2-} + 2Fe^{3+} + 2\,H_2O \rightarrow 2Fe^{2+} + 2SO_3^{2-} + 4H^+ \qquad [1]$$

Because sulfate is eventually produced, extracting treated aquifers after dithionite injection is sometimes used if secondary drinking water limits are a concern (site dependent). Once the aquifer solids are reduced, subsequent oxidation of the adsorbed and structural ferrous iron in the reduced zone (i.e., redox barrier) occurs passively by the inflow of dissolved oxygen and additionally by contaminants that can serve as electron acceptors (i.e., RDX, Cr(VI), TCE). The longevity of the reduced sediment barrier is dependent on the flux of electron acceptors. In relatively uncontaminated aquifers, dissolved oxygen in water is the dominant oxidant. Although oxidation of Fe(II) occurs relatively quickly at alkaline pH, slower rates of oxidation are likely for surface Fe(II) phases (Szecsody et al. 2004).

Considerable research on ISRM has been conducted with chlorinated solvents and Cr(VI) but only recently has this technology been investigated for high explosives. ISRM is currently being considered at the Pantex Plant for treatment of the RDX-contaminated perched aquifer. Initial testing by Pacific Northwest National Laboratory (PNNL) showed that RDX was quickly degraded (i.e., minutes) in batch and column studies by dithionite-treated Pantex sediments. As observed with Fe^0 treatment of RDX (Singh *et al.* 1998b), dithionite-reduced sediments also produced nitroso derivatives of RDX but these degradates were further reduced into ring fragments that were not strongly adsorbed (based on ^{14}C data, Szecsody *et al.* 2001). Subsequent biodegradation studies of the transformed products showed that the RDX degradates produced from the reduced sediments were readily biodegradable under aerobic conditions, with approximately 50% of the initial ^{14}C recovered as $^{14}CO_2$ after 100 d (Adams *et al.* 2005). Consequently, abiotic reduction of RDX by a redox barrier followed by biodegradation of the transformed products may result in a viable treatment scenario for ground water contaminated with RDX.

Field applicability of ISRM however, is also dependent on geochemical (redox capacity) and hydro geological considerations as well as injection design. Dithionite treatment of the perched aquifer material at Pantex yielded a high redox capacity (0.4% Fe(II)/g), which is equal or greater than other sites in which field-scale remediation is in progress or being considered (Szecsody *et al.* 2001). While 100% reduction of aquifer solids may never be achieved in the field (greater reduction near injection well and less reduction further away), it is plausible that partially reduced Pantex sediments (<0.4%Fe(II)/g) will also be able to reduce RDX, as can dithionite itself. Column studies conducted at PNNL indicate that reduced Pantex sediments are capable of treating several hundred pore volumes of ground water. Considering the hydrological characteristics of the Pantex site, this could relate to lifetime of 30 years or more for the redox barrier (Aquifer Solutions, Inc. 2002).

Electrochemical Reduction of RDX in Aqueous Solutions

Electrolysis, the use of electrical energy to drive an otherwise unfavorable chemical reaction, is a developing technology that has been used to remediate industrial wastes and recently applied to explosives for wastewater treatment (Meenakshisundaram *et al.* 1999, Rodgers and Bunce 2001a, Doppalapudi *et al.* 2002, Bonin *et al.* 2004). Some potential advantages of an electrochemical treatment include the low cost of electricity compared with the cost of chemical treatments, relatively low capital costs, modular design, operations under ambient conditions, and the possibility of higher energy efficiency than thermal or photolysis treatments (Rodgers and Bunce 2001a). Rodgers and Bunce (2001a) demonstrated electrochemical reduction of 2,4,6-trinitrotoluene (TNT) at a reticulated vitreous carbon cathode while Bonin *et al.* (2004) utilized a cascade of divided flow through reactors and showed that an aqueous solution of RDX (48 mg L^{-1}) was completely degraded by a current of 10 mA after flowing through three reactors. The major degradation pathway involved reduction of RDX to MNX followed by ring cleavage to yield formaldehyde and methylenedinitramine, which underwent further reduction and/or hydrolysis (Bonin *et al.* 2004). Doppalapudi *et al.* (2002) also demonstrated that RDX (10 mg L^{-1}) could be degraded under anoxic and oxic conditions by electrolysis while Meenakshisundaram *et al.* (1999) found that RDX degradation increased with increasing current (~20 - 50 mA) and stir rate (630 -2040 rpm).

Chemical Oxidation for In Situ Remediation of Soils and Ground Water

In situ chemical oxidation (ISCO) is class of remediation technologies that delivers oxidants on-site and in-place to ground water or the vadose zone. While municipal and industrial companies have routinely used chemicals to oxidize organic contaminants in drinking and wastewater, it is the ability to treat contaminated field sites that has fueled ISCO popularity, especially when bioremediation is inadequate or where treatment time is considered a factor (Environmental Protection Agency 1998). Increased interest and research in ISCO has caused significant developments and application changes over the last five years with numerous site demonstrations (Vance 2002).

Much of the groundwork for ISCO applications can be traced back to the 20 years or more of research conducted on advanced oxidative processes (AOPs), which employ reactive oxidizing agents such as H_2O_2 or ozone, with or without additional catalysts or photolysis, to generate short-lived chemical species of high oxidation power. Past studies specific to the treatment of explosives include oxidative systems such as H_2O_2/ozone, H_2O_2/ultraviolet light (UV), ozone/UV, or Fenton's reagent for rapid destruction of nitroaromatic and nitramine compounds (see review by Rodgers and Bunce 2001b). Examples of this research include Ho (1986) who demonstrated that photooxidation of 2,4-dinitrotoluene by an H_2O_2/UV system resulted in a side-chain oxidation converting 2,4-DNT to 1,3-dinitrobenzene followed by hydroxylation and cleavage of the benzene ring to produce carboxylic acids and aldehydes. Fleming *et al.* (1997) used a 1:1 mixture of ozone and H_2O_2 at pH>7 (peroxone) to generate hydroxyl radicals and reported that RDX, HMX, and several nitroaromatics in ground water from the Cornhusker Army Ammunitions Plant were degraded by ≥64%, with a destruction efficiency of 90% for RDX. Bose *et al.*

(1998a, b) also conducted a detailed evaluation of oxidative treatments for RDX using a combination of ozone, UV and hydrogen peroxide and showed that that side-chain oxidation and elimination of nitro radicals or nitrous oxide equivalents occurred, followed by cleavage of the heterocyclic ring that resulted in the formation of urea and formamide. A pilot-scale assessment of UV/photolysis is currently on-going at the Nebraska Ordnance plant for treatment of the RDX plume. In this test, a ground water circulation well (a combination of traditional pump and treat with an *in situ* treatment) is being used to extract and treat the water below surface grade. Results from this trial have shown RDX concentrations (5-78 ug L^{-1}) were typically reduced below 5 ug L^{-1} (Elmore and Graff 2001).

Successfully implementing ISCO requires that the oxidant react with the contaminants of concern and that an effective means of dispersing the oxidant to the subsurface is achieved. Technology advances in this regard include delivery processes such as deep soil mixing, hydraulic fracturing, mulit-point vertical lancing, horizontal well recirculation, and vertical well recirculation (U.S. Department of Energy 1999). Because of their high oxidation potential, the three oxidants commonly employed include hydrogen peroxide (H_2O_2, 1.78 V) either alone or in the form of the Fenton's reagent ($H_2O_2 + Fe^{2+}$), ozone (O_3, 2.07 V), and permanganate (MnO_4^-, 1.68 V). Specific examples illustrating the use of these oxidants for treating RDX/HMX-contaminated soil and water follow.

Permanganate. Chemical oxidation using permanganate has been widely used for treatment of pollutants in drinking water and wastewater for more than 50 years (U.S. Department of Energy 1999). In 2001, more than 100 field applications involving permanganate had been completed or planned (Siegrist *et al.* 2001). Like most of the abiotic treatments discussed thus far, these applications have focused on chlorinated solvents and only recently has permanganate treatments been directed toward treating explosives. Site specific issues are always a concern and the Office of Environmental Management concluded that ISCO using $KMnO_4$ is applicable for the destruction of dissolved organic compounds in saturated permeable zones with hydraulic conductivities $> 10^{-4}$ cm/s, low organic carbon contents ($<0.5\%$) and a pH range between 3 and 10 (optimum, 7-8) (U.S. Department of Energy 1999).

Commonly manufactured and sold as a solid ($KMnO_4$) or liquid ($NaMnO_4$), permanganate is an oxidizing agent with a strong affinity for organic compounds containing carbon-carbon double bonds, aldehyde groups, or hydroxyl groups. Research with chlorinated solvents has shown that permanganate is attracted to the negative charge associated with the π electrons of chlorinated alkenes such as tetrachlroethene, trichloroethene, dichloroethene, and vinyl chloride (Oberle and Schroder 2000). Although the chemical structure of RDX and HMX does not readily lend itself to reaction with permanganate, IT and Stroller Corporation (2000) initially demonstrated effective RDX destruction by $KMnO_4$ treatment. Based on favorable laboratory results, a single-well push-pull test was also conducted at the Pantex site. In this test, permanganate was injected (push) into a single well, allowed to react, and then extracted (pull). Significant degradation of all HE compounds was observed with RDX half-life estimated at ~7d at a $KMnO_4$ concentration of 7000 mg L-1 (Siegrist et al. 2001).

In a follow up to the observations of IT and Stroller (2000), Adam et al. (2004) subsequently used ^{14}C-RDX with $KMnO_4$ and found that a 2.8 mg L^{-1} solution of RDX solution treated with 20,000 mg L^{-1} $KMnO_4$ decreased to 0.1 mg L^{-1} within 11 d with cumulative mineralization continuing for 14 d until 87% of the labeled carbon was trapped as [14]CO_2. Moreover, they showed lower $KMnO_4$ concentrations (1000-4000 mg L^{-1}) also produced slow (weeks) but sustainable RDX destruction (Fig. 10) (Adam et al. 2004). Treatment parameters such as initial RDX concentration (1.3-10.4 mg L^{-1}) or pH (4.1-11.3) had no significant effects on reaction rates. Microcosm studies also demonstrated that RDX products produced by permanganate were more biodegradable than parent RDX. While Adam et al. (2004) hypothesized that permanganate may be facilitating hydrolysis of RDX and that 4-nitro-2,4-diaza-butanal (4-NDAB) may be an intermediate product of the reaction, more detailed studies are needed to determine destructive mechanisms. Nevertheless, the high destructive and mineralization rates observed combined with the ability of permanganate to remain active in the subsurface for weeks to months and allow wider injection well spacing (IT and Stoller 2000) lends supports for use of permanganate in treating RDX/ HMX plumes.

Fenton Reaction. The Fenton reaction (Fenton 1894) is recognized as one of the oldest and most powerful oxidizing reactions available. This reaction has been used to decompose a wide range of refractory synthesized or natural organic compounds (Sedlak and Andren 1991, Watts et al. 1991). The Fenton reagent is a mixture of hydrogen peroxide (H_2O_2) and ferrous iron (Fe^{2+}), which produces OH radicals (Haber and Weiss, 1934).

$$H_2O_2 + Fe^{2+} \rightarrow Fe^{3+} + \bullet OH + OH^- \qquad [2]$$

Although several propagating reactions can occur (Walling, 1975), Tomita et al. (1994) provided strong experimental evidence that the OH radical is the primary oxidizing species formed by Fe(II)-catalyzed decomposition of H_2O_2 in the absence of a iron chelator. The hydroxyl radical is second only to fluorine as an oxidizing agent and is capable of nonspecific oxidation of many organic compounds. If a sufficient

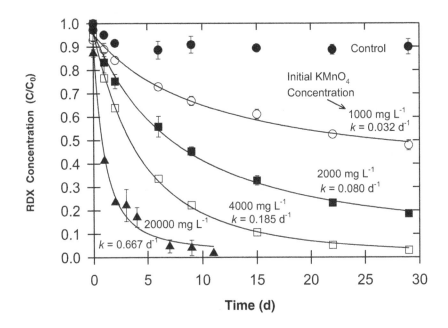

Figure 10. Loss of RDX from an aquifer slurry treated with varying $KMnO_4$ concentrations. Bars on symbols represent standard deviations of means (n=4); where absent, bars fall within symbols. Originally printed in *Journal of Environmental Quality* 33:in press.

concentrations of •OH are generated, the reaction can continue to completion, ultimately oxidizing organic compounds to CO_2, H_2O and low molecular weight mono- or di-carboxylic acids.

The Fenton reaction has been effective in treating volatile organic carbons (VOCs), light and dense non-aqueous phase liquids (LNAPL, DNAPL), petroleum hydrocarbons, PCBs, and high explosives. A significant advantage of using the Fenton reaction for treatment of RDX/ HMX is that destruction is rapid. Zoh and Stenstrom (2002) investigated Fenton treatment of both RDX and HMX and reported 90% removal of RDX from a solution within 70 min, with HMX removal one-third as rapid. Most researchers have found that reaction works best between pH 3 and 5, but destruction has been observed across a wider pH range (3-7). High subsurface pH can limit the effectiveness of the reaction, especially when free-radical scavengers are present, such as carbonate (Siegrist *et al.* 2001).

Bier *et al.* (1999) found that Fenton's reagent readily oxidized RDX under a wide range of conditions. They performed experiments with baseline RDX concentrations ranging from 4.4 to 28 mg L^{-1} and controlled reaction variables such as pH (2.0-7.5), ferrous iron concentrations (0-320 mg L^{-1}) and hydrogen peroxide (0-4%). Results showed a 100% transformation of all baseline RDX concentrations was achieved at pH 3 with hydrogen peroxide concentrations $\geq 0.5\%$ and ferrous iron ≥ 8.2 mg L^{-1}. More relevant to aquifer treatments, Bier *et al.* (1999) also showed 80% transformation of RDX at pH 7.5. Bier *et al.* (1999) found formic acid, nitrate, ammonium were formed as intermediate or final oxidation products and presented evidence that methylenedinitramine might also be a product of Fenton oxidation of RDX. A nitrogen mass balance indicated that 80% of the nitrogen from RDX was accounted for by nitrate and ammonium. Recently, Liou *et al.* (2003) investigated Fenton and photo-Fenton processes for treatment of a wide variety of explosives in wastewater. Their results showed that RDX and HMX were more difficult to destroy than TNT but oxidation rates significantly increased with increasing Fe(II) concentrations and illumination with UV.

Although the majority of research with the Fenton reaction has been directed at treating wastewaters, examples of soil treatments are available (Gauger *et al.* 1991, Li *et al.* 1997a, b, c, Pignatello and Day 1996, Ravikumar and Gurol 1992, Tyre *et al.* 1991, Watts *et al.* 1990, 1991, 1993). Many soils contain enough iron to initiate the Fenton's reactions, but those with insufficient iron require the additional step of adding a source of Fe^{2+} (Gates-Anderson *et al.* 2001). Bier *et al.* (1999) also conducted oxidation tests with soil slurries, using soils from the Nebraska Ordnance Plant that had RDX concentrations >900 mg kg^{-1}. In one set of experiments, contaminated soil was washed with water and the wash solutions treated with Fenton's reagent. RDX in the wash solution was oxidized but not as rapidly as pure aqueous solutions due to the scavenging effects of soil organic matter, carbonates, or other oxidizable materials.

Soil washing combined with Fenton oxidation was also considered at the Massachusetts Military Reservation (MMR). Under this scenario, full scale soil washing equipment, like that used by Brice Environmental (Fairbanks, AK) (Fig. 11), would be used to remove lead-based bullets (also an environmental concern at training ranges) and reduce the volume of contaminated soil by concentrating the fine soil fraction, which contains the highest percentage of contaminants. This technology was adapted from the mining industry and essentially offers a physical approach to treating contaminated soils but when in combination with a Fenton oxidation treatment offers an innovative physical-chemical approach. Under this scenario, the Fenton's reagent would be added during the washing procedure or as post treatment for the wash water. Laboratory studies conducted by the University of Nebraska showed that a 15 min treatment of a MMR soil slurry (7% by weight) with 1% H_2O_2 and 80 mg Fe^{2+} L^{-1}

Figure 11. Photograph of soil washing equipment used for physical and chemical treatment of contaminated soils. Photo courtesy of Brice Environmental Services Corporation (Fairbanks, AK).

significantly reduced RDX and HMX concentrations as well as other explosives and degradation products in the aqueous phase (Table 2) but solvent extractable soil concentrations did not meet the stringent soil remediation goals for the MMR site (RDX: 0.12 mg kg^{-1}; HMX: 0.25 mg kg^{-1}) unless other treatments were used (i.e., Fe^0).

Commercial applications of the Fenton reaction to treat soils include pilot and field-scale studies sponsored by the Gas Research Institute (Chicago, IL), where Fenton oxidation of contaminated soil slurries (primarily associated with manufactured gas plants) has been combined with biodegradation. Another unique example is the commercial remediation services provided by H&H Eco Systems (North Bonneville, WA) where their patented soil mixer (Microenfractionator®; Fig. 8) has been successfully used to spray and coat well-mixed soil with 50% H$_2$O$_2$ while self propelling itself through windrows, thus providing an ex situ treatment for unsaturated soils (Horn and Funk 1998).

A few commercial firms also specialize in using the Fenton reaction as well as other chemical oxidants to treat contaminated ground water (e.g., Geo-Cleansen International, Kenilworth, NJ; ORIN Remediation Techno-

Table 2. Aqueous solution concentrations of explosives and degradation products following Fenton oxidation of MMR soil slurry.

	Treatment							
	HMX	RDX	MNX	DNX	TNX	TNT	2ADNT	TNB
	ug L^{-1}							
Control	6.47	3.13	1.05	0.31	1.38	0.39	0.69	<0.30
Fenton	1.41	1.00	<0.30	<0.30	<0.30	0.34	<0.30	<0.30

[1]Slurry agitated for 15 min followed by 30 min settling time before decanting water. Fenton treatment was 1% H_2O_2 + 80 mg Fe^{2+} L^{-1}.

[2]Abbreviations not previously defined: 2ADNT, 2-aminodinitrotoluene, TNB, trinitrobenzene.

[3]Values <0.3 ug L^{-1} indicate concentrations below reporting limits (Cassada et al. 1999).

[4]Source: Final report prepared for AMEC Earth and Environmental, Inc. entitled: Massachusetts Military Reservation, Innovative Technology Evaluation, Physical Treatment and Chemical Oxidation/Reduction Laboratory Results, March 30, 2001, by Brice Environmental Services Corporation and University of Nebraska-Lincoln.

logies, McFarland, WI). While specific applications are site dependent, ISCO treatments using Fenton's reagent typically include H_2O_2 concentrations between 5 to 50% (v/v) and where native iron is lacking or unavailable, ferrous sulfate is commonly added in mM concentrations (Siegrist et al. 2001). In some cases, acetic or mineral acids are added to reduce the pH. Potassium phosphate (KH_2PO_4) is sometimes added to prevent premature decomposition of hydrogen peroxide in soil systems (Tarr 2003a). Delivery systems have included common ground water wells or specialized injectors with compressed air or deep soil mixing equipment (Siegrist et al. 2001). Not all sites are appropriate for ISCO treatment with the Fenton's reagent. Suitable ground water characteristics for ISCO treatment using Fenton reagent typically include: pH <7.8; alkalinity ≥400 mg L^{-1} (as $CaCO_3$), depth to ground water >5ft below grade; and hydraulic conductivity >10-6 cm sec^{-1}.

Examples of Fenton treatment of RDX/HMX plumes in the field are limited but Geo-Cleanse International conducted a test program at a former munitions production facility that had contaminated ground water (Pueblo Chemical Depot, Pueblo, CO). In this field test program, 1,100 gallons of 50% H_2O_2 was injected with catalysts into a test plot (40 x 40 x 13 ft) over two days. After 26 d, HMX was completely removed and RDX concentrations had decreased by 60%. Decreases in the nitroaromatic compounds also present decreased by 72 to 100 % (http://www.geocleanse.com).

Ozone. Ozone (O_3) was first discovered in 1840 and used as a disinfectant at the end of the 19th century (Beltrán 2003). Commonly used in treating drinking water, ozone has been more recently applied to treat organic contaminants in ground water and the vadose zone. Chemically, ozone can be represented as a hybrid of four resonance structures that present negative and positively charged oxygen atoms. This allows ozone to react through two different mechanisms, namely direct and indirect ozonation. Direct ozonation can be through electrophillic substitution while indirect results in the formation of hydroxide radicals. Ozone is also similar to permanganate in that it has a strong affinity for organic compounds containing carbon-carbon double bonds by forming unstable ozonide intermediates.

Slightly soluble in water, ozone is a very reactive reagent in both air and water. Ozone is a gas that is highly reactive and must be produced on-site. It can be vented into a soil profile for remediation purposes and has been studied as an alternative for unsaturated soils contaminated with compounds resistant to soil vapor extraction (Masten and Davies 1997, Hsu and Masten 1997, Choi et al. 2001, Kim and Choi 2002). Although the fate and reaction mechanisms of ozone in porous geologic media is not completely understood, it is probable that hydroxyl radical production will occur in the vadose zone through catalytic reactions of O_3 with iron oxides and organic material. The OH radicals produced should in turn be able to transform RDX present in the soil. While much is known regarding the destructive mechanisms of ozone on chlorinated solvents and HE in groundwater, far less is known regarding how ozone attacks and breaks down RDX in unsaturated soils (vadose zone).

Ozonation is being considered for the treatment of RDX in vadose zone at the Pantex site. In a preliminary feasibility study, we obtained vadose zone soil from the Pantex site (~20-30 ft deep) for treatment of soil columns with ozone under varing soil water contents. Soils initially had background concentrations of RDX (~1-2 mg kg^{-1}) but were augmented with ^{14}C-RDX to quantify mineralization. Ozone generated from O_2 was then passed through the soil columns (26-30 mg L^{-1}) at ~125 mL min^{-1} and subsequently through two midget bubblers containing 0.5 M NaOH to trap emitted $^{14}CO_2$. Initial experiments showed that ozonation was highly effective in mineralizing RDX with >80% of the initial ^{14}C recovered as $^{14}CO_2$; small differences were observed between columns that had different initial soil water contents (Fig. 12).

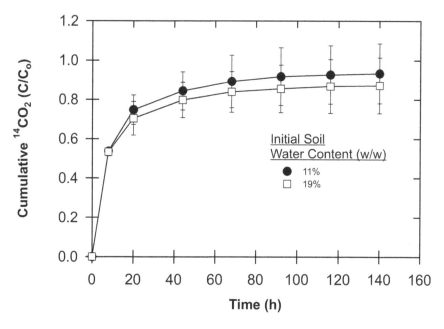

Figure 12. Cumulative $^{14}CO_2$ recovered following ozonation of columns packed with vadose zone soils from Pantex Plant (Amarillo, TX). C_o represents initial ^{14}C spiked into soil columns as RDX.

SUMMARY

Considerable progress has been made in developing and refining new and innovative abiotic approaches for remediating RDX/HMX contaminated soil and water. Although this chapter summarized some of these approaches, those reported should not be considered an inclusive list of treatments. Rather, other effective methods could have been mentioned (such as, supercritical water oxidation (Hawthorne *et al.* 2000), alkaline hydrolysis, (Bakakrishnan *et al.* 2003), solvated electron reduction, and thermal treatments (Tarr 2003b). Given the numerous treatments developed for remediating RDX/HMX contamination, it is perhaps noteworthy that when the EPA compiled all government-sponsored new and innovative field-scale technologies for treating contaminated soil, sediments and ground water (Environmental Protection Agency 2000), only 17 of the 601 reported projects dealt with explosives and more than two-thirds of those were biological approaches. Consequently, despite the considerable research put forth on developing techniques for remediating munitions contamination, this work has not yet progressed into multiple field-scale demonstrations. While a number of government-based programs are in place for implementing field-scale technologies to meet DoD's most urgent environmental needs (e.g., Environmental Security Technology Certification Program, ESTCP), more aggressive efforts will be needed in the future to field test and document the performance of these abiotic approaches.

Acknowledgement

Sincere appreciation is expressed to my former and current graduate students for generating and synthesizing much of the data presented in this chapter. A contribution of Agric. Res. Div. Project NEB-40-002.

REFERENCES

Achtnich, C., U. Sieglen, H-J. Knackmuss, and H. Lenke. 1999. Irreversible binding of biologically reduced 2,4,6-trinitrotoluene to soil. *Environ. Toxicol. Chem.* 18: 2416-2423.

Adam, M.L., S.D. Comfort, M.C. Morley, and D.D. Snow. 2004. Remediating RDX-contaminated ground water with permanganate: laboratory investigations for the Pantex perched aquifer. *J. Environ. Qual.* 33: 2165-2173.

Adam, M.L., S.D. Comfort, T.C. Zhang, and M.C. Morley. 2005. Evaluating biodegradation as a primary and secondary treatment for removing RDX (Hexahydro-1,3,5-trinitro-1,3,5-traizine) from a perched aquifer. *Bioremed. J.* 9:1-11.

Agrawal, A., and P.G. Tratnyek. 1996. Reduction of nitro aromatic compounds by zero-valent iron metal. *Environ. Sci. Technol.* 30: 153-160.

Akhavan, J. 1998. *The Chemistry of Explosives*, The Royal Society of Chemistry, Cambridge, UK.

Alowitz, M.J., and M.M. Scherer. 2002. Kinetics of nitrate, nitrite and Cr(VI) reduction by iron metal. *Environ. Sci. Technol.* 36: 299-306.

Amonette, J.E., D.J. Workman, D.W. Kennedy, J.S. Fruchter, and Y.A. Gorby. 2000. Dechlorination of carbon tetrachloride by Fe(II) associated with goethite. *Environ. Sci. Technol.* 34: 4606-4613.

Aquifer Solutions, Inc. 2002. Conceptual deployment scenarios for *in situ* remediation of the southeast perched aquifer plume Pantex Plant, Amarillo, Texas. Aquifer Solutions, Inc. Evergreen, Colorado.

ATSDR. 1996. ToxFAQs for RDX. Agency for toxic substances and disease registry. Available from http://www.atsdr.cec.gov/tfacts78html. August, 2004.

ATSDR. 1997. ToxFAQs for HDX. Agency for toxic substances and disease registry. Available from http://www.atsdr.cec.gov/tfacts98html. August, 2004.

Bachmann, W.E., and J.C. Sheehan. 1949. A new method of preparing the high explosive RDX. *J. Amer. Chem. Soc.* 71: 1842-1845.

Balakrishnan, V.K., A. Halasz, and J. Hawari. 2003. Alkaline hydrolysis of the cyclic nitramine explosives RDX, HMX, and CL-20: New insights into degradation pathways obtained by the observation of novel intermediates. *Environ. Sci. Technol.* 37: 1838-1843.

Balakrishnan, V.K., F. Monteil-Rivera, M.A. Gautier, and J. Hawari. 2004. Sorption and stability of the polycyclic nitramine explosive CL-20 in soil. *J. Environ. Qual.* 33: 1362-1368.

Banerjee, S., S.H. Yalkowsky, and S.C. Valvani. 1980. Water solubility and octanol/water partition coefficients of organics: Limitations of the solubility - partition coefficient correlation. *Environ. Sci. Technol.* 14: 1277-1279.

Bartha, R., and T. Hsu. 1974. Interaction of pesticide-derived chloroaniline residues with soil organic matter. *Soil Sci.* 115: 444-453.

Beltrán, F.J. 2003. Ozone-UV radiation-hydrogen peroxide oxidation technologies. Pages 1-75 in *Chemical degradation methods for wastes and pollutants. Environmental and Industrial Applications,* M.A. Tarr, ed., Marcel Dekker, New York.

Bentley, R.E., J.W. Dean, S.J. Ells, T.A. Hollister, G.A. LeBlanc, S. Sauter, and B.H. Sleight III. 1977a. Laboratory evaluation of the toxicity of RDX to aquatic organisms. Final Report, U.S. Arm Medical Research and Development Command, AD-A061730.

Bentley, R.E., G.A. LeBlanc, T.A. Hollister, and B.H. Sleight III. 1977b. Acute toxicity of 1,3,5,7-tetranitrooctahydro-1,3,5,7-tetrazocine (HMX) to aquatic organisms. EG&G Bionomics, Wareham, Massachusetts, AD-A054981.

Bier, E.L., J. Singh, Z. Li, S.D. Comfort, and P. Shea. 1999. Remediating hexahydro-1,3,5-1,3,5-triazine-contaminated water and soil by Fenton oxidation. *Environ. Toxicol. Chem.* 18: 1078-1084.

Bollag, J.-M. 1992. Decontaminating soil with enzymes. An *in situ* method using phenolic and anilinic compounds. *Environ. Sci. Technol.* 26: 1876-1881.

Bollag, J.-M., R.D. Minard, and S.-Y. Liu. 1983. Cross-linkage between anilines and phenolic humus constituents. *Environ. Sci. Technol.* 17: 72-80.

Bonin, P.M.L., D Bejan, D. Bejan, L. Schutt, J. Hawari, and N.J. Bunce. 2004. Electrochemical reduction of hexahydro-1,3,5-trinitro-1,3,5-triazine in aqueous solutions. *Environ. Sci. Technol.* 38: 1595-1599.

Bose, P., W.H. Glaze, and D.S. Maddox. 1998a. Degradation of RDX by various advanced oxidation processes: I. Reaction rates. *Water Res.* 32: 997-1004.

Bose, P., W.H. Glaze, and D.S. Maddox. 1998b. Degradation of RDX by various advanced oxidation processes: II. Organic by-products. *Water Res.* 32: 1005-1018.

Bradley, P.M., and F.H. Chapelle. 1995. Factors affecting microbial 2,4,6-trinitrotoluene mineralization in contaminated soil. *Environ. Sci. Technol.* 29: 802-806.

Brannon, J.M., D.D. Adrian, J.C. Pennington, T.E. Myers, and C.A. Hayes. 1992. Slow release of PCB, TNT, and RDX from soils and sediments. Technical Report EL-92-38. U.S. Army Engineer Waterways Experiment Station, Vicksburg, Mississippi.

Brannon, J.M., P.N. Deliman, J.A. Gerald, C.E. Ruiz, C.B. Price, C. Hayes, S. Yost, and M. Qasim. 1999. Conceptual model and process descriptor formulations for fate and transport of UXO. Technical Report IRRP-99-1. U.S. Army Engineer Waterways Experiment Station, Vicksburg, Mississippi.

Brannon, J.M., and J.C. Pennington. 2002. Environmental fate and transport process descriptors for explosives. ERDC/EL TR-02-10, U.S. Army Engineer Research and Development Center, Vicksburg, Mississippi.

Bruns-Nagel, D. J. Breitung, E. von Low, K. Steinbach, T. Gorontzy, M. Kahl, K.-H. Blotevogel, and D. Gemas. 1996. Microbial transformation of 2,4,6-trinitrotoluene in aerobic soil columns. *Appl. Environ. Microbiol.* 62: 2651-2656.

Butler, L.C., D.C. Staiff, G.W. Sovocool, and J.E. Davis. 1981. Field disposal of methyl parathion using acidified powdered zinc. *J. Environ. Sci. Health* B16: 49-58.

Cassada, D.A., S.J. Monson, D.D. Snow, and R.F. Spalding. 1999. Sensitive determination of RDX, nitroso-RDX metabolites and other munitions in ground water by solid-phase extraction and isotope dilution liquid chromatography-atmospheric pressure chemical ionization mass spectrometry. *J. Chromatogr.* A 844: 87-96.

Chilakapati, A., M. Williams, S. Yabusaki, C. Cole, and J. Szecsody. 2000. Optimal design of an *in situ* Fe(II) barrier: transport limited reoxidation. *Environ. Sci. Technol.* 34: 5215-5221.

Choi, H., H.N. Lim, J.Y. Kim, and J. Cho. 2001. Oxidation of polycyclic aromatic hydrocarbons by ozone in the presence of sand. *Water Sci. Technol.* 43: 349-356.

Coleman, N.V., D.R. Nelson, and T. Duxbury. 1998. Aerobic biodegradation of hexahydro-1,3,5-trinitro-1,3,5-triazine (RDX) as a nitrogen source by a *Rhodococcus* sp., Strain DN22. *Soil Biol. Biochem.* 30: 1159-1167.

Comfort, S.D., P.J. Shea, L.S. Hundal, Z. Li, B.L. Woodbury, J.L. Martin, and W.L. Powers. 1995. TNT transport and fate in contaminated soil. *J. Environ. Qual.* 24: 1174-1182.

Comfort, S.D., P.J. Shea, T.A. Machacek, H. Gaber, and B.-T. Oh. 2001. Field-scale remediation of a metolachlor-contaminated spill site using zerovalent iron. *J. Environ. Qual.* 30: 1636-1643.

Comfort, S.D., P.J. Shea, T.A. Machacek, and T. Satapanajaru. 2003. Pilot-scale treatment of RDX-contaminated soil with zerovalent iron. *J. Environ. Qual.* 32: 1636-1643.

Crockett, A.B., H.D. Craig, T.F. Jenkins, and W.E. Sisk. 1996. Field sampling and selecting on-site analytical methods for explosives in soil. EPA/540/S-97/501. National Exposure Research Laboratory, Las Vegas, NV.

Cuttshall, E.R., G. Felling, S.D. Scott, and G.S. Tottle. 1993. Method and apparatus for treating PCB-containing soil. U.S. Patent no. 5 197 823. Date Issued: 30 March.

Defense Science Board. 1998. Defense Science Board Task Force on Unexploded Ordnance Clearance, Active Range UXO Clearance, and Explosive Ordnance Disposal Programs. 1998. Report of the Defense Science Board Task Force on Unexploded Ordnance (UXO) Clearance, Active Range UXO Clearance, and Explosive Ordnance Disposal (EOD) Programs; Office of the Under Secretary of Defense for Acquisition and Technology, U.S. Government Printing Office: Washington, DC.

Dickel, O., W. Haug, and H.-J. Knackmuss. 1993. Biodegradation of a nitrobenzene by a sequential anaerobic-aerobic process. *Biodegradation* 4: 187-194.

Doppalapudi, R.B., G.A. Sorial, and S.W. Maloney. 2002. Electrochemical reduction of simulated munitions wastewater in a bench-scale batch reactor. *Environ. Eng. Sci.* 19: 115-130.

Eary, L.E., and D. Rai. 1988. Chromate removal from aqueous wastes by reduction with ferrous ion. *Environ. Sci. Technol.* 22: 972-977.

Elmore, A.C., and T. Graff. 2001. Groundwater circulation wells using innovative treatment systems. Proceedings of the 2001 International Containment & Remediation Technology Conference. Orlando, Florida. Available from http://www.containment.fsu.edu/ce/content/index.htm, July, 2004.

EnviroMetal Technologies, Inc. 2004. Available from http://eti.ca, June, 2004.

Environmental Protection Agency. 1998. Field applications of in situ remediation technologies: chemical oxidation. EPA 5421-R-98-008. United States Environmental Protection Agency, Office of Solid Waste and Emergency Response, Technology Innovation Office. Washington, DC.

Environmental Protection Agency. 2000. Innovative remediation technologies: field-scale demonstration projects in North America, 2nd Edition. Year 2000 Report. EPA 542-B-00-004. United States Environmental Protection Agency, Office of Solid Waste and Emergency Response, Technology Innovation Office. Washington, DC.

Etnier, E.L. 1989. Water quality criteria for hexahydro-1,3,5-trinitro-1,3,5-triazine (RDX). *Regulatory Toxicol. Pharmacol.* 9: 147-157.

Etnier, E., and W.R. Hartley. 1990. Comparison of water quality criterion and lifetime health advisory for hexahydro-1,3,5-trinitro-1,3,5-triazine (RDX). *Regulatory Toxicol. Pharmacol.* 11: 118-122.

Fedoroff, B.T., and O.E. Sheffield. 1966. Encyclopedia of explosives and related items. Volume 3. Picatinny Arsenal, Dover, NJ.

Fenton, H.J.H. 1894. Oxidation of tartaric acid in presence of iron. *J. Chem. Soc.* 65: 899-910.

Fleming, E.C., M.E. Zappi, J. Miller, R. Hernandez, and E. Toro. 1997. Evaluation of peroxone oxidation techniques for removal of explosives from Cornhusker Army Ammunition plant waters. Technical Report SERDP-97-2. U.S. Army Engineer Waterways Experiment Station, Vicksburg, Mississippi.

Funk, S., D. Roberts, D. Crawford, and R. Crawford. 1993. Initial-phase optimization for bioremediation of munitions compound contaminated soils. *Appl. Environ. Microbiol.* 59: 2171-2177.

Gaber, H.M., S.D. Comfort, P.J. Shea, and T.A. Machacek. 2002. Metolachlor dechlorination by zerovalent iron during unsaturated transport. *J. Environ. Qual.* 31: 962-969.

Gates-Anderson, D.D., R.L. Siegrist, and S.R. Cline. 2001. Comparison of potassium permanganate and hydrogen peroxide as chemical oxidants for organically contaminated soils. *J. Environ. Eng.* 127: 337-347.

Gauger, W.K., V.J. Srivastava, T.D. Hayes, and D.G. Linz. 1991. Enhanced biodegradation of polycyclic aromatic hydrocarbons in manufactured gas plant wastes. Pages 75-92 in *Gas, Oil, Coal, and Environmental Biotechnology III*, C. Akin and J. Smith, eds., Institute of Gas Technology, Chicago, Illinois.

George, S.E., G. Huggins-Clark, and L.R. Brooks. 2001. Use of Salmonella microsuspensions bioassay to detect the mutagenicity of munitions compounds at low concentrations. *Mutations Res.* 490: 45-56.

Gibb, C., T. Satapanajaru, S.D. Comfort, and P.J. Shea. 2004. Remediating dicamba-contaminated water with zerovalent iron. *Chemosphere* 54: 841-848.

Gilcrease, P.C., and V.G. Murphy. 1995. Bioconversion of 2,4-diamino-6-nitrotoluene to a novel metabolite under anoxic and aerobic conditions. *Appl. Environ. Microbiol.* 61: 4209-4214.

Glover, D.J., and J.C. Hoffsommer. 1973. Thin-layer chromatographic analysis of HMX in water. *Bull. Environ. Contam. Toxicol.* 10: 302-304.

Gong, P., J. Hawari, S. Thiboutot, G. Ampleman, G.I. Sunahara. 2001a. Ecotoxicological effects of hexahydro-1,3,5-trinitro-1,3,5-triazine on soil microbial activities. *Environ. Toxicol. Chem.* 20: 947-951.

Gong, P., J. Hawari, S. Thiboutot, G. Ampleman, G.I. Sunahara. 2001b. Toxicity of octahydro-1,3,5,7-tetranitro-1,3,5,7-tetrazocine (HMX) to soil microbes. *Bull. Environ. Contam. Toxicol.* 69: 97-103.

Gorontzy, T., O. Drzyzga, M.W. Kahl., D. Bruns-Nagel, J. Breitung, E.V. Loew, and K.H. Blotevogel. 1994. Microbial degradation of explosives and related compounds. *Crit. Rev. Microbiol.* 20: 265-284.

Gregory, K.B., P. Larese-Cassanova, G.F. Parkin, and M.M. Scherer. 2004. Abiotic transformation of hexahydro-1,3,5-trinitro-1,3,5-triazine by Fe^{II} bound to magnetite. *Environ. Sci. Technol.* 38: 1408-1414.

Haber, F., and J. Weiss. 1934. The catalytic decomposition of hydrogen peroxide by iron salts. *Proc. Roy. Soc. Lond. Series A.* 147: 332-351.

Hawari, J. 2000. Biodegradation of RDX and HMX: From basic research to field application. Pages 277-310 in *Biodegradation of Nitroaromatic Compounds and Explosives,* J.C. Spain, J.B. Hughes, and H. Knackmuss, eds., Lewis Publ., Boca Raton, Florida.

Hawari, J., S. Baudet, A. Halasz, S. Thiboutot, and G. Ampleman. 2000. Microbial degradation of explosives: biotransformation versus mineralization. *Appl. Microbiol. Biotechnol.* 54: 605-618.

Hawthorne, S.B., A.J.M. Lagadec, D. Kalderis, A.V. Lilke, and D.J. Miller. 2000. Pilot-scale destruction of TNT, RDX, and HMX on contaminated soils using subcritical water. *Environ. Sci. Technol.* 34: 3224-3228.

Heijman, C.G., C. Holliger, M.A. Glaus, R.P. Schwarzenbach, and J. Zeyer. 1993. Abiotic reduction of 4-chloronitrobenzene to 4-chloroaniline in a dissimilatory iron-reducing enrichment culture. *J. Appl. Environ. Microbiol.* 59: 4350-4353.

Heijman, C.G., E. Grieder, C. Holliger, and R.P. Schwarzenbach. 1995. Reduction of ntiroaromatic compounds coupled to microbial iron reduction in laboratory aquifer columns. *Environ. Sci. Technol.* 29: 775-783.

Ho, P.C., 1986. Photoxidation of 2,4-dinitrotoluene in aqueous solution in the presence of hydrogen peroxide. *Environ. Sci. Technol.* 20: 260-267.

Horn, T., and S. Funk. 1998. Seeking and finding the quickest solution. *Soil Groundwater Cleanup.* November, 1998: 6-9.

Hundal, L.S., P.J. Shea, S.D. Comfort, W.L. Powers, and J. Singh. 1997a. Long-term TNT sorption and bound residue formation in soil. *J. Environ. Qual.* 26: 869-904.

Hundal, L., J. Singh, E.L. Bier, P.J. Shea, S.D. Comfort, and W.L. Powers. 1997b. Removal of TNT and RDX from water and soil using iron metal. *Environ. Poll.* 97: 55-64.

Hsu, I., and S.J. Masten. 1997. The kinetics of the reaction of ozone with phenanthrene in unsaturated soils. *Env. Eng. Sci.* 14: 207-217.

Hsu, T.S., and H. Bartha. 1976. Hydrolyzable and nonhydrolyzable 3,4-dichloroaniline-humus complexes and their respective rates of degradation. *J. Agric. Food. Chem.* 24: 118-122.

Isbister, J.D., R.C. Doyle, and J.F. Kitchens. 1980. Engineering and development support of general decontamination technology for the DARCOM installation restoration program. Task 6-adapted/mutant biological treatment. Phase 1-literature review. U.S. Defense Technical Information Center, Alexandria, Virginia, Rep. No. DRXTHIS-CR-80132.

Island Pyrochemical Industries (IPI). 2004. Available from http://www.islandgroup.com/ExplosiveChemistry.html, August, 2004.

IT Corporation, and S.M. Stoller Corporation. 2000. Implementation report of remediation technology screening and treatability testing of possible remediation technologies for the Pantex perched aquifer. Pantex Environmental Restoration Department. U.S. Department of Energy Pantex Plant, Amarillo, Texas.

Jenkins, T.F., J.C. Pennington, T.A. Ranney, T.E. Berry, Jr., P.H. Miyares, M.E. Walsh, A.D. Hewitt, N.M.Perron, L.V. Parker, C.A. Hayes, and E.G. Wahlgren. 2001. Characterization of explosive contamination at military firing ranges. U.S. Army Corps of Eng. ERDC TR-01-5.

Joo, S.H., A.J. Feitz, and T.D. Waite. (2004) Oxidative degradation of the carbothioate herbicide, molinate, using nanoscale zero-valent iron. *Environ. Sci. Technol.* 38: 2242-2247.

Kaplan, A.S., C.F. Berghout, and L.A. Peczenik. 1965. Human intoxication from RDX. *Arch. Environ. Health* 10: 877-883.

Kim, J., and H. Choi. 2002. Modeling in situ ozonation for the remediation of nonvolatile PAH-contaminated unsaturated soils. *J. Contam. Hydrol.* 55: 261-285.

Klausen, J., S.P. Tröber, S.B. Haderlein, and R.P. Schwarzenbach. 1995. Reduction of substituted nitrobenzenes by Fe(II) in aqueous mineral suspensions. *Environ. Sci. Technol.* 29: 2396-2404.

Kreslavski, V.D., G.K. Vasilyeva, S.D. Comfort, R.A. Drijber, and P.J. Shea. 1999. Accelerated transformation and binding of 2,4,6-trinitrotoluene in rhizosphere soil. *Bioremediation* 3: 59-67.

Lechner, C. 1993. Incineration of soils and sludges. Approaches for the remediation of federal facility sites contaminated with explosives or radiation wastes. EPA/625/R-93/013. pp. 30-33.

Li, Z.M., S.D. Comfort, and P.J. Shea. 1997a. Destruction of 2,4,6-trinitrotoluene (TNT) by Fenton oxidation. *J. Environ. Qual.* 26: 480-487.

Li, Z.M., M.M. Peterson, S.D. Comfort, G.L. Horst, P.J. Shea, and B.T. Oh. 1997b. Remediating TNT-contaminated soil by soil washing and Fenton oxidation. *Sci. Tot. Environ.* 204: 107-115.

Li, Z.M. P.J. Shea, and S.D. Comfort. 1997c. Fenton oxidation of 2,4,6-trinitrotoluene in contaminated soil slurries. *Environ. Eng. Sci.* 14: 55-66.

Liou, M.-J., M.-C. Lu, and J.-N. Chen. 2003. Oxidation of explosives by Fenton and photo-Fenton processes. *Water Res.* 37: 3172-3179.

Lotufo, G.R., J.D. Farrar, L.S. Inouye, T.S. Bridges, and D.B. Ringelberg. 2001. Toxicity of sediment-associated nitroaromatic and cyclonitramine compounds to bethnic invertebrates. *Environ. Toxciol. Chem.* 20: 1762-1771.

MacDonald, J.A. 2001. Cleaning up unexploded ordnance. *Environ. Sci. Technol.* 35: 373A-376A.

Marvin-Sikkema, F.D., and J.A.M. de Bont. 1994. Degradation of nitroaromatic compounds by microorganisms. *Appl. Microbiol. Biotechnol.* 42: 499-507.

Masten, S.J., and S.H.R. Davies. 1997. Efficacy of *in-situ* ozonation for the remediation of PAH-contaminated soils. *J. Contam. Hydrol.* 28: 327-335.

Matheson, L.J., and P.G. Tratnyek. 1994. Reductive dehalogenation of chlorinated methanes by iron metal. *Environ. Sci. Technol.* 28: 2045-2053.

Meenakshisundaram, D.M. Mehta, S. Pehkonen, and S.W. Maloney. 1999. Electrochemical reduction of nitro-aromatic compounds. Report TR 99/85 U.S. Army Construction Engineering Research Laboratory, Champaign, Illinois.

McCormick, N.G., F.E. Feeherry, and H.S. Levinson. 1976. Microbial transformation of 2,4,6-trinitrotoluene and other nitroaromatic compounds. *Appl. Environ. Microbiol.* 31: 949-958.

McCormick, N.G., J.H. Cornell, and A.M. Kaplan. 1981. Biodegradation of hexahydro-1,3,5-trinitro-1,3,5-triazine. *Appl. Environ. Microbiol.* 42: 817-823.

McGrath, C.J. 1995. Review of formulations for processes affecting the subsurface transport of explosives. Technical Report IRRP-95-2. U.S. Army Engineer Waterways Experiment Station, Vicksburg, Mississippi.

McLellan W.L., W.R. Hartly, and M.E. Brower. 1988a. Health advisory for hexahydro-1,3,5-tetranitro-1,3,5-triazine (RDX). Technical Report No. PB90-273533. US Army Medical Research and Development Command, Fort Detrick, Maryland.

McLellan W.L., W.R. Hartly, and M.E. Brower. 1988b. Health advisory for octahydro-1,3,5,7-tetranitro-1,3,5,7-tetrazocine (HMX). Technical Report No. PB90-273525. US Army Medical Research and Development Command, Fort Detrick, Maryland.

Miehr, R., J.Z. Bandstra, R. Po, and P.G. Tratnyek. 2003. Remediation of 2,4,6-trinitrotoluene (TNT) by iron metal: kinetic controls on product distributions in batch and column experiments. 225th National Meeting, New Orleans, Louisiana, American Chemical Society. 43: 1.

Mirvish, S.S., P. Issenberg, and H.C. Sornson. 1976. Air-water and ether-water distribution of N-nitroso compounds: Implications for laboratory safety, analytical methodology, and carcinogenicity for the rat esophagus, nose, and liver. *J. Natl. Cancer Inst.* 56: 1125-1129.

Monteil-Rivera, F., C. Groom, and J. Hawari. 2003. Sorption and degradation of octahydro-1,3,5,7-tetranitro-1,3,5,7-tetrazocine in soil. *Environ. Sci. Technol.* 37: 3878-3884.

Myers, T.E. , J.M. Brannon, J.C. Pennington, D.M. Townsend, W.M. Davis, M.K. Ochman, C.A. Hayes, and K.F. Myers. 1998. Laboratory studies of soil sorption/transformation of TNT, RDX, and HMX. Technical Report IRRP-98-8. U.S. Army Engineer Waterways Experiment Station, Vicksburg, Mississippi.

Oberle, D.W., and D.L. Schroder. 2000. Design considerations for *in situ* chemical oxidation. Pages 91-99 in *Chemical Oxidation and Reactive Barriers: Remediation of Chlorinated and Recalcitrant Compounds*, G.B.

Wickramanayake, A.R. Gavaskar, and A.S.C. Chen, eds., Battelle Press, Columbus, Ohio.

Oh, B.-T., C.L. Just, and P.J.J. Alvarez. 2001. Hexahydro-1,3,5-trinitro-1,3,5-triazine mineralization by zerovalent iron and mixed anaerobic cultures. *Environ. Sci. Technol.* 35: 4341-4346.

Oh, S.-Y., D.K. Cha, B.-J. Kim, and P.C. Chiu. 2002. Effect of adsorption to elemental iron on the transformation of 2,4,6-trinitrotoluene and hexahydro-1,3,5-trinitro-1,3,5-triaizine in solution. *Environ. Toxicol. Chem.* 21: 1384-1389.

Oh, B.-T., and P.J.J. Alvarez. 2002. Hexahydro-1,3,5-trinitro-1,3,5-triazine (RDX) degradation in biologically-active iron columns. *Water Air Soil Pollution* 141: 325-335.

Park, J., S.D. Comfort, P.J. Shea, and T.A. Machacek. 2004. Remediating muntions-contaminated soil with zerovalent iron and cationic surfactants. *J. Environ. Qual.* 33: 1305-1313.

Bonin, P.M.L., D. Bejan, L. Schutt, J. Hawari, and N.J. Bunce. 2004. Electrochemical reduction of hexahydro-1,3,5-trinitro-1,3,5-triazine in aqueous solutions. *Environ. Sci. Technol.* 38: 1595-1599.

Pasti-Grigsby, M.B., T.A. Lewis, D.L. Crawford, and R.L. Crawford. 1996. Transformation of 2,4,6-trinitrotoluene (TNT) by actinomycetes isolated from TNT-contaminated and uncontaminated environments. *Appl. Environ. Microbiol.* 62: 1120-1123.

Pignatello, J.J., and M. Day. 1996. Mineralization of methyl parathion insecticide in soil by hydrogen peroxide activated with iron(III)-NTA or -HEIDA complexes. *Hazard. Waste Hazard. Mater.* 13:137-143.

Peterson, M.M., G.L. Horst, P.J. Shea, S.D. Comfort, and R.K.D. Peterson. 1996. TNT and 4-amino-2,6-dinitrotoluene influence on germination and early seeding development of tall fescue. *Environ. Pollution* 93: 57-62.

Peterson, M.M., G.L. Horst, P.J. Shea, and S.D. Comfort. 1998. Germination and seeding development of switchgrass and smooth bromegrass exposed to 2,4,6-trinitrotoluene. *Environ. Pollution* 99: 53-59.

Price, C.B., J.M. Brannon, S.L. Yost, and C.A. Hayes. 2001. Relationship between redox potential and pH on RDX transformation in soil-water slurries. *J. Environ. Eng.* 127: 26-31.

Ravikumar, J.X., and M.D. Gurol. 1992. Fenton reagent as a chemical oxidant for soil contaminants. Pages 206-229 in, *Chemical Oxidation Technologies for the Nineties*, W.W. Eckenfelder, A.R. Bowers, and J.A. Roth eds., Technomic Pub. Co., Lancaster, Pennsylvania.

Reynolds, G.W., J.T. Hoff, and R.W. Gillham. 1990. Sampling bias caused by materials used to monitor halocarbons in groundwater. *Environ. Sci. Technol.* 24: 135-142.

Rodgers, J.D., and N.J. Bunce. 2001a. Electrochemical treatment of 2,4,6-trinitrotoluene and related compounds. *Environ. Sci. Technol.* 35: 406-410.

Rodgers, J.D., and N.J. Bunce. 2001b. Treatment methods for the remediation of nitroaromatic explosives. *Water Res.* 35: 2101-2111.

Robidoux, P.Y., J. Hawari, G. Bardai, L. Paquet, G. Ampleman, S. Thiboutot, and G.I. Sunahara. 2002. TNT, RDX, and HMX decrease earthworm (*Eisenia andrei*) life-cycle responses in a spiked natural forest soil. *Arch. Environ. Contam. Toxicol.* 43: 379-388.

Rosser, S.J., A. Basran, E.R. Travis, C.E. French, and N.C. Bruce. 2001. Microbial transformation of explosives. *Adv. Appl. Microbiol.* 49: 1-35.

Rueda, E.H., M.C. Ballesteros, R.L. Grassi, and M.A. Blesa. 1992. Dithionite as a dissolving reagent for goethite in the presence of EDTA and citrate. Application to soil analysis. *Clays and Clay Minerals.* 40: 575-585.

Satapanajaru, T., S.D. Comfort, and P.J. Shea. 2003a. Enhancing metolachlor destruction rates with aluminum and iron salts during zerovalent iron treatment. *J. Environ. Qual.* 32: 1726-1734.

Satapanajaru, T., P.J. Shea, S.D. Comfort, and Y. Roh. 2003b. Green rust and iron oxide formation influences metolachlor dechlorination during zerovalent iron treatment. *Environ. Sci. Technol.* 37: 5219-5227.

Schackmann, A., and R. Müller. 1991. Reduction of nitroaromatic compounds by different *Pseudomonas* sp. under aerobic conditions. *Appl. Microbiol. Biotechnol.* 34: 809-813.

Scherer, M.M., B.A. Balko, and P.G. Tratnyek. 1999. The role of oxides in reduction reactions at the metal-water interface. Pages 301-322 in *Mineral-Water Interfacial Reactions: Kinetics and Mechanisms*, D.L. Sparks, and T.J. Grundl, eds., American Chemical Society, Washington, DC.

Schrader, P.S., and T.F. Hess. 2004. Coupled abiotic-biotic mineralization of 2,4,6-trinitrotoluene (TNT) in soil slurry. *J. Environ. Qual.* 33: 1202-1209.

Sedlak, D.L., and A.W. Andren. 1991. Oxidation of chlorobenzene with Fenton reagent. *Environ. Sci. Technol.* 25: 777-782.

Senzaki, T., and Y. Kumagai. 1988. Removal of chlorinated organic compounds from wastewater by reduction process: Treatment of 1,1,2,2-tetrachloroethane with iron powder. *Kogyo Yosui.* 357: 2-7.

Senzaki, T., and Y. Kumagai. 1989. Removal of chlorinated organic compounds from wastewater by reduction process: II. Treatment of trichloroethylene iron powder. *Kogyo Yosui.* 369: 19-25.

Shen, C.F., J. Hawari, G. Ampleman, S. Thiboutot, and S.R. Guiot. 2000. Enhanced biodegradation and fate of hexahydro-1,3,5-trinitro-1,3,5-triazine (RDX) and octahydro-1,3,5,7-tetranitro-1,3,5,7-tetrazocine (HMX) in anaerobic soil bioslurry process. *Biorem. J.* 41: 27-39.

Sheremata, T.W., An. Halasz, L. Paquet, S. Thiboutot, G. Ampleman, and J. Hawari. 2001. The fate of the cyclic nitramine explosive RDX in natural soil. *Environ. Sci. Technol.* 35: 1037-1040.

Siegel, L. 1998. A stakeholder's guide to the cleanup of federal facilities. Center for Public Environmental Oversight. San Francisco, CA.

Siegrist, R.L., M.A. Urynowicz, O.R. West, M.L. Crimi, and K.S. Lowe. 2001. *Principles and Practices of in Situ Chemical Oxidation Using Permanganate.* Battelle Press, Columbus, Ohio.

Sikka, H.C., S. Banerjee, E.J. Pack, and H.T. Appelton. 1980. Environmental fate of RDX and TNT. Technical Report 81538. U.S. Army Medical Research and Development Command, Fort Detrick, Maryland.

Singh, J., S.D. Comfort, L.S. Hundal, and P.J. Shea. 1998a. Long-term RDX sorption and fate in soil. *J. Environ. Qual.* 27: 572-577.

Singh, J., S.D. Comfort, and P.J. Shea. 1998b. Remediating RDX-contaminated water and soil using zero-valent iron. *J. Environ. Qual.* 27: 1240-1245.

Singh, J. P.J. Shea. L.S. Hundal, S.D. Comfort, T.C. Zhang, and D.S. Hage. 1998c. Iron-enhanced remediation of water and soil containing atrazine. *Weed Sci.* 46: 381-388.

Singh, J., S.D. Comfort, and P.J. Shea. 1999. Optimizing Eh/pH for iron-mediated remediation of RDX-contaminated water and soil. *Environ. Sci. Technol.* 33: 1488-1494.

Sisk, W. 1993. Granuar activated carbon. Approaches for the remediation of federal facility sites contaminated with explosives or radiation wastes. EPA/625/R-93/013. pp. 38-39.

Smith-Simon, C., and S. Goldhaber. 1995. Toxicological profile for RDX. Agency for Toxic Substances and Disease Registry. Atlanta, Georgia.

Spain, J.C. 1995. *Biodegradation of Nitroaromatic Compounds.* Plenum Press, New York.

Spain, J.B., J.B. Hughes, and H.J. Knackmus. 2000. *Biodegradation of Nitroaromatic Compounds and Explosives.* Lewis Publ., Boca Raton, Florida.

Spalding, R.F., and J.W. Fulton. 1988. Groundwater munitions residues and nitrate near Grand Island, Nebraska, U.S.A., *J. Contam. Hydrol.* 2: 139-153.

Spanggord, R.J., R.W. Mabey, T.W. Chou, D.L. Haynes, P.L. Alfernese, D.S. Tse, and T. Mill. 1982. Environmental fate studies of HMX, screening studies, final report, Phase I - Laboratory study. SRI Project LSU-4412, SRI International, Menlo Park, California. U.S. Army Medical and Research and Development Command, Fort Detrick, Maryland.

Spanggord, R.J., R.W. Mabey, T. Mill, T.W. Chou, J.H. Smith, S. Lee, and D. Roberts. 1983. Environmental fate studies on certain munitions wastewater constituents: Phase IV - Lagoon model studies. AD-A082372. SRI International, Menlo Park, CA. U.S. Medical Research and Development Command, Fort Detrick, Maryland.

Spanggord, R.J., T. Mill, T.W. Chou, R.W. Mabey, W.H. Smith, and S. Lee. 1980. Environmental fate studies on certain munition watewater constituents, final report part II - laboratory study. AD A099256. SRI International, Menlo Park, CA. U.S. Army Medical Research and Development Command, Fort Detrick, Maryland.

Staiff, D.C., L.C. Butler, and J.E. Davis. 1977. Field disposal of DDT: Effectiveness of acidified powdered zinc on reduction of DDT in soil. *J. Environ. Sci. Health* B12: 1-13.

Steevens, J.A., B.M. Duke, G.R. Lotufo, and T.S. Bridges. 2002. Toxicity of the explosives 2,4,6-trinitrotoluene, hexahydro-1,3,5-trinitro-1,3,5-triazine, and octahydro-1,3,5,7-tetrazocine in sediments to *Chironomus tentans*, and *Hyalella azteca*: low-dose hormesis and high-dose mortality. *Environ. Toxicol. Chem.* 21: 1475-1482.

Sunahara, G.I., S. Dodard, M. Sarrazin, L. Paquet, G. Ampleman, S. Thiboutot, J. Hawari, and A.Y. Renoux. 1998. Development of a soil extraction procedure for ecotoxicity characterization of energetic compounds. *Ecotoxicol. Environ. Saf.* 39: 185-194.

Sweeny, K.H. 1979. Reductive degradation treatment of industrial and municipal wastewaters. *Am. Water Works Assoc. Res. Found.* 2: 1487-1497.

Sweeny, K.H. 1981. The reductive treatment of industrial wastewaters. *Am. Inst. Chem. Eng. Sym. Ser.* 77: 72-78.

Szecsody, J.E., D.C. Girvin, B.J. Devary, and J.A. Campbell. 2004. Sorption and oxic degradation of the explosive CL-20 during transport in subsurface sediments. *Chemosphere* 56: 593-610.

Szecsody, J., J. Fruchter, M.A. McKinley, C.T. Reach, and T.J. Gillmore. 2001. Feasibility of *in-situ* redox manipulation of subsurface sediments for RDX remediation at Pantex. Pacific Northwest National Laboratory. PNNL-13746. Richland, Washington.

Szecsody, J., J. Fruchter, M.D. Williams, V.R. Vermeul, and D. Sklarew. 2004. In situ reduction of aquifer sediments: Enhancement of reactive iron phases and TCE dechlorination. *Environ. Sci. Technol.* 38: 4656-4663.

Talmage, S.S., D.M. Opresko, C.J. Maxwell, C.J. Welsh, F.M. Cretella, P.H. Reno, and F.B. Daniel. 1999. Nitroaromatic munitions compounds: environmental effects and screening values. *Rev. Environ. Contam. Toxicol.* 161: 1-156.

Tarr, M.A., 2003a. Fenton and modified Fenton methods for pollutant degradation. Pages 165-200 in *Chemical Degradation Methods for Wastes and Pollutants. Environmental and Industrial Applications*, M.A. Tarr, ed., Marcel Dekker, Inc. New York.

Tarr, M.A., 2003b. Chemical degradation methods for wastes and pollutants. Environmental and Industrial Applications. Marcel Dekker, Inc., New York.

Thiboutot, S., G. Ampleman, A. Gagnon, A. Marois, T.F. Jenkins, M.E. Walsh, P.G. Thorne, and T.A. Ranney. 1998. Characterization of antitank firing ranges at CFB Valcartier, WATC Wainwright and CFAD Dundrun. Defense Research Establishment Valcartier, Quebec, Report #DREV-R-9809.

Tomita, M., T. Okuyama, S. Watanabe, and H. Watanabe. 1994. Quantitation of the hydroxyl radical adducts of salicylic acid by micellar electrokinetic capillary chromatography: oxidizing species formed by Fenton reaction.

Arch. Toxicol. 68: 428-433.

Tucker, W.A., I.E.V. Dose, and G.J. Gensheimer. 1985. Evaluation of critical parameters affecting contaminant migration through soils, final report. Report No. AMXTH-TE-85030. U.S. Army Toxic and Hazardous Materials Agency, Aberdeen Proving Ground, Maryland.

Turner, W.B. 1971. *Fungal Metabolites.* Academic Press, London.

Tratnyek, P.G., T.L. Johnson, and A. Schattauer. 1995. Interfacial phenomena affecting contaminant remediation with zero-valent iron metal. Emerging Technologies in Hazardous Waste Management VII. Atlanta, GA. American Chemical Society. pp 589-592.

Tratnyek, P.G., M.M. Scherer, T.J. Johnson, and L.J. Matheson. 2003. Permeable reactive barriers of iron and other zero-valent metals. Pages 371-421 in *Chemical Degradation Methods for Wastes and Pollutants: Environmental and Industrial Applications,* M.A.Tarr, ed., Marcel Dekker, New York.

Tyre, B.T., R.J. Watts, and G.C. Miller. 1991. Treatment of four biorefractory contaminants in soils using catalyzed hydrogen peroxide. *J. Environ. Qual.* 20: 832-837.

Urbanski, T. 1984. *Chemistry and Technology of Explosives.* Pergamon Press, Oxford.

U.S. Department of Energy. 1999. In situ chemical oxidation using potassium permanganate. Innovative Technology Summary Report DOE/EM-0496. U.S. Department of Energy, Washington, DC.

Van Aken, B., and S.N. Agathos. 2001. Biodegradation of nitrosubstituted explosives by white-rot fungi: a mechanistic approach. *Adv. Appl. Microbiol.* 48: 1-70.

Vance, D.B. 2002. A review of chemical oxidation technology. Available from http://2the4.net/html/chemoxwp.htm. July, 2004.

Walker, J.E., and D.L. Kaplan. 1992. Biological degradation of explosives and chemical agents. *Biodegradation* 3: 369-385.

Walling, C. 1975. Fenton's reagent revisited. *Acc. Chem. Res.* 8: 125-131.

Walsh, M.E., T.F. Jenkins, P.S. Schnitker, J.W. Elwell, and M.H. Stutz. 1993. USA Cold Regions Research and Engineering Laboratory CRREL Special Report 93-5. Hanover, New Hampshire, pp. 1-17.

Watts, R.J., M.D. Udell, and R.M. Monsen. 1993. Use of iron minerals in optimizing the peroxide treatment of contaminated soils. *Water Environ. Res.* 65: 839-844.

Watts, R.J., M.D. Udell, P.A. Rauch, and S.W. Leung. 1990. Treatment of pentachlorophenol contaminated soils using Fenton reagent. *Hazard. Waste Hazard. Mater.* 7: 335-345.

Watts, R.J., M.D. Udell, and S.W. Leung. 1991. Treatment of contaminated soils using catalyzed hydrogen peroxide. Page 37-50 in *Chemical Oxidation Technologies for the Nineties,* W.W. Eckenfelder, A.R. Bowers, and J.A. Roth, eds., Technomic Pub. Co., Lancaster, Pennsylvania.

Weber, E.J. 1996. Iron-mediated reductive transformations: Investigation of reaction mechanism. *Environ. Sci. Technol.* 30: 716-719.

Widman, M.J., and P.J.J. Alvarez. 2001. RDX degradation using an integrated Fe(0)-microbial treatment approach. *Water Sci. Technol.* 43: 25-33.

Wilson, E.K. 1995. Zero-valent metals provide possible solution to groundwater problems. *Chem. Eng. News* 73: 19-22.

Yinon, J. 1990. *Toxicity and Metabolism of Explosives.* CRC press, Ann Arbor, MI.

Zoh, K., and K.D. Stenstrom. 2002. Fenton oxidation of hexahydro-1,3,5-trinitro-1,3,5-triazine (RDX) and octahydro-1,3,5,7-tetranitro-1,3,5,7-tetrazocine (HMX). *Water Res.* 36: 1331-1341.

Microbial Surfactants and Their Use in Soil Remediation

Nick Christofi[1] and Irena Ivshina[2]

[1]Pollution Research Unit, School of Life Sciences, Napier University, 10 Colinton Road, Edinburgh, EH10 5DT, Scotland, UK

[2]Alkanotrophic Bacteria Laboratory, Institute of Ecology and Genetics of Microorganisms, Russian Academy of Sciences, 13 Golev Street, Perm 614081, Russian Federation

Introduction

In the remediation of organic and inorganic pollutants in soil, a number of physical, chemical and biological treatments are utilised, including excavation and removal, thermal evaporation, flushing, vapour extraction and bioremediation. In *in situ* and *ex situ* bioremediation and soil washing techniques the use of surfactants can be beneficial. Many environmental pollutants, particularly organics such as polycyclic aromatic hydrocarbons (PAH), polychlorinated biphenyls (PCB), and many petroleum hydrocarbons and biocides, are hydrophobic with low solubility and dissolution in aqueous media. This often reduces their removal from soils as their biodegradation may depend, for example, on their mass transfer into the water phase (Van Loosdrecht *et al.* 1990, Weissenfels *et al.* 1992). There is, however, evidence that microorganisms can overcome mass transfer limitations by producing biofilms on pollutant-coated surfaces as shown, for example, by Johnsen and Karlson (2004) for poorly soluble PAH. The use of surfactants in increasing the availability of hydrophobic pollutants in soils and other environments for bioremediation is a fairly recent consideration (Vigon and Rubin 1989). Surfactants produced naturally by organisms or chemically synthesised have been shown to be capable of increasing the solubility and dispersion of hydrophobic organic pollutants from particulates (Zang *et al.* 1997) and have been utilised in the solubilisation of water immiscible substances. Most work to date has utilised synthetic (petrochemically-derived) surfactants to enhance the solubility and removal of organic and inorganic (heavy metal) pollutants from soils (Christofi and Ivshina 2002). Both synthetic and natural surfactants can be expensive to manufacture and use, and the former are often toxic, affecting bioremediation processes. Although microbially produced biosurfactants can be expensive to manufacture, some are being commercially prepared at relatively low cost.

Biosurfactants

Biosurfactants are biological compounds produced by microorganisms, plants and animals that exhibit high surface-active properties (Georgiou *et al.* 1992). They can have a low or high molecular weight and their size affects their properties and role in pollutant solubilisation and dispersion. Similar to synthetic surfactants, they are amphiphilic and have a hydrophilic and a hydrophobic/lipophilic (non-polar) portion in the molecule (Fiechter 1992, Haig 1996). The hydrophilic part may include amino acids (or peptides), anions or cations, or mono-, di- or poly-saccharides (Banat 1995). Fatty acids or peptides form the hydrophobic portion of the amphiphile. The majority of surfactants used in soil remediation are synthetic but an advantage in using biosurfactants is that they are potentially less toxic and more biodegradable than petrochemical types (Torrens *et al.* 1998, Banat *et al.* 2000, Christofi and Ivshina, 2002). Biosurfactants have a wide range of industrial applications (see Kosaric 2001), and some, rhamnolipids (from *Pseudomonas*), Surfactin (from *Bacillus*) and Emulsan (from *Acinetobacter*), are produced on a large scale and have been evaluated for environmental use. Figure 1 shows the structure of one type of rhamnolipid (a dirhamnolipid) produced by *Pseudomonas aeruginosa*.

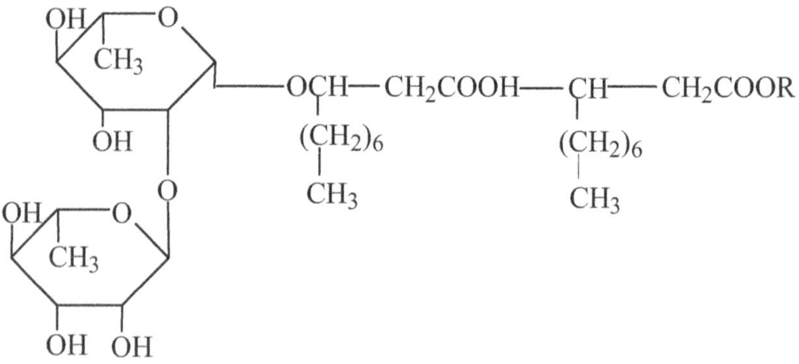

Figure 1. Structure of *Pseudomonas* dirhamnolipid. R = H and R = CH_3 for acid and methyl dirhamnolipids respectively.

An increase in the concentration of hydrophobic compounds in the water phase (solubilisation) is achieved by the formation of micelles (see Fig. 2). These structures can be spherical, ellipsoidal and/or cylindrical with varying size distribution depending on the pH of the solution (see e.g., Knoblich *et al.* 1995 for the biosurfactant Surfactin). Micelle formation occurs at a concentration above what is referred to as the Critical Micelle Concentration (CMC) where biosurfactant molecules aggregate to form spherical structures as the hydrocarbon moiety of the surfactant becomes situated in the centre with the hydrophilic part in contact with water (Haigh 1996). Enhancement of biodegradation is normally pronounced only at biosurfactant concentrations above the critical micelle concentration (CMC) (Robinson *et al.* 1996). It may be that in some studies where the opposite is observed, the use of a concentration of surfactants needed to achieve solubilisation in aqueous systems may not be appropriate in soil systems and sorption onto soils may be a factor affecting efficacy. It is known that the CMC relies on the structure of the surfactant (size of the hydrophobic moiety), ionic concentration of the solution and other factors such as temperature (see Cserháti *et al.* 2002).

There are contrasting reports on the role of biosurfactants in bioavailability (Volkering *et al.* 1998). Some research has shown that the surfactants can enhance solubilisation while others indicate no change or

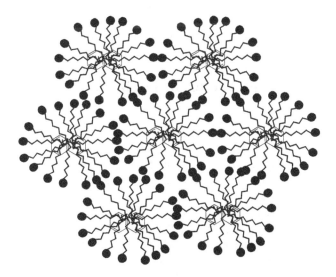

Figure 2. Biosurfactant micelles with core filled with hydrophobic pollutant (PAH).

enhanced sorption (Bruheim 1997). Hua *et al.* (2003) demonstrated that the biosurfactant BS-UC produced by *Candida antarctica* enhanced the biodegradation of a number of *n*-alkanes and had the ability to modify the hydrophobicity and zeta potential of the cell surface. This enabled the microbial cell to attach to the hydrophobic substrate more easily. Although biosurfactants are considered less toxic than synthetic types, toxicity (many biosurfactants are associated with antimicrobial activity) may compromise remediation. In addition, a lack of remediation in their presence may be a function of preferential degradation of biosurfactants added over the organic pollutant (Maslin and Maier 2000).

The role of biosurfactants in solubilisation and enhancement of bioremediation may not be realized in soils if there is reliance of surfactant production *in situ*. The natural production of effective concentrations of biosurfactant may require a large microbial density as induction of the agent has been shown to involve quorum sensing in some organisms. Production should exceed the CMC for effective solubilisation of hydrophobic pollutants. This may not be possible in real systems where sufficient populations throughout a contaminated site may not occur and where emulsified pollutants may not be stable and be easily dispersed in an open system (Ron and Rosenberg 2002). Processes such as surfactant enhancement of microbial cell hydrophobicity and localized sorption (biofilm formation) of surfactant leading to surface solubilisation of pollutants may predominate in natural systems. Some work suggests that some microorganisms can modify their cell walls to attach to hydrophobic surfaces by removal of lipopolysaccharide but detachment is also possible (Rosenberg *et al.* 1983, Zang and Miller 1994).

Biosurfactant Production

Microorganisms involved

Many microorganisms are surfactant producers with concomitant ability to solubilise and degrade hydrophobic pollutants. Figure 3 shows one possible mechanism of hydrocarbon availability facilitated by biosurfactants. Alkanotrophic (oil-degrading) bacteria of the genus *Rhodococcus*, for example, produce surfactants with excellent properties (Christofi and Ivshina 2002, Philp *et al.* 2002). Table 1 provides information on some of the many microbial surfactants. Few studies have been carried out to determine the distribution and abundance in natural environments such as soils (Bodour *et al.* 2003); but scientists continue to isolate new biosurfactant-producing microorganisms and new surfactants (see e.g., Philp *et al.* 2002, Tuleva *et al.* 2002). An examination of contaminated and

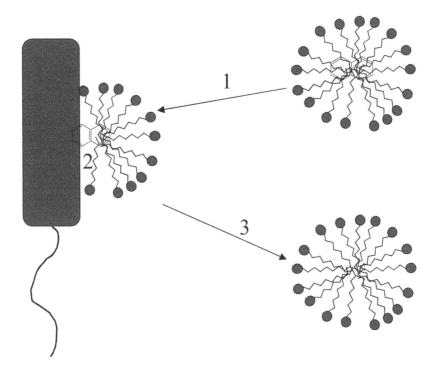

Figure 3. Mechanism of bioavailability of the contaminant partitioned in the micellar phase of non-ionic surfactant to a bacterial cell. 1. Transfer of micelle phase contaminant to hemi-micellesformed on cells; 2. Diffusion of contaminant into cell- biodegradation; 3. Empty micelles are exchanged with new filled micelles.

uncontaminated soils shows their presence and the study by Bodour *et al* (2003) indicated that Gram-positive surfactant-producing bacteria predominate in metal-contaminated or uncontaminated soils whereas soils contaminated or co-contaminated with organic pollutants contained a predominance of Gram-negative surfactant producers.

Mass (cost-effective) production

The most widely studied biosurfactants are two types of rhamnolipids (mono- and di-rhamnolipids) produced by *P. aeruginosa* (Maier and Soberón-Chávez 2000). These can reduce surface tension to ~29 mN m^{-1} (see Christofi and Ivshina 2002) and are currently being produced on a commercial scale. A problem with mass production of microbial

Table 1. A selection of microbial surfactants and their characteristics (see also Desai and Banat 1997, Lang 2002, Vardar-Sukar and Kosaric 2000).

MICRORGANISMS	SURFACTANTS TYPES
BACTERIA	
Acinetobacter spp.	**Emulsan** (heterolipopolysaccharide); whole cell lipopeptide (acylpeptide); fatty acids; mono- and di-glycerides
Acinetobacter radioresistens	**Alasan,** alanine-containing polysaccharide-protein complex.
Alcanivorax borkumensis	Glycolipids
Arthrobacter spp.	Glycolipids; Lipopeptides; heteropolysaccharides
Bacillus licheniformis and other spp.	**Surfactin; Lichenysis;** rhamnose lipids; hydrocarboprotein complex; polymyxin, gramicidin antibiotics
Brevibacterium	Acylpolyols
Clostridium spp.	Neutral lipids
Corynebacterium spp.	Acylpolyols; polysaccharide-protein complex; phospholipids; corynemycolic acids; fatty acids
Flavobacterium spp.	Flavolipids
Micobacterium spp.	Glycoglycerolipid
Mycolata (mycolic acid producing bacteria)- *Nocardia, Rhodococcus, Mycobacterium*	Glycolipids; whole cell de-emulsifiers; neutral lipids and fatty acids; trehalose dimycolates and dicorynomycolates; polysaccharide
Pseudomonas spp.	**Viscosin; Ornithin;** Glycolipids; cyclic lipopeptides **(Arthrofactin)**
Serratia spp.	**Rubiwettin; Serrawettin;** glycolipids
Thiobacillus spp.	Phospholipids
FUNGI	
Candida bombicola, Pseudozyma (Candida) antarctica	**Liposan** (mainly carbohydrate); Glycolipids- sophorolipids, mannosylerythritol lipids; peptidolipid; polysaccharide-fatty acid complex
Torulopsis	Glycolipids; proteins
Ustilago	Cellobiolipids

surfactants, in general, relates to their low level of synthesis in culture and the high costs of recovery of these amphiphilic substances (Georgiou et al. 1992). Synthetic surfactants are primarily used in various applications and this is due to the expense of producing them from biological materials. The use of low-cost substrates including renewable substances, such as waste frying oils (Haba et al. 2000), olive oil mill effluents (Mercadé et al. 1993), potato processing wastes (Fox and Bala 2000) and various urban and agro-industrial wastes (Makkar and Cameotra 1999) to manufacture biosurfactants, can reduce production costs for these important compounds. Thompson et al. (2000) examined the production of surfactin by *Bacillus subtilis* using high- and low-solids potato processing waste and concluded that the latter waste materials could be used to produce biosurfactant, under non-sterile conditions, that can be cost effective in oil recovery and the remediation of organic-contaminated environments.

In some microorganisms the surfactants are released extracellularly, but in others they are part of cell constituents and require costly solvent extraction (Lang and Philp 1998). The use of solvents, many of which are toxic and environmentally unfriendly, needs to be carefully considered. In the extraction of biosurfactants from *Rhodococcus ruber*, Philp et al. (2002) used MTBE (methyl tertiary-butyl ether) that has reduced toxicity, and is less likely to explode or produce peroxides (Rosenkrantz and Klopman 1991, Gupta and Lin 1995).

It is evident that for some microorganisms, high surfactant yields can be obtained (Kuyukina et al. 2001). Cheap cultivation of the micro-organisms, and, cost effective and safe recovery of biosurfactants may enable wide scale replacement of synthetic surfactants in the future.

Laboratory and Field Studies

Most studies utilizing biosurfactants for soil remediation do so in laboratory studies possibly because of the indicated costs of mass production. In laboratory studies they have been used to treat all forms of organic and metal pollutants (Christofi and Ivshina 2002).

There are many studies which have shown that biosurfactants can solubilise and mobilise organics sorbed onto soil constituents (Bai et al. 1997, Brusseau et al. 1995, Ghosh et al. 1995, Ivshina et al. 1998, Noordman et al. 2002, Page et al. 1999, Park et al. 1998, Scheibenbogen et al. 1994, Zang and Miller 1992, Zang et al. 1997). Biosurfactants may be involved in different processes affecting pollutant availability and removal from soil environments including dispersion, displacement and solubilisation. They have the ability of lowering surface and interfacial tensions of liquids and an important factor affecting degradation and removal of soil contaminants is their availability within the matrix, either for microbial degradation or extraction (particularly in the case for metals). A recent study using a *lux*-marked bacterial biosensor which is supposed to monitor the toxicity of PAH indicates that rhamnolipid biosurfactant, used to extract a model PAH, phenanthrene, was found to enhance the transfer of phenanthrene into the aqueous phase leading to higher toxicity of the biosensing organisms (Gu and Chang 2001). Al-Tahhan *et al.* (2000) have shown that the lipopolysaccharide cellular component of a hexadecane-utilising pseudomonad was extracted by rhamnolipid biosurfactant and that this enabled the microbial cells (with increased hydrophobicity) to attach to droplets enhancing hexadecane degradation. Jordan *et al.* (1999) hypothesised that the sorption of biosurfactants at the solid-liquid interface increases bioavailability of adsorped substances.

Christofi and Ivshina (2002) advocated the promotion of biosurfactant production in natural systems as the most cost-effective method of affording organic pollutant bioavailability and increased degradation but this requires and understanding of the factors affecting gene activation in natural systems. Moran *et al.* (2000) indicated *in situ* stimulation of biosurfactant production in contaminated sites and that these can be recovered and recycled. The study by Holden *et al.* (2002) attempted to determine indigenous production and showed that biosurfactants were likely produced in sand cultures but enhanced degradation was not observed as surfactant-producing bacterial cells exhibited direct hexadecane contact. In liquid culture hexadecane degradation was shown to be advantageous as bacteria partitioned at the hexadecane-water interfaces in the presence of biosurfactants.

Petroleum hydrocarbons in soils have been shown to be degraded at faster rates in the presence of rhamnolipid biosurfactants (from a *Pseudomonas* sp.) in *ex situ* treatment systems utilizing nutrient supplements, bioaugmentation with a consortium of oil degrading bacteria and bulking with coir pith and poultry litter to increase aeration (Rahman *et al.* 2002). Treatment combinations without biosurfactant exhibited reduced hydrocarbon degradation suggesting a role in enhancing bioavailability of the contaminant. Noordman and Janssen (2002) presented data indicating that an energy-dependent system is present in biosurfactant-producing *P. aeruginosa* (strain UG2) which mediates fast uptake of hydrophobic compounds in the presence of rhamnolipid. Exogenous addition of this biosurfactant may enhance hydrocarbon degradation. Soil columns have been used to test the efficacy of surfactant foams (Triton X100 and a natural *Pseudomonas* rhamnolipid) in bioremediation of PCP (Mulligan and Eftekhari 2003). The increased removal of PCP (higher with Triton X100 than rhamnolipid) was found to be due to volatilization and increased degradation in foams containing a larger air content. Reduced removal was evident in liquid remediation columns. Sophorolipid biosurfactant produced by the yeast *Candida bombicola* has been used in soil suspension cultures with increased phenanthrene removal through enhanced pollutant availability (Schippers *et al.* 2000).

Heavy metal contamination of soils represents a significant problem requiring solution and unlike organic pollutants, microorganisms cannot degrade these substances to end-products that can leave the soil ecosystem. Soil washing using biosurfactants may be a possibility. Surfactants have been shown to remove metals from surfaces and facilitate their solubilisation by a number of processes. These include metal contact and desorption, complexation reactions leading to removal of metals from surfaces by the Le Chatelier Principle (Miller 1995) and the reduction of metal-particulate interaction (in the case of cationic surfactants) by competition for some, but not all, negatively charged surfaces (Beveridge and Pickering 1983). Metals associated with surfactants in the aqueous phase would require separation and removal. Tan *et al.* (1994) carried out a study on the formation of monorhamnolipid biosurfactant-metal complexes. They showed that rapid and stable surfactant-metal combinations were produced. Miller (1995), using the same biosurfactant, showed that the formation of metal complexes were similar with complexes formed by a range of polymers released by microorganisms. Associations such as these should permit the partitioning of the complex in the aqueous phase and subsequent removal from the soil in washing processes.

Biosurfactants have been used to facilitate removal in soil batch wash systems (Mulligan and Yong 1997, Mulligan *et al.* 1999, 2001a, b, c). Neilson *et al.* (2003) showed that lead can be extracted in its various forms from contaminated soils using rhamnolipid biosurfactant but complete removal was not possible. Mulligan and Yong (1997) used biosurfactants extracted from *Bacillus subtilis* (surfactin), *Pseudomonas aeruginosa* (rhamnolipids) and *Torulopsis bombicola* (sophorolipids) to remove metals from oil-contaminated soils. Soil washing was carried out with different concentrations of surfactant under different pH conditions. Negligible metal extraction was achieved with water but appreciable removal of copper (~37%) and zinc (~20%) was obtained in systems containing different combinations of biosurfactants and HCl/NaOH. Yong *et al.* (1993) used a sequential metal extraction technique to determine the partitioning of metals within the organic, carbonate, oxide, exchangeable and residual contaminant fractions of a particular soil. Mulligan and Yong (1997) produced data showing that copper, zinc and lead found in contaminated soil partition differently. Copper removal from the organic fraction was achieved by surfactin and rhamnolipid; zinc was removed from the oxide fraction by the same surfactants and an acid/sophorolipid mix removed zinc from the carbonate fraction. This preliminary study indicated that metal extraction is possible using anionic surfactants even under conditions of low exchangeable metal fractions. More recent studies by Mulligan *et al.* (1999a, b) used surfactin to treat soil and sediments contaminated with oil and grease, Zn, Cu and Cd,. It was found that the heavy metals were associated with carbonate, oxides and organic fractions in the contaminated material and that these could be removed using a combination of surfactin and NaOH. It was suggested that sequential extraction could be developed to enhance soil-washing procedures. Metal removal involved attachment of surfactin to the soil interface leading to lowering of the interfacial tension and micellar complexation (Mulligan *et al.* 1999a, b).

Many metals are toxic to soil microorganisms and are often associated with organic pollutions (Balrich and Stotsky 1985). Indeed, even organics have an inhibitory effect on microorganisms at certain concentrations (Huesemann 1994). In order to carry out bioremediation of the latter, the metal toxicity requires attenuation. Studies have been made to determine whether metal-rhamnolipid complexes could alleviate metal toxicity and enhance organic degradation by a *Burkholderia* sp. Sandrin *et al.* (2000) showed that rhamnolipid eliminated cadmium toxicity when added at a 10-fold greater concentrations than cadmium (890 mM) There was a decrease in toxicity when added at an equimolar concentrations (89 mM) but no effect at a 10-fold lower concentration (8.9 mM). Reduced toxicity was considered to be a combined function of rhamnolipid complexation with cadmium and the biosurfactant interacting with the cell surface affecting cadmium uptake.

Field studies utilising microbial biosurfactants are not common. Shoreline field experiments have been carried out using proprietary nutrient formulations (BIOREN 1 and 2) to clean oil-contaminated sediments (Le Floch *et al.* 1999). BIOREN 1 contains a biosurfactant which was shown, initially, to produce enhanced oil degradation but, ultimately, differences in using the formulation without surfactant were not obvious. *Ex situ* biopile (composting-type) remediation systems have been used in field experiments on bioremediation of oil contaminated agricultural soils following an accidental oil-spill in the Perm Oblast in Russia (Christofi *et al.* 1998). The biopiles (Plate 1) provided an environment of increased oxygen transfer with biodegradation further enhanced by combination of nutrient additions, bulking with straw and inoculations of *Rhodococcus*-biosurfactant complexes (Ivshina *et al.* 2001). An active surfactant producer *R. ruber* AC 235 isolated from oil-contaminated sites was used to prepare biosurfactant complexes. Organisms were grown in hexadecane and surfactants produced by the organism were extracted using methyl tertiary-butyl ether (MTBE) using sonication (Kuyukina *et al.* 2001). The complexes were able to reduce the surface and interfacial tension of water to values of 26.8 and 0.9 mN m^{-1}. It was shown that up to 57.1% crude oil removal was achieved depending on the combinations of treatments and additives used. The inclusion of biosurfactants, increased ventilation and nutrient addition, lead to more effective remediation in composting systems (Christofi and Ivshina 2002). In the Russian sites examined, it was shown that representatives of *R. erythropolis*, *R. opacus* and *R. "longus"* predominated in soils contaminated with crude oil. The two species *R. ruber* and *R. rhodochrous* were dominant within the subsurface bacterial populations of oil and gas deposits and represented 90-100% of hydrocarbon degraders.

The aim of other work carried out in crude-oil contaminated sites in Russia was to study the ecological behaviour and competitive ability of biosurfactant-producing *Rhodococcus* bacteria introduced into crude oil contaminated soil, prospects for their survival, reproduction, environmental effects on the development of the introduced rhodococcal populations and the estimation of introduction of *Rhodococcus* species into the open ecosystem. Studies have been carried out utilizing *Rhodococcus* biosurfactant complexes to stimulate indigenous crude oil degrading microflora to facilitate bioremediation of crude oil contaminated soil. The introduction of the surfactant complex resulted in increased oil degradation and the increases in the hydrocarbon degrading bacteria. It was evident from the results that the abundance of *Rhodococcus* species is enhanced by surfactant and that the oil degrading populations provide an indicator of the potential for bioremediation of crude oil contaminated soils (Christofi and Ivshina 2002). The study showed that the number of hydrocarbon degraders in the control soil was reduced by 6.7 times following addition of crude oil at a concentration of 4.5 % (w/w) and that enhancing biosurfactant production by the manipulation of the soil matrix can be important in oil degradation. Bioaugmentation was also tested in these trials. The introduction of rhodococci, able to degrade aliphatic and aromatic hydrocarbons, into oil-contaminated soil accelerated bioremediation process by 20-25%. Simultaneous introduction of both *R. erythropolis* and *R. ruber* proved to be most efficient and resulted in 75.5% decrease in the oil content within three months.

Recently field experiments were done on soils heavily contaminated with crude oil at concentration of up to 200 g kg^{-1} total recoverable petroleum hydrocarbons (TRPH). The fractions consisted of aliphatics (64%), aromatics (25%), heterocyclics (8%) and tars/asphaltenes (3%).

Ex situ bioslurry and land farming techniques were utilized in sequence to remediate the soil. In both treatment systems, an oleophilic biofertiliser based on *Rhodococcus* biosurfactant complexes (Ivshina *et al.* 2001) was used. The slurry-based reactor biotreatment lead to an 88% reduction in soil oil content after 2 months. Following transfer of the reactor content of ~25 g Kg^{-1} TRPH into landfarming cells receiving oleophilic biofertilizer, watering tilling and bulking with woodchips, the contamination decreased to 1-1.5 g kg^{-1} TRPH after 5-7 weeks. This latter treatment facilitated the removal of 0.3-0.6 g Kg^{-1} day^{-1} TRPH. Tertiary soil management involving phytoremediation was also used in the study (Kuyukina *et al.* 2003).

Summary

Most studies utilizing biosurfactants in soil pollutant removal and remediation have used laboratory scale systems. Few field scale remediation programmes have been initiated. Generally for *in situ* bioremediation, the use of biosurfactants poses problems similar to supplementing with nutrients in that the substances are difficult to distribute to the contaminated sites for effective removal processes to take place. Also, biosurfactants may participate in a number of reactions in soils leading to positive, negative and no effect outcomes to pollutant remediation. More research is still needed on their role in real system remediation prior to their wide scale use.

Acknowledgements

We acknowledge support for aspects of this work over the years from The Royal Society, London; NATO; LUKOIL, Russia and the European Commission.

REFERENCES

Al-Tahhan, R., T.R. Sandrin, A.A. Bodour, and R.M. Maier. 2000. Rhamnolipid-induced removal of lipopolysaccharide from *Pseudomonas aeruginosa*: affect on cell surface properties and interaction with hydrophobic substrates. *J. Biotechnol.* 94: 195-212.

Bai, G., M.L. Brusseau, and R.M. Miller. 1997. Biosurfactant-enhanced removal of residual hydrocarbon from soil. *J. Contam. Hydrol.* 25: 157-170.

Balrich, H., and G. Stotsky. 1985. Heavy metal toxicity to microbe-mediated ecological processes: a review and potential application to regulatory process. *Environ. Res.* 36: 111-137.

Banat, I.M. 1995. Biosurfactant production and possible uses in microbial enhanced oil recovery and oil pollution remediation: A review. *Biores. Technol.* 51: 1-12.

Banat, I.M., R.S. Makkar, and S.S. Cameotra. 2000. Potential commercial applications of microbial surfactants. *Appl. Microbiol. Biotechnol.* 53: 495-508.

Beveridge, A., and W.F. Pickering. 1983. The influence of surfactants on the adsorption of heavy metal ions to clays. *Water Res.* 17: 215-225.

Bodour, A.A., K.P. Drees, and R.M. Maier. 2003. Distribution of biosurfactant-producing bacteria in undisturbed and contaminated arid southwestern soils. *Appl. Environ. Microbiol.* 69: 3280-3287.

Bruheim, P. 1997. Bacterial degradation of emulsified crude oil and the effect of various surfactants. *Can. J. Microbiol.* 43: 17-22.

Brusseau, M.L., R.M. Miller, Y. Zhang, X. Wang, and G.-Y. Bai. 1995. Biosurfactant- and cosolvent-enhanced remediation of contaminated media. Pages 82-94 in *Microbial Processes for Remediation*, R.E. Hinchee, F.J. Brockman, and C.M. Vogel., eds., Battelle Press, Columbus, Ohio.

Christofi, N., and I.B. Ivshina. 2002. Microbial surfactants and their use in remediation of contaminated soils. *J. App. Microbiol.* 93: 915-929.

Christofi, N., I.B. Ivshina, M.S. Kuyukina, and J.C. Philp. 1998. Biological treatment of crude oil contaminated soil in Russia. Pages 45-51 in *Contaminated Land and Groundwater: Future Directions*, D.N. Lerner, and N. R.G. Walton, eds., London: Geological Society- Engineering Geology Special Publication, 14.

Cserháti, T., E. Forgás, and G. Oros. 2002. Biological activity and environmental impact of anionic surfactants. *Environ. Internat.* 28: 337-348.

Desai, J.D., and I.M. Banat. 1997. Microbial production of surfactants and their commercial potential. *Microbiol. Mol. Biol. Rev.* 61: 47-64.

Fiechter, A. 1992. Biosurfactants: moving towards industrial application. *TIBTECH* 10: 208-217.

Fox, S.L., and G.A. Bala. 2000. Production of surfactant from *Bacillus* ATCC 21332 using potato substrates. *Biores. Technol.* 75: 235-240.

Georgiou, G., S.-C. Lin, and M.M. Sharma. 1992. Surface-active compounds from microorganisms. *Biotechnology* 10: 60-65.

Ghosh, M.M., I.T. Yeom, Z. Shi, C.D. Cox, and K.G. Robinson. 1995. Surface-enhanced bioremediation of PAH-and PCB-contaminated soil. Pages 15-23 in *Microbial Processes for Remediation*, R.E. Hinchee, F.J. Brockman, and C.M. Vogel, eds., Battelle Press, Columbus, Ohio.

Gu, M.B., and S.K. Chang. 2001. Soil biosensor for the detection of PAH toxicity using an immobilized recombinant bacterium and a biosensor. *Biosens. Bioelectron.* 16: 667-674.

Gupta, G., and Y.J. Lin. 1995. Toxicity of methyl tertiary butyl ether to *Daphnia magna* and *Photobacterium phosphoreum. Bull. Environ. Contam. Toxicol.* 55: 618-620.

Haba, E., M.J. Espuny, M. Busquets, and A. Manresa. 2000. Screening and production of rhamnolipids by *Pseudomonas aeruginosa* 47T2 NCIB 40044 from waste frying oils. *J. Appl. Microbiol.* 88: 379-387.

Haigh, S.D. 1996. A review of the interaction of surfactants with organic contaminants in soil. *Sci. Total Environ.* 185: 161-170.

Holden, P.A., M.G. LaMontagne, A.K. Bruce, W.G. Miller, and S.E. Lindow. 2002. Assessing the role of *Pseudomonas aeruginosa* surface-active gene expression in hexadecane biodegradation in sand. *Appl. Envir. Microbiol.* 68: 2509-2518.

Hua, Z., J. Chen, S. Lun, and X. Wang. 2003. Influence of biosurfactants produced by *Candida antarctica* on surface properties of microorganism and biodegradation of n-alkanes. *Water Res.* 37: 4143-4150.

Huesemann, M.H. 1994. Guidelines for land-treating petroleum hydrocarbon contaminated soils. *J. Soil Contam.* 3: 299-318.

Ivshina, I.B., M.S. Kuyukina, J.C. Philp, and N. Christofi. 1998. Oil desorption from mineral and organic materials using biosurfactant complexes produced by *Rhodococcus* species. *World J. Microbiol. Biotechnol.* 14: 307-312.

Ivshina, I.B., M.S. Kuyukina, M.I. Ritchkova, J.C. Philp, C.J. Cunningham, and N. Christofi. 2001. Oleophilic biofertilizer based on a *Rhodococcus* surfactant complex for the bioremediation of crude oil-contaminated soil. *AEHS Contaminated Soil Sediment and Water: International Issue,* August 2001, pp. 20-24.

Johnsen, A.R., and U. Karlson. 2004. Evaluation of bacterial strategies to promote the bioavailability of polycyclic aromatic hydrocarbons. *Appl. Microbiol. Biotechnol.* 63: 452-459.

Jordan, R.N., E.P. Nichols, and A.B. Cunningham. 1999. The role of (bio)surfactant sorption in promoting the bioavailability of nutrients localized at the solid-water interface. *Water Sci. Technol.* 39: 91-98.

Knoblich, A., M. Matsumoto, R. Ishiguro, K. Murata, Y. Fujiyoshi, Y. Ishigami, and M. Osman. 1995. Electron cryo-microscopic studies on micellar shape and size of surfactin, an anionic lipopeptide. *Colloids and Surfaces B: Biointerfaces* 5: 43-48.

Kosaric, N. 2001. Biosurfactants and their application for soil remediation. *Food Technol. Biotechnol.* 39: 295-304.

Kuyukina, M.S., I.B. Ivshina, J.C. Philp, N. Christofi, S.A. Dunbar, and M.I. Ritchkova. 2001. Recovery of *Rhodococcus* biosurfactants using methyl tertiary-butyl ether extraction. *J. Microbiol. Methods* 46: 149-156.

Kuyukina, M.S., I.B. Ivshina, M.I. Ritchkova, J.C. Philp, C.J. Cunningham, and N. Christofi. 2003. Bioremediation of crude oil-contaminated soil using slurry-phase biological treatment and land farming techniques. *Soil Sed. Contam.* 12: 85-99.

Lang, S. 2002. Biological amphiphiles (microbial biosurfactants). *Cur. Opinion Coll. Interface Sci.* 7: 12-20.

Lang, S., and J.C. Philp. 1998. Surface-active lipids in rhodococci. *Antonie Leeuwenhoek* 74: 59-70.

Le Floch, S., F.X. Merlin, M. Guillerme, C. Dalmazzone, and P. Le Corre. 1999. A field experimentation on bioremediation: BIOREN. *Environ. Technol.* 20: 897-907.

Maier, R.M., and G. Soberón-Chávez. 2000. *Pseudomonas aeruginosa* rhamnolipids: biosynthesis and potential applications. *Appl. Microbiol. Biotechnol.* 54: 625-633.

Makkar, R.S., and S.S. Cameotra. 1999. Biosurfactant production by microorganisms on unconventional carbon sources. *J. Surf. Deter.* 2: 237-241.

Maslin, P., and R.M. Maier. 2000. Rhamnolipid enhanced mineralization of phenanthrene by indigenous microbial populations in organic-metal contaminated soils. *Biorem. J.* 4: 295-308.

Mercadé, M.E., M.A. Manresa, M. Robert, M.J. Espury, C de Andrés, and J. Guinea. 1993. Olive oil mill effluent (OOME). New substrate for biosurfactant production. *Biores. Technol.* 43: 1-6.

Miller, R.M. 1995. Biosurfactant-facilitated remediation of metal-contaminated soils. *Environ. Health Persp.* 103: 59-62.

Moran, A.C., N. Olivera, M. Commendatore, J.L. Esteves, and F. Sineriz. 2000. Enhancement of hydrocarbon waste biodegradation by addition of a biosurfactant from *Bacillus subtilis* O9. *Biodegradation* 11: 65-71.

Mulligan, C.N., and F. Eftekhari. 2003. Remediation of surfactant foam of PCP-contaminated soil. *Eng. Geol.* 2179: 1-11.

Mulligan, C.N., and R.N. Yong. 1997. The use of biosurfactants in the removal of metals from oil-contaminated soil. *Proceedings of the Geoenviromental Conference on Contaminated Ground: Fate of Pollutants and Remediation,* Cardiff, UK, page 461-466.

Mulligan, C.N., R.N. Yong, and B.F. Gibbs. 1999. On the use of biosurfactants for the removal of heavy metals from oil-contaminated soil. *Environm. Prog.* 18: 50-54.

Mulligan, C.N., R.N. Yong, and B.F. Gibbs. 1999a. Removal of heavy metals from contaminated soil and sediments using the biosurfactant surfactin. *J. Soil Contam.* 8: 231-254.

Mulligan, C.N., R.N. Yong, and B.F. Gibbs, S. James, and H.P.J. Bennett. 1999b. Metal removal from contaminated soil and sediments by the biosurfactant surfactin. *Environ. Sci. Technol.* 33: 3812-3820.

Mulligan, C.N., R.N. Yong, and B.F. Gibbs. 2001a. Heavy metal removal from sediments by biosurfactants. *J. Hazard. Mater.* 85: 111-125.

Mulligan, C.N., R.N. Yong, and B.F. Gibbs. 2001b. Remediation technologies for metal-contaminated soils and groundwater: an evaluation. *Eng. Geol.* 60: 193-207.

Mulligan, C.N., R.N. Yong, and B.F. Gibbs. 2001c. Surfactant remediation of contaminated soil: a review. *Eng. Geol.* 60: 371-380.

Neilson, J.W., J.F. Artiola, and R.M. Maier. 2003. Characterization of lead removal from contaminated soils by non-toxic soil washing agents. *J. Environ. Qual.* 32: 899-908.

Noordman, W.H., and D.B. Janssen. 2002. Rhamnolipid stimulates uptake of hydrophobic compounds by *Pseudomonas aeruginosa*. *Appl. Environ. Microbiol.* 68: 4502-4508.

Noordman, W.H., J.H.J. Wachter, G.J. de Boer, and D.B. Janssen. 2002. The enhancement of surfactants of hexadecane degradation by *Pseudomonas aeruginosa* varies with substrate availability. *J. Biotechnol.* 94: 195-212.

Page, C.A., J.S. Bonner, S.A. Kanga, M.A. Mills, and R.L. Autenrieth. 1999. Biosurfactant solubilization of PAHs. *Environ. Eng. Sci.* 16: 465-474.

Park, A.J., D.K. Cha, and M. Holsen. 1998. Enhancing solubilization of sparingly soluble organic compounds by biosurfactants produced by *Nocardia erythropolis*. *Water Environ. Res.* 70: 351-355.

Philp, J.C., M.S. Kuyukina, I.B. Ivshina, S.A. Dunbar, N. Christofi, S. Lang, and V. Wray. 2002. Alkanotrophic *Rhodococcus* as a biosurfactant producer. *Appl. Microbiol. Biotechnol.* 59: 318-324.

Rahman, K.S.M., I.M. Banat, J. Thahira, T. Thayumanavan, and P. Lakshmanaperumalsamy. 2002. Bioremediation of gasoline contaminated soil by a bacterial consortium amended with poultry litter, coir pith and rhamnolipid biosurfactant. *Biores. Technol.* 81: 25-32.

Robinson, K.G., M.M. Ghosh, and Z. Shi. 1996. Mineralization enhancement of non-aqueous phase and soil-bound PCB using biosurfactant. *Water Sci. Technol.* 34: 303-309.

Ron, E.Z., and E. Rosenberg. 2002. Biosurfactants and oil bioremediation. *Curr. Opinion Biotechnol.* 13: 249-252.

Rosenberg, E., A. Gottlieb, and M. Rosenberg. 1983. Inhibition of bacterial adherence to hydrocarbons and epithelial cells by emulsan. *Infect. Immun.* 39: 1024-1028.

Rosenkrantz, H.S., and G. Klopman. 1991. Predictions of the lack of genotoxicity and carcinogenicity in rodents of two gasoline additives: methyl- and ethyl-*t*-butyl ethers. *In Vitro Toxicol.* 4: 49-54.

Sandrin, T.R., A.M. Chech, and R.M. Maier. 2000. A rhamnolipid biosurfactant reduces cadmium toxicity during naphthalene degradation. *Appl. Environ. Microbiol.* 66: 4585-4588.

Scheibenbogen, K., R.G. Zytner, H. Lee, and J.T. Trevors. 1994. Enhanced removal of selected hydrocarbons from soil by *Pseudomonas aeruginosa* UG2 biosurfactants and some chemical surfactants. *J. Chem. Technol. Biotechnol.* 59: 53-59.

Schippers, C., K. Gessner, T. Müller, and T. Scheper. 2000. Microbial degradation of phenanthrene by addition of a sophorolipid mixture. *J. Biotechnol.* 83: 189-198.

Tan, H., J.T. Champion, J.F. Artiola, M.L. Brusseau, and R.M. Miller. 1994. Complexation of cadmium by a rhamnolipid biosurfactant. *Environ. Sci. Technol.* 28: 2402-2406.

Thompson, D.N, S.L. Fox, and G.A. Bala. 2000. Biosurfactants from potato process effluents. *Appl. Biochem. Biotechnol.* 84-86: 917-930.

Torrens, J.L., D.C. Herman, and R.M. Miller-Maier. 1998. Biosurfactant (rhamnolipid) sorption and the impact on rhamnolipid-facilitated removal of cadmium from various soils under saturated flow conditions. *Environ. Sci. Technol.* 32: 776-781.

Tuleva, B.K., G.R. Ivanov, and N.E. Christova. 2002. Biosurfactant production by a new *Pseudomonas putida* strain. *Z. Naturforsch.* 57c: 356-360.

Van Loosdrecht, M.C.M., J. Lycklema, W. Norde, and A.J.B. Zehnder, 1990. Influence of interfaces on microbial activity. *Microbiol Rev.* 54: 75-87.

Vardar-Sukar, F., and N. Kosaric. 2000. Biosurfactants. Pages 618-635 in *Encyclopedia of Microbiology* (Second Edition, volume 1), J. Lederberg, ed., Academic Press, New York.

Vigon, B.W., and A.J. Rubin. 1989. Practical considerations in the surfactant aided mobilization of contaminants in aquifers. *J. Water Pollut. Cont. Fed.* 61: 1233-1240.

Volkering, F., A.M. Breure, and W.H. Rulkens. 1998. Microbiological aspects of surfactant use for biological soil remediation. *Biodegradation* 8: 401-417.

Weissenfels, W.D., H.J. Klewer, and J. Langhoff. 1992. Adsorption of polycyclic aromatic hydrocarbons (PAHs) by soil particles: influence on biodegradability and biotoxicity. *Appl. Microbiol. Biotechnol.* 36: 689-696.

Yong, R.N., R. Galvez-Cloutier, and Y. Phadungchewit. 1993. Selective extraction analysis of heavy metal retention in soil. *Can. Geotechnic. J.* 28: 378-387.

Zang, Y., W.J. Maier, and R.M. Miller. 1997. Effects of rhamnolipids on the dissolution, bioavailability and biodegradation of phenanthrene. *Environ. Sci. Technol.* 31: 2211-2217.

Zang, Y., and R.M. Miller. 1992. Enhanced octadecane dispersion and biodegradation by *Pseudomonas* rhamnolipid surfactant (biosurfactant). *Appl. Environ. Microbiol.* 58: 3276-3282.

Zang, Y., and R.M. Miller. 1994. Effect of a *Pseudomonas* rhamnolipid biosurfactant on cell hydrophobicity and biodegradation of octadecane. *Appl. Environ. Microbiol.* 60: 2101-2106.

Phytoremediation Using Constructed Treatment Wetlands: An Overview

Alex J. Horne and Maia Fleming-Singer

Ecological Engineering Group, Department of Civil and Environmental Engineering, University of California, Berkeley, California 94720, USA

Introduction

Water pollution is endemic to human activities, ranging from urban industrial and municipal point source discharges to non-point source discharges from agriculture, logging, and mining. Until recently, only point sources have been subject to serious water quality regulation, which mandates cleanup to levels that are not expected to harm the environment. Regulation for non-point source pollution has been limited by the vast scale of the pollution and an appreciation for the economic ramifications of this scale when ordering cleanup. Ecological engineering, with its emphasis on sustainable energy sources and acceleration of pollution amelioration using natural treatment systems, offers a relatively low cost alternative for large-scale treatment of non-point source pollution in particular, but also that of point sources. The flagship of ecological engineering is phytoremediation using constructed wetlands. Terrestrial phytoremediation plays a similar role in brownfield cleanup.

The use of solar energy to power photosynthesis has enhanced the general topic of phytoremediation in both its terrestrial and aquatic forms. *Phytoremediation* is the use of photosynthetic plants or other autotrophic organisms to clean up and manage hazardous and non-hazardous pollutants (McCutcheon and Schnoor 2003, USEPA 1998, 1999). Cleanup and manage-ment of a pollutant refers to its destruction, inactivation, or immobilization in a form that is neither directly nor indirectly harmful to the environment. The above definition includes the use of living green plants for cleanup of a steady flow of wastewater, in contrast to earlier definitions of phytoremediation (USEPA 1998, 1999, Cronk and Fennessey 2001) which involved only *in situ* remediation of soil, sludge, sediments, and groundwater that had been contaminated by past use.

In aquatic phytoremediation, wetlands have come to play the dominant role. Phytoremediation in wetlands occurs via the well-identified functioning of these ecosystems as biogeochemical transformers in the greater landscape (Kadlec and Knight 1996, Mitsch and Gosselink 2000). A wide range of wastes can be transformed in wetlands. These wastes include nutrients (e.g., nitrogen, phosphorous, Mitsch 2000), metals (e.g., chromium, copper, selenium), trace organic compounds (tri-nitrotoluene and pesticides such as atrazine, chlorpyrifos, endosulfan, Rodgers and Dunn 1992, Alvord and Kadlec 1996, Moore 2000, Schulz and Peall 2001) and pathogens (total and fecal coliform, bacteriophages, protozoans, Kadlec and Knight 1996, Karpiscak *et al.* 1996, Quiñónez *et al.*1997).

Wetlands are ecotones, transitional environments between upland terrestrial and deep aquatic ecosystems. However, if ecologists had described wetlands first, it is likely that terrestrial and lake ecosystems would have been defined as fringes of wetlands. Either way, wetlands are a very distinct habitat. Technically, jurisdictional wetlands are defined by three common components: 1) shallow water coverage for at least a few weeks during the year, 2) permanent or temporarily anoxic soils, and 3) characteristic vegetation possessing morphological adaptations for coping with life in anoxic soils (i.e., no roots or roots that can survive anoxia, Lyon 1993). Although this definition includes small lakes or ponds surrounded by a margin of aquatic macrophytes, efficient phytoremediation requires systems having at least a 50% aerial cover of submerged or emergent macrophytes or attached algae. Lakes and ponds are poor at pollutant remediation relative to wetlands primarily because the aquatic macrophytes and a few large algae species are absent in deeper, open lake waters. These plants provide the reduced organic carbon and biofilm substrate required for wetland phytoremediation. In addition, open water in lakes and ponds short-circuits to the discharge point, reducing the residence time for pollution treatment.

The innate characteristics of wetland ecosystems vary across a wide continuum, much as meadow ecosystems differ dramatically from forest ecosystems. The particular type of wetland regulates its phytoremediation potential. Wetlands are usually divided into four groups based either on their hydraulic regime or on the type of vegetation. *Marshes* are dominated by emergent macrophytes such as the cattail (*Typha* sp.), bulrush (*Scirpus* sp.) and common reed (*Phragmites australis*). Marshes are the main engine of constructed wetlands phytoremediation. *Swamps* are characterized by large trees, such as cypress and tupelo, that are too slow growing to be much used in wetlands phytoremediation although some existing ones are used for waste treatment. *Fens* or *alkaline mires* are colonized by mosses, sedges and grasses; and *bogs* contain low growing plants, typically the acidophilic moss *Sphagnum*. Bogs and fens are also too slow growing to be created easily and are little used in phytoremediation apart from some metals removal in existing sites.

Depending on the water depth and degree of shading, many wetlands contain submerged macrophytes in addition to characteristic emergent species. Along with the dead wetland litter layer, submerged plant leaves and stems often support an abundant periphyton community. This biofilm community contains attached bacteria, algae and protozoa, which take up nutrients or transform them in oxidation-reduction reactions (Cronk and Fennessey 2001). Together, the periphyton community on submerged surfaces and the microbial biofilms present in the relatively shallow rhizospheres of wetland plants are responsible for the majority of microbial processing that occurs in wetlands (Nichols 1983, Brix 1997). However, the very slow kinetics of transport of contaminants through soils relative to that in free water and the lack of labile carbon limits the role of root and soils for wetlands phytoremediation. The aquatic biofilm among the plant stems and litter is the engine of aquatic phytoremediation.

Wetland phytoremediation relative to conventional wastewater treatment technology

Traditional remediation of wastes has a long history with many false starts. In 1357 King Edward III attempted to clean up some of the River Thames in England but it took another five centuries before sewage ponds were invented to actually treat waste by using bioprocesses. Over the past two decades in the United States, many new advanced technological methods have been developed based on the need for large-scale cleanup of USEPA Superfund and other lesser-polluted sites (Mineral Policy Center 1997). Conventional wastewater treatment for municipal and industrial sewage involves highly mechanized, energy-intensive processes including oxygenated activated sludge, trickling filters, and high rate oxidation ponds to treat primarily biochemical oxygen demand (BOD), total suspended solids (TSS), fecal coliform bacteria, pH, nutrients, and oil and grease (Metcalf and Eddy 1991). Flocculation, sorption and coagulation, ion exchange and membrane filtration, reverse osmosis, and tertiary treatment for nutrient polishing are additional examples of conventional treatment technologies which rely on a highly mechanized, centralized collection approach and which have a realistic service life of 25-30 years. Conventional treatment systems are dependent on electricity and contribute to (1) depletion of nonrenewable fossil fuel sources and (2) the environmental degradation that occurs from extraction of nonrenewable resources, and also from the byproducts and/or final products of these technologies, such as biosolids and sludge (Sundaravadivel and Vigneswaran 2001).

In contrast to municipal and industrial wastewater, agricultural and storm runoff are usually produced in such high volumes that treatment of any kind is rare. For this type of treatment problem, pollutant source control by best management practices (BMPs), usually involving soil conservation and detention basins but also including wetlands, has been tried with moderate success (Meade and Parker 1985). A more recent regulatory tool, the total maximum daily load (TMDL), is being implemented to provide the quantitative tool lacking in previous BMP programs. Unfortunately, no BMPs match the needs of the TMDL concept, particularly for soluble contaminants. Constructed phytoremediation wetlands could be such a tool (IRWD 2003).

Industrial and mining wastes have been conventionally remediated using typical physical-chemical methods including addition of bases such as limestone or metals such as iron that will neutralize and precipitate soluble toxic metals such as copper and zinc. Groundwater bioremediation can involve infusion of nutrients and a microbial consortium to metabolize the toxicant *in situ*. Groundwater extraction from confined aquifers, alone or following additions of steam or solvents has also been practiced for cleanup of industrial wastes such as dense non-aqueous phase liquids (DNAPL). When remediation is not economical, containment by grout walls or other impermeable barriers, including on-site burial, can be used. However, the scale of many of these problems is too large for such methods.

Advanced technological methods are expensive, relying on electricity, pumping, or oxygen additions, and often require large concrete or steel vessels. In contrast, wetland phytoremediation harnesses ambient solar energy and requires no sophisticated containment system. The usual design is a shallow depression in the ground surrounded by earthen berms from the excavation. Simple, renewable technologies are particularly appropriate in locations lacking infrastructure support for conventional wastewater treatment, such as developing countries. However, many of the most advanced and prosperous regions such as Orange County, California, are leaders in developing phytoremediation wetlands. Additionally, no specific design life period is generally prescribed for treatment wetlands (Sundaravadivel and Vigneswaran 2001), meaning expensive overhauls or equipment replacement is not an obstacle for long-term use. In some cases, phytoremediation wetlands can be relatively tolerant to shock hydraulic and pollutant loads, allowing for reliable treatment quality. These systems can also provide indirect benefits such as green space, wildlife habitat, and recreational and educational areas.

Phytoremediation in wetlands was first reported in 1952, with the possibility of decreasing over-fertilization, pollution, and silting up of inland waters through appropriate plant mediation (Seidel and Kickuth of the Max Planck Institute in Plon, Germany; Brix 1994a). Engineering of wetland systems for pollutant treatment has advanced over the past five decades, creating constructed treatment wetlands that can reproduce the range of biogeochemical transformations occurring in natural wetlands (Kadlec and Knight 1996, Sundaravadivel and Vigneswaran 2001). Compared with the other two types of natural treatment systems currently in use (terrestrial or land application systems and aquatic or pond/lagoon systems), phytoremediation in treatment wetlands offers design simplicity, as well as relatively low installation, operation, and maintenance costs.

The anoxic conditions and aquatic milieu that characterize wetland ecosystems means that successful phytoremediation can occur for those reactions requiring low REDOX (reduction/oxidation) potential and for both dissolved or particulate pollution. Wetlands can remove biochemical oxygen demand (BOD) and total suspended solids (TSS) from wastewater streams. However as characteristically low-oxygen environments they are best reserved for (1) *polishing* of already partially treated (oxidized) industrial or municipal waste; and (2) *removal of specific pollutants,* such as nutrients, metals, trace organics, and pathogens. Wetlands are capable of treating large volumes of contaminated water, although they perform best when contaminant concentrations are low or moderate. These conditions are often precisely those that are most costly to treat using conventional technologies. Examples of pollutants removed by phytoremediation in wetlands are shown in Table 1.

Despite the aforementioned benefits of phytoremediation using treatment wetlands, there are some commonly cited limitations of these systems. The following list summarizes these limitations (Sundaravadivel and Vigneswaran 2001) and offers updated perspectives by the present authors in italics.

— Large land areas are required for the same or lower level of treatment produced by conventional systems, making them unsuitable for large, centralized wastewater sources such as cities. *If combined with parks or wildlife areas, even quite large wetlands (> 150 ha or 300 acres) are often welcomed in cities. Almost all large conventional treatment systems are surrounded by land easements where homes are not permitted for reasons of odor and safety. These easements can easily be developed as wetland parks, incorporating wildlife areas, providing public access, and acting as a visual screen for the concrete infrastructure of conventional treatment plants (see IRWD example later in this paper).*

Table 1. Summary of known uses of phytoremediation wetlands. Phytoremediation using wetlands ranges more widely than terrestrial phytoremediation since drinking water supplies as well as streams and rivers are targets for clean up. Types of wetlands used for this purpose range from acid *Sphagnum* bogs for acid-mine drainage to cattail and duckweed marshes for denitrification and pesticide removal.

Pollutant or toxicant	Human problem	Environmental problem
Biological oxygen demand	Impaired drinking water quality, malodors	Fish kills, slime production
Nitrate	Methemoglobenemia, impaired lake use	Eutrophication, avian botulism, blue-green algae toxins to birds & mammals
Particulate-N/P	Impaired lake use	Decreased water clarity
Phosphorus	Impaired lake use	Eutrophication
[1]Heavy metals (Cu, Pb, acid mine drainage, storm runoff)	Impacted drinking water standards	Toxicity
Metalloid ([1]Se from agriculture, copiers, taillight production)	Toxicity to livestock (blind staggers)	Bird embryo deformities, skeletal deformation in fish
Pesticides	Food chain toxicity, cancers	Non-target organism deaths
Trace organics (chlorinated organics, estrogen mimics)	Major long-term objection to human water reuse, long term heath concern	Subtle toxic effects, feminization of males
Bacterial & viral pathogens	Common microbial diseases	None?
Protozoan pathogens	Hard to treat some "spores"	None?

[1] Wetlands phytoremediation will not work for strongly chelated metals such as nickel. Selenium, mercury and arsenic need special treatment.

— There is a long equilibration period, typically two to three growing seasons, during which treatment efficiencies may be greater or lesser than during the subsequent stable phase.

In warmer climates, most of the pollutant removal potential can be expressed within a year. Some benefits, such as increasing water clarity, can occur within weeks.

— Process dynamics in these (wetland) systems are yet to be clearly understood, leading to imprecise design and operating criteria.

In general, much remains to be understood about conventional treatment processes as well as natural treatment systems. Recent work on wetlands has lead to a much better understanding of internal processes and thus better design. However, much remains to be accomplished; fortunately many of the discoveries for conventional treatment microbial kinetics can be applied to wetland systems.

— (Wetlands are) outdoor systems spread over a large area and are highly susceptible to variability in performance due to temperature variations, storm, wind, and floods.

It is not uncommon for modern conventional treatment systems to experience partial and occasional total failures to meet discharge standards. Infiltration of storm water into sewers often overwhelms conventional plants too. Hurricanes are more of a threat to the large complex structures and electricity dependent conventional plants than to wetlands. Recent wetland designs have improved reliability by expansion of the wetland area to compensate for a lower processing rate during colder months or to capture storm floods.

— Pest control is necessary due to mosquitoes and other insects or pests that may use these systems as a breeding ground.

Even in dry and desert areas no mosquitoes or other pests need occur, so long as the proper biological controls are in place. In particular, the use of mosquitofish or bacterial pathogen of mosquitoes (Bti) has proven successful. Substantial mosquito problems have been found to be a result of wastewater inflows that deterred or killed mosquitofish. Ammonia present in wastewater at levels >5-20 mg/L seems to have been the main culprit (Horne 2002). For other views see Russell (1999).

— Steep topography and high water table may limit application of these systems in certain regions.

Treatment wetland designs are highly flexible and several cascade type phytoremediation wetlands have been proposed for steep topography (Horne 2003c). However, construction costs can be relatively high for such cascades. High water table at some sites may be an insoluble problem, but it can be partially overcome by building berms and raising the wetland above the local water table.

Natural wetlands compared with constructed treatment wetlands for phytoremediation

Natural wetlands are not very efficient nor are they reliable at pollutant removal. Short-circuiting of flows decreases the typical retention time for water in the wetland and hence decreases the ability of the system to treat pollutants. The annual mass balance for nutrients in natural wetlands often shows seasonal effects of nutrient cycling but no net loss (Elder 1985). In contrast, although many features of large natural wetlands are uncontrollable, the hydraulic regime, types of plants and animals, and drying cycles of most constructed treatment wetland can be modified to maximize treatment potential and reduce unwanted effects.

The most important difference between constructed and natural wetlands is the isolation of the hydraulic regime from its natural pattern. Relative constancy in hydraulic loadings for constructed treatment wetlands begets predictability and allows for the application of simplified mathematical constructs to model system performance and design for required removal efficiencies. Simple reactor models of treatment wetlands use inlet and outlet parameters and the assumption of steady state behavior. Although more complex versions of the reactor models can be applied under non-steady state conditions (Kadlec and Knight 1996), accurate knowledge of flow and volume fluctuations is required for all modeling periods; a requirement that is often not met for either natural or constructed systems. Pollutant loading to the wetland is ultimately controlled by how much flow is entering and leaving, which then contributes to controls on removal efficiency.

Additionally, regulation of flows and volume fluctuations in a wetland can control phytoremediation efficiency by controlling the type of plants that grow in the system. For example, many wetland plants, such as mosses and water primrose, will not grow in water more a few centimeters deep, and even cattail (*Typha* sp.) and large bulrushes (*Scirpus* sp.) do not grow well in water over 1.5 m deep. Similarly, the natural cycle of seasonal wetlands includes drying during summer that will kill many larger plant species and desiccate the biofilm. Drying in natural wetlands increases diversity since the seeds of small annuals dominate early the next year. The initial bottom contouring, flooding depth, and hydroperiod of the constructed wetland can control the general kind of plants in the system, and plant type will have an effect on pollutant removal.

Treatment wetland design: the evolution of sequential or unit process phytoremediation

Constructed treatment wetlands are classified into two broad categories, depending on the level of water column with respect to the substrate bed. In *surface flow wetlands* (SF), the substrate bed is densely vegetated and the water column is well above the soil surface of the bed. No special treatment of the soil is required. Various aquatic plants are planted on the soil with depth of water column ranging from 10-75 cm, typically less than 40 cm

(Kadlec and Knight 1995). In *sub-surface flow wetlands* (SSF) the water level is maintained below the surface of the substrate bed. The substrate medium in SSF wetlands is made of imported gravel or soil, and these systems can be either the horizontal flow type (depth commonly less than 0.6 m) or vertical flow type (depth ranges 2-3 m). Constructed treatment wetlands are also classified on the basis of plant habitat. Thus, they can be dominated by either:

- Floating macrophytes (e.g. water hyacinth, duckweed, *Lemna*);
- Submerged macrophytes (e.g., pondweeds, *Potomogeton* spp., *Chara*.
- Rooted emergent macrophytes (e.g., cattail, bulrush, common reed, water grasses).

Free surface flow (SF) wetlands are the main topic of this review since, by definition, phytoremediation treatment processes are dominated by plants. In wetlands plants provide both a carbon source and a physical structure for microbial transformations as they grow and die in the wetland. SF wetlands are also highly attractive to wildlife and can be designed with curved edges, open water areas and decorative planting to be aesthetically

Figure 1. Photograph of a large phytoremediation wetland in Southern California at the Irvine Ranch Water District's San Joaquin Marsh. This 200 ha wetland has reversed eutrophication in the downstream Newport Bay and is also designed to provide good wildlife habitat (Horne and Fleming-Singer 2003). Photograph by A.J. Horne.

enhancing (Fig. 1). Because the main processes in SSF wetlands are carried out underground, and the main carbon source is usually added or part of the inflow, SSF are best reserved for relatively smaller sites where the nature of the waste is not compatible for wildlife or human access. They are ideal for small contaminated flows from landfills or individual septic systems, while SF wetlands are better for larger flows or where beauty and wildlife are desired for the overall result.

Initially constructed SF wetlands had a simple design; they were created to mimic natural systems. Since holding a large amount of water was important they tended to be quite deep (> 2 m), too deep for higher plants. Thus they consisted of ponds having a fringe of emergent macrophytes. The pond-type design does not provide enough contact time between pollutants and the biofilm attached to plants. Natural SF wetlands used for water treatment were true marshes but usually had a central channel draining a large vegetated area. The natural SF marsh may have sufficient biofilm area, but provides insufficient contact time in the channel and excessive contact time in the rest of the wetland. Despite those inherent limitations, the early ponds and canalized marshes were often constructed as a series of several individual ponds. The ponds-in-series design substitutes for the desired plug-flow conditions in the system as a whole and approximates biofilm contact when there are more than five ponds in series. This is the case even if hydraulic short-circuiting occurs in individual ponds. Thus, early phytoremediation wetlands worked reasonably well, even if they operated well short of their optimum.

The evolution of wetland phytoremediation science and practice can be summarized as follows, where "reed" is a general descriptor for emergent macrophytes of many kinds:

- Some reed-covered islands within reed-fringed ponds of variable depth and incidental wildlife habitat.
- Dense reed beds in shallow water interspersed with a few deeper pools for wildlife habitat.
- Series of dense reed beds with ~ 5 units in the series to give the equivalent of plug flow hydraulics.
- Secondary design for specific groups of birds (e.g., shorebirds, mallard). For example a swan requires much more open water "runway" for takeoff than a mallard or other "puddle ducks".
- Series of unit-process reactors with plant species designed to carry out general groups of processes (e.g., bacterial treatment, physical sorption, suspended sediment deposition).
- Series of unit-process reactors with plant and animal species designed for specific chemical or physical treatments (e.g., denitrification, selenium removal, pathogen removal).

The developments in treatment wetland design reflect a general progression of phytoremediation knowledge and a need for performance improvements as the technology becomes more widely accepted. In particular, agencies and individuals who had not considered wetlands as phytoremediation or treatment options are now considering them seriously. Their consideration is often due to pressure from citizens or environmental groups looking for greener and more sustainable solutions for water and wastewater cleanup. In addition, most regulatory agencies such as the US Environmental Protection Agency and its state equivalents look favorably on wetlands in general.

Sequential wetlands in a unit process design

Unit processes are typical in conventional water and wastewater treatment (Tchobanoglous and Schroeder 1985) and are based on a mixture of as many as seven separate tanks (units) using different physical and chemical-biological processes for each step of the treatment. For wastewater, conventional wastewater treatment begins with an initial screen that removes large objects and a sedimentation basin removes the grit and smaller particles. Remaining organic matter and liquid is then oxygenated to convert organic compounds such as fat and protein to inorganic molecules such as carbon dioxide, phosphate, and ammonia. The bacterial-fungal association that carries out the activated sludge process also takes up some metals and refractory organic compounds. The next stages involve particle settling to remove the bacterial floc and produce a clear liquid, phosphate removal, nitrification (possibly denitrification), disinfection and finally discharge usually with a dispersion unit into the receiving water. Different unit processes are used in the treatment of drinking water, including various flocculants and coagulants which are added for specific purposes along the treatment train.

In conventional sewage treatment tanks, the water is fully mixed. In early wetlands phytoremediation design, all processes were initially assumed to be similar and go on in any part of the wetland at any time. However, wetlands are physically complex systems having considerable spatial and temporal heterogeneity, e.g., open water areas, thick reed beds, shallow areas of high dissolved oxygen, deeper anoxic zones. Thus, the oversimplified assumptions acted to decrease potential treatment efficiency of the wetland system. Use of sequential unit processes can be applied to wetland treatment systems as well as conventional wastewater treatment facilities, as shown in the following example (Fig. 2).

- Unit process #1. **Detention basin** - larger sediment removal step
- Unit process # 2. **Cattail wetland** - main bio-processing step

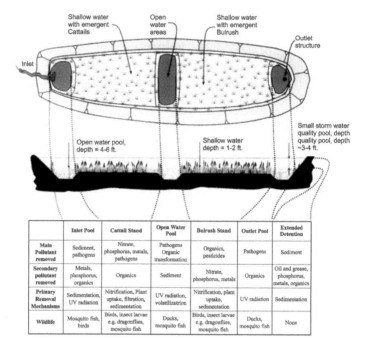

Figure 2. Generic diagram of a unit process phytoremediation wetland proposed for a watershed-scale (320 km², 120 sq miles). Designed by A.J. Horne for IRWD 2003 EIR.

- Unit process # 3. **Deep open pond** - UV and free-oxygen radical organic degradation step
- Unit process # 4. **Bulrush wetland** - main organic contaminant absorption step
- Unit Process #5. **Very shallow outlet pond** - final bacterial destruction using UV and free oxygen radicals in a shallow water layer.
- Unit process # 6. **Final clean up,** mixed emergent stands or cattails to filter out organics and particulate matter formed in unit # 5.
- Unit process # 7. **Optional sand filter** to back up process # 6
- Unit process # 8. **Disinfection step** if needed

Unit process #1- the detention pond for sediment removal and optional phytoremediation. The first unit process is an inlet pond, serving primarily as a sedimentation basin to remove silt and pathogens often attached to particles. Sediment is frequently removed by excavation and that does not lend time for plant growth. In addition, conventional detention basins in arid climates are dry for most of the year so the role of aquatic phytoremediation is small or absent. In wetter climates detention ponds are often permanent with a fringe of reeds, but phytoremediation appears also to be small since there is little contact between inflow and plants.

Detention ponds are required in most cities to prevent flooding so will continue to be a common urban feature. They are not beautiful but it is possible to combine the use of aquatic phytoremediation with aesthetic improvement in detention ponds. The main need, as always, is to provide some plant component that would enhance the detention basin's function. Two ways are possible; increased baffling and coagulation potential. Some detention ponds have concrete baffles to slow down the water and increase sedimentation rates. The stiff two or three meter high cattail or bulrush stems serve a similar purpose in natural marshes and do the same in detention ponds so long as the water runs between and not around them. Appropriate contouring of the basin can ensure the flow path is through the vegetation stand. The organic matter in wetlands, especially if there is a wet biofilm, increases the amount of flocculation and setting since microbes excrete many organic compounds such as mucopolysaccharides.

In arid climates detention ponds are dry most of the year and require a summer water source. One way to maintain a wetland in such conditions is to divert summer "nuisance" runoff from landscape irrigation overflow or driveway car washing. An innovative combination in Orange County, California that treats summer runoff in a detention basin solves the problem of sediment excavation by an initial internal rock berm that holds back most of the heavy silt from the wetland section. The summer flows are much smaller than the winter storms so the wetlands are confined to a series of small marshes set into the larger detention basin.

Unit process #2: the cattail marsh and its microbial treatment system. The second unit process is the main biological treatment system, usually a cattail unit. Cattails are hardy, rapidly-growing emergent macrophytes that are easy to grow, are resistant to overgrowth by most other plants and are large enough to provide a lot of biomass each year. Most importantly for biological treatment, cattails have a relatively large amount of labile carbon relative to lignin in their tissues (Hume *et al.* 2002). Thus, when cattails die and fall into the water, they provide an excellent carbon substrate for bacteria as well as a physical surface for the microbial biofilm. Microbial respiration depletes the dissolved oxygen supply in the water column creating anoxic sediment and lower water zones in the wetland treatment facility. Primary removal mechanisms in the cattail unit include 1) active microbial processes, including nitrification-denitrification, transformation of soluble ionic heavy metals to insoluble sulfide precipitates and uptake of phosphate into the biofilm; 2) plant uptake, filtration and sedimentation;

and 3) pathogen removal by active consumption, passive coagulation and settling (see later section for details).

A secondary role of the cattail unit is to treat organics by providing the carbon source and substrate habitat for microbial action to destroy or partially degrade many organic contaminants. Atrazine is a good example of a pesticide that is fully metabolized and destroyed in the cattail zone, while the common explosive TNT is an example of partial destruction to DNT and additional degradation products (Zoh and Horne 1999). Some organic degradation products are comparable to or even more toxic than the original contaminant, so care must be taken not to make such treatment systems attractive to wildlife. More complex organic contaminants such as multi-ring compounds, like PAHs and PCBs, are not likely to be degraded in the cattail marsh but can be removed intact in the later bulrush-peat unit.

The cattail unit is not ideal as a habitat for larger wildlife because cattails do not have large nutritious seeds. The leaves and stems are a poor food source for birds, mammals and insects. However, there are certain moths that eat cattail leaves, blackbirds will nest in the dense thickets, and muskrats will eat cattail rhizomes when the ponds are drawn down. More promisingly, down in the water column invertebrate larvae thrive and provide food for fish, ducks, and wading birds, particularly along the edges of the dense vegetation.

Unit process #3 - algae and UV pond. The third unit process is the deep, open-water pond that provides algal phytoremediation and UV/oxygen free radical destruction of organics and pathogens.. Deep water prevents encroachment of adjacent cattail and bulrush plants into the open-water pond. Another kind of phytoremediation takes place in the open-water pond since at this stage in the treatment process there are often still sufficient nutrients for substantial algal growth. While algae rapidly take up many pollutants, their short life cycle relative to other plants in the ecosystem means that re-cycling of pollutants can occur rather than permanent immobilization. However, algae increase dissolved oxygen in the water column as a byproduct of photosynthesis, and during warm afternoons it is not uncommon to find 20 mg/L or about 200% saturation. Combined with the strong UV of a sunny day, it is likely that free oxygen radicals are present in the water which will assist UV in pathogen destruction. In addition, high levels of photosynthesis can increase pH to greater than 9.5, adding to the discomfort of pathogens. In addition, direct photo-destruction of some organics such as pharmacologically active substances including birth control drugs can occur in this high UV/high oxygen environment. Finally, the open water provides the main open water habitat for birds and good wildlife viewing opportunities for humans.

Unit process #4- bulrush-peat marsh for absorption of organics. The fourth unit process is the bulrush stand that provides peat which absorbs organic contaminants. Bulrush are stiff, upright plants because they contain relatively high amounts of lignin although consequently less labile carbon for bacteria (Hume *et al.* 2002). The high lignin content resists decay in the anoxic conditions in flooded sediments. Planting a wetland with bulrush as the dominant emergent macrophyte thus increases long-term peat in the wetland sediments. Peat contains large molecules of humic substances that bind many organic contaminants including PAHs and pesticides as well as some metals. While total destruction of organic contaminants by phytoremediation is the ideal goal, it may not be possible with all compounds, particularly those containing recalcitrant benzene rings or other aromatic hydrocarbons (Coates *et al.* 1997). With a few exceptions, only fungi, especially the class of white rot fungi, possess the lignin peroxidase enzymes that can break the ring compounds present in aromatic hydrocarbons (Srebotnik *et al.* 1994). Fungi are obligate aerobes and do not grow well in the ever-present anoxia of permanent wetlands. They require oxygen for metabolism; however they also require damp conditions and so are a part of the wetland ecosystem only at the air-water interface in association with decaying plant material. Coordination between the new field of fungal or mycoremediation and phytoremediation in wetlands (see end of this paper) offers promise for the successful destruction of many aromatic hydrocarbon contaminants. Until this new science advances, sorption to humic substances is the primary design removal mechanism for natural treatment systems. Crompton and his students at the University of Iowa, Ames, have shown that pesticides absorb rapidly to humic substances such as peat in wetlands. Further, this work has shown that the attachment becomes stronger with time (Crompton, pers. comm.), presumably due to partitioning of the pesticide into the humic material, partial degradation, rearrangement, and recombination of the original molecules present in the humic matrix with those of the pesticide. Drying or other wetland manipulations apparently do not re-release absorbed pesticide. The removal of heavy metal by absorption onto refractory carbon is discussed later in this review.

Unit process #5 -intense algal and UV treatment for xenobiotics and pathogens. The fifth unit process is the shallow exit unit. Water only 10 cm deep and clear of turbidity due to prior treatment in the wetland is ideal for further UV destruction of pathogens and large organic molecules. By lining the exit site with concrete, most plants are excluded and shading does not occur. However, algae will grow on the bottom and in the summer months their dark color can increase the water temperature to over $30°C$, the pH to 10, and the dissolved oxygen concentration to 25 mg/L. Even more so than in unit process #3 (the deep pond), these conditions are ideal for pathogen and organic destruction by UVB and free oxygen radicals. A unit of this design has never been constructed and is yet to be tested for quantitative performance. The concrete base can be cleaned of algae by truck-based rubber blades and brushes, as needed.

Unit process #6 - algae and organic fragment removals with cattail & bulrushes. The sixth unit process is required to filter out and degrade the lower molecular weight compounds produced during the destruction of dissolved organic pollutants in the 5th unit process. In addition, any algae sloughed from the surface of the concrete bottom need to be filtered out. In particular, a turbidity of < 2 NTU is usually required for discharge to many surface waters and is needed for conventional disinfection steps. Additionally, some degradation products from unit process #5 may be harmful to wildlife and should be treated rather than discharged. A bulrush-peat wetland or a mixed cattail-bulrush wetland is appropriate for the final plant-based stage.

Unit processes #7&8. The 7th and 8th unit processes include a sand filter to ensure low turbidity (< 2 NTU) and a disinfection step as needed. These are not plant-based steps and will not be considered further here. They may not be needed for most wetlands but are prudent considerations given the present state of wetlands phytoremediation science and practice.

Summary of unit processes

Not all phytoremediation treatment wetlands require or are designed to incorporate as many or the same unit processes as described in the multi-use example above. For example, if nitrate is the only pollutant of concern in a particular water, then a cattail wetland alone will suffice (Philips and Crumpton 1994, Bachand and Horne 2000b). Similarly, if the waste is clear but contains ammonia, a nitrification step (often a sand bed) is useful (Reed *et al.* 1995). Where only refractory pesticides and other xenobiotic organic contaminants must be removed a bulrush wetland may be all that is needed. In a complex test, eight combinations of unit processes consisting of mixture of cattails or common reed beds were used in combination with other beds of sand, fine and coarse gravels (Cerezo *et al.* 2001). This Spanish study confirmed that the choice of unit processes is dependent on the kind of waste inflow and the legal standards that must be met for outflowing water.

However, in most wastewater treatment applications the water will contain a variety of contaminants, from nutrients to metals to pesticides and PAHs. Figure 2 suggests a guide for general wastes. Wetland phytoremediation using unit processes may be combined with conventional wastewater treatment, as in the recent case involving the City Council of

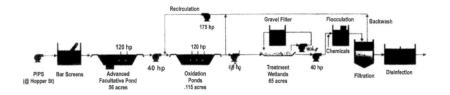

Figure 3. A phytoremediation free surface treatment wetland in a unit process train proposed in 1999 as one alternative for the City of Petaluma, California. Corollo Engineers, Walnut Creek, California.

Petaluma, in Northern California (Fig. 3). They specifically requested a "green sustainable solution" to be incorporated into wastewater treatment plant expansion plans. Based on this, a 35 ha wetland was added to the treatment train to reduce algal growth in the shallow oxidation pond, and is expected to reduce the need for the sand filtration step needed in the conventional treatment train. Some polishing of organics and metals is expected from this system but is not essential.

Examples of phytoremediation wetlands

Nutrient Removal

Nitrogen removal in the San Joaquin Wildlife Sanctuary, Irvine, CA.

The San Joaquin Wildlife Sanctuary (SJWS) is a 32 ha series of 6 shallow ponds owned and operated by the Irvine Ranch Water District (IRWD). The marsh was created to maximize nitrogen removal rates while still maintaining 90% open water and episodically exposed shoreline for waterfowl, shorebird, and wading bird habitat. These avian design elements created non-ideal denitrification conditions in the marsh by diminishing an important source of organic carbon (emergent vegetation) and increasing sediment exposure to oxygen. A novel phytoremediation strategy was used in the SJWS to enhance organic carbon and related denitrification potential by seasonally planting barnyard grass (*Echinochloa crusgalli*) in two of the largest ponds in the system. The grass was intended to serve both as a carbon amendment for denitrification and as a physical surface for microbial attachment within the water column. Use of barnyard grass was based on a 1999 study which compared the denitrification enhancement potential of several carbon amendments including barnyard grass (*E. crusgalli*), disked-in wheat straw (*Triticeae* sp.), purple three-awn grass (*Aristida purpurea*), molasses, and the native soil and bulrush (Hume 2000, Horne *et al.* 1999). During the 1999 study, planted *E. crusgalli* enhanced denitrification by providing labile carbon and a greater surface area for the attachment of denitrifying bacteria.

Recently, SJWS aqueous nitrogen and avian data for the non-winter months of 1999-2002 were analyzed to determine whether design and operating conditions allowed for simultaneous nitrogen removal and diverse, abundant avian habitat (Horne and Fleming-Singer 2004a). Marsh management practices currently involve draw-down of Ponds 1 and 2 once per season for *E. crusgalli* seeding, and Ponds 3 and 4 approximately bi-weekly throughout the year in order to provide foraging sites for shorebirds. Thus, on-going pond volume perturbations occur on a roughly two-week cycle with an additional 8-week cycle occurring during the summer months. Four-week running averages of hydraulic and water quality parameters (e.g., flow, residence time, nitrogen, temperature) were used to account for information about system dynamics without being overwhelmed by the extremes of changing pond volumes occurring on smaller time scales. Denitrification rates were estimated using inlet and outlet parameters and avian species diversity and abundance were analyzed and compared with similar systems in Northern California (Fleming-Singer and Horne 2004b).

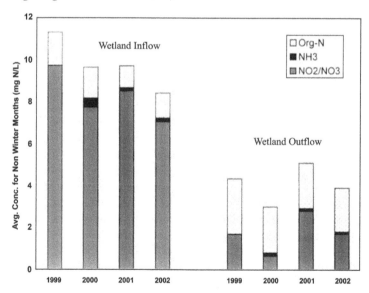

Figure 4. Removal of nitrate fractions in San Joaquin Marsh, a phytoremediation system with both algal and bulrush components. The open water decreases overall efficiency but increases bird use (see Horne and Fleming-Singer 2003).

Overall, avian design features did not appear to inhibit high rates of denitrification in the SJWS during 1999-2002 (Fig. 4). The highest aerial nitrate removal rates occurred during April-May (350-500 mg/m^2/d) and September-October (250-425 mg/m^2/d) of each year, corresponding to the highest loading periods for inorganic nitrogen in the marsh (Horne and Fleming-Singer 2003). These rates are comparable to denitrification rates in other constructed treatment wetlands systems (Horne 1995). First order rate constants ranged 0.05-0.25 d^{-1}. There was no discernable difference in nitrate removal when comparing carbon amended and non-amended conditions, which may be because data averaging obscured a small, localized enhancement signal.

For 2001-2002, the average combined bird density was 46 birds ha^{-1} and the total number of bird species observed was 156. The number of bird species observed there is higher than that of other constructed wetland systems while average combined bird density at SJWS indicated that species abundance was roughly mid-way between reported abundance levels of other constructed wetland systems (Fleming-Singer and Horne 2004b). Thus, the SJWS appears to be successfully removing nitrate and providing habitat for a large variety of bird species. Low levels of organic-N were produced in the SJWS (mean = 1 mg/L) and based on chlorophyll a measurements, roughly 40% of it was present as algae, while the remaining 60% was likely leaving the system as dissolved organic matter (DOM). Algal-N production was greatest relative to TN-removed in July and August of each year.

Phosphorous removal in Florida using a triple unit process

Another unit process that has promise for solving one of the more intractable but also important contaminant problems is the use of a wide range of aquatic plant types set in series to gradually remove phosphorus from water. As mentioned earlier, phosphorus (P) is normally only temporarily retained in wetlands and is usually swiftly recycled. Often less than 5% of added P is permanently retained and Richardson et al. (1997) suggest that only 1 g P/m^2 can be removed in the long term in wetlands. In contrast, as shown above, up to 200 g N/m^2 can be removed by wetlands phytoremediation. Since phosphorus is an important stimulator in the eutrophication of lakes, its removal is desirable, especially in areas where land development or farming has increased nitrate loadings. A low total phosphorus (TP) standard of 10 ug/L has been set for the protection of the Florida Everglades for storm water entering from the agricultural and small urban areas to the north. Given that the storm water volume is very large ~ 1 x 10^9 m^3 (~800,000 af) and the TP concentration is 70-220 ug/L, the TP levels are up to 20 times the desired standard. Phosphorus removal by wetlands phytoremediation seems impossible.

Using the unit process concept in studies sponsored by the South Florida Water Management District, a group of wetland engineers and scientists came up with an ingenious solution to the Everglades TP requirements. A series of three phytoremediation cells were linked in series to gradually lower the TP level to the required level. The stages were a typical cattail wetland similar to that described as unit process # 2 above, followed by a submerged aquatic vegetation (SAV) wetland, and finally a periphyton-based stormwater treatment area (PSTA) wetland (CH2M-Hill 2001). In this system, the SAV consisted of various macrophyte species mixtures dominated by *Najas* and *Ceratophyllum* with lesser amounts of *Potamogeton* and *Hydrilla*. The periphyton cells were planted with sparse stands of spikerush (*Eleocharis cellulosa*) and bladderwort (*Utricularia* spp.), since these are not invasive macrophytes and are good substrates for attached algae. Of the over 300 species of periphyton that grew on the plants and soil, there was and even split between diatoms, green algae and blue-green algae.

The ingenious part of the system is that neither the SAV nor PSTA could survive at higher TP than the inflows provided by the upstream cattail wetland. Cattails would rapidly overgrow the other species. The concept was partially inspired by the observations in the natural Everglades wetlands, the famous "river of grass" named in the 1940s by Marjory Stoneman Douglas (Douglas 1988). Sawgrass, more accurately a sedge marsh (*Cladium jamaicense*), wetlands with the natural supply of rainwater are low in phosphorus. They are indeed dominated with sawgrass but are pocketed with beautiful clear pools of shallow water containing water lilies and bladderwort intertwined with blue-green algae (cyanobacteria) buoyed up by the bladderwort's air sacs. The key point is that among the blue-green algae mats are whitish precipitates of insoluble calcium phosphate (apatite) which permanently immobilizes the phosphorus (Fig. 5). The high pH produced during the day by the photosynthesis of the algae and high concentration of calcium ions in Florida waters combine to form the phosphorus-rich precipitate.

Using the polluted stormwater supply at a site upstream of the Everglades in various kinds and sizes of mesocosms, the cattail wetland reduced TP from over 200 to 40-50 ug/L. The submerged plants dropped TP to about 24 ug/L and the periphyton treatment further reduced TP to 8-14 ug/L under the best conditions (SFWMD 2002, Fig. 6). These experiments were run over a few years so the removals are assumed to be in equilibrium with any recycling and releases of phosphorus. The overall phytoremediation system thus depends on a chain of phytoremediation starting with large robust plants such as cattails then large submerged plants and finally encouraging an apatite-precipitation reaction with small and delicate

Figure 5. Natural wetlands phytoremediation in the Everglades. The whitish areas on the water surface are precipitates of calcium phosphate (apatite) produced by an interaction of blue-green algae buoyed up and supported by bladderwort (*Utricularia*). Apatite production is the only common long-term way to immobilize the normally readily recycled-P.   Sawgrass and lilies are also present. Photograph by A.J. Horne.

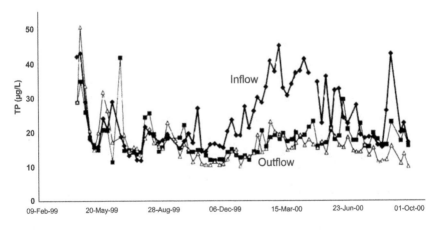

Figure 6. Reduction of total phosphorus to very low levels by a sequential unit process wetland series. The figure shows the periphyton- small submerged aquatic vegetation final step that often reaches the target of 10 ug/L (SFWMD 2002).

periphyton algae. A very careful ecological balance is needed to achieve this very difficult goal and keeping the plant groups separated in a unit process train is essential. In some cases, SAV and periphyton can be considered together since one often grows on the other. Details of a combined SAV-periphyton system as used in other parts of Florida are provided by Knight *et al.* (2003).

Metals Removal

Metals removal in phytoremediation wetlands occurs by several mechanisms, including straightforward sedimentation of particles as would occur in conventional detention ponds. In terms of phytoremediation, metal removal occurs via the (1) creation of insoluble forms such as sulfides (powered by decaying plant carbon and mediated by bacteria), (2) direct absorption to dead plant or animal material and (3) accumulation in plant tissue. The order given is the probable order of importance in wetlands. Metal sulfides, and other complex compounds such as basic copper carbonates, are similar or identical to the ore from which the metals were originally mined. The green color on church roofs is due to basic copper carbonate and provides visual confirmation that the copper is insoluble even under the acid rain found in cities. The biogeochemical process of metal removal via insoluble sulfide compounds depends on sulfur reducing bacteria (SRB) which are present under anoxic conditions. Metals sulfides and other ores are quite stable even when dried out, but can be broken down if water, oxygen, and SRB occur together. The threefold combination is found in open mine spoil heaps where rain provides the water, the loose rocks allow free access of air, and sulfur oxidizing bacteria are quite common in nature. However, the wetland sediments present a very different picture since flooded soils are always anoxic. Thus wetlands can provide a long-term removal of metal if they are transformed into the sulfide form or a similar ore-like chemical.

Heavy metals removal from highway runoff

Most work on metals removals in wetlands has concerned storm or mine runoff and this biases the data base considerably. The use of wetland phytoremediation for the treatment of metals from highway runoff has been documented in the US (Kadlec and Knight 1996) and in the UK (Shutes *et al.* 2001). An experimental highway runoff treatment system in England was tested for removal of several metals, including lead, zinc, copper, cadmium, chromium, nickel, vanadium, molybdenum, platinum, and palladium (Shutes *et al.* 2001). The treatment system incorporated two different types of wetlands; a SSF treatment wetland planted with cattails (*Typha latifolia*) and the common reed (*Phragmites australis*), and a pond planted with a mixture of floating, emergent, and submergent species. Both systems were preceded by oil separators and silt traps, and a settling pond was also placed ahead of the SSF wetland. Overall aqueous metal concentrations were low, presumably because the highway was relatively new and metals may have been retained in the roadway matrix. However, despite this, several metals were consistently removed from runoff (maximum efficiencies of 68% copper, 70% chromium, 60% zinc, and 87% nickel), particularly during summer months. Although not mentioned specifically, almost all of the metals would be present as sediment bound fractions or other particulates (e. g., copper particles from brake linings). As in most other cases of storm water wetlands, it is hard to determine how much of the metal removal was phytoremediation and how much was detention by simple sedimentation or filtration by the underground gravel beds.

Metal accumulation in root/rhizome tissues was indicated for both plant types in the gravel bed sub-surface wetlands, particularly for zinc, nickel, and copper in *Phragmites* root/rhizomes and vanadium in *Typha* root/rhizomes, but also in *Phragmites* tissues. Nonetheless, the fraction of metal retained in vegetation is tiny (Mitch and Wise 1998). The most likely mechanism for rhizome heavy metal accumulation is on the outer surface via absorption to ferric hydroxide flocs that form where oxygen leaks out from the root hairs into the surrounding soil (Horne 2000). Lead, molybdenum, chromium, platinum, cadmium, and palladium concentrations were low in both plant root/ rhizome and leaf tissues, reflecting the generally low concentration of bioavailable forms of these metals in the rhizosphere. The wetland was a source of copper during the only two storms sampled during the study period. This was explained by release of organically bound copper following aquatic plant senescence. Such imbalances are temporary since over the long term wetlands cannot be a source of metals unless they receive aerial fallout, as occurs in the case of mercury and lead.

By directly measuring the sedimentation of storm water metals attached to particles in a large wetland Walker and Hurl (2002), found the magnitude of phytoremediation by difference. In this Australian study in a large pond-like wetland ($A = \sim 0.1 \times 10^6 \, m^2$) it was found that about half of the wetland removal of zinc and copper and almost three-quarters of the inflow of lead was due to phytoremediation. They defined the phytoremediation metal removal as "filtration by plants, adsorption, biological assimilation, decomposition, chemical transformation and volatilization." A Canadian study on metals removals from stormwater in a young wetland, where plant uptake would not be in equilibrium, showed that first order removal kinetics did not apply for most of the year for most metals (Goulet *et al.* 2001). They concluded that "biological and well as hydrological variables" control metal removals, especially in cold climates.

In contrast to terrestrial phytoremediation the accumulation of heavy metals in plant tissue is less used in wetland phytoremediation since the other methods are easier and less costly. In addition, the production of edible wetland plants high in heavy metals could easily lead to the deaths of birds or insects as was found for Kesterson Reservoir marsh (see below). In the English example discussed above (Shutes *et al.* 2001), most removal occurred by passive sedimentation although the authors did find some un-quantified level of rhizosphere removal in the sub-surface wetlands. Work by others suggests that these root coatings range from 32-93% of the total metal concentration in the roots (Mays and Edwards 2001). However, the cost of constructing gravel beds limits such rhizosphere activity to small wetlands. In addition, the flow of water through gravel beds is much slower than through open water and reeds. The amount of metals taken up in cattail-bulrush-juncus marshes was a small fraction (1 or 2%) of the total metal load for iron, manganese, copper, and nickel (Eger and Lapakko 1988, Mays and Edwards 2001). In a recent review of the topic it was concluded that typical treatment wetlands store less than 0.1% of the annual inflow of iron in above ground vegetation (Mitch and Wise 1998). Adding the below-ground average amounts might increase this value to 0.3% of total annual load. The small quantities emphasize the general conclusion that aquatic metal removal is independent of vegetation uptake and dependent on vegetation as a site for precipitation or, more likely, enhanced sedimentation in various ways. These conclusions, however, are based on particulate metal phytoremediation in wetlands. The now pressing topic of removal of soluble metals in wetland may produce a different role for vegetation.

Role of carbon in dissolved metals removal in wetlands

When dissolved metals come into contact with organic matter they can be chelated and rendered non-toxic but still soluble. They can also be bound more firmly and removed from solution. In treatment wetlands, the abundance of plant material allows the direct ligand-binding potential of organic matter to play a role (Kerndorff and Schnitzer 1980). The removal occurs when the metal comes in contact with functional groups of the organic matter such as carboxyl (COOH), phenolic (OH), quinine and ketonic carboxyl (C=O), amino (NH_2 and R-NH), and sulfhydryl groups (Sposito 1986). The sulfhydryl groups promote the strongest bond with metals but are not the most common functional group on degraded organic matter. Humic and fulvic acids are major products of plant decomposition and are present in the peat or leaf litter in wetlands. Humic acids have a complex mobile structure and change in response to pH and salt concentrations as well as age. A Portuguese study on the composted sewage and natural soils showed that the fluvic acid fraction formed "quite stable complexes with divalent metal ions" of copper, lead, and cadmium (Esteves *et al.* 2002) at least in near neutral conditions. Thus it is possible to harness both methods for metals removal (microbially mediated and physical adsorption). In the example shown in Table 2, it is probable that a combination of sulfide reduction and direct metal absorption occurred, although the slow kinetics would appear to favor eventual sulfide formation.

Table 2. Removal of dissolved copper from mine runoff by various kinds of carbon substrates. All samples contained about 4 g C per microcosm experiment. Leached cattails were used as the substrate for the growth of bacteria and were leached to remove labile carbons. The wetlands plants alone were slower acting than the combination of cattails and other carbon sources (modified from Hauri 2001).

Substrate	Dissolved copper ppm		
	Start	*After 60 days*	*After 120 days*
Control	8	8	8
Cattails alone	8	6	4
Leached cattails + redwood leaves	8	0	0
Leached cattails + sawdust	8	2	0.6
Leached cattails + molasses	8	2	0.5

Mercury and selenium as potential threats to wetland phytoremediation

Under typical anoxic conditions in wetlands, most metals can be removed from solution. However, not all transition rapidly and safely from their initial form to the final immobile and non-toxic form. Two metals, mercury (Hg) and selenium (Se), are particular problems in wetlands. Mercury is transformed from the biologically unavailable forms of inorganic mercury (Hg^0, Hg^{2+}) to biologically available and toxic methymercury ($HgCH_3$) in wetlands. Bioaccumulation of mercury refers to the net incorporation of mercury in an organism from its environment, which typically results in biota concentrations that are orders of magnitude greater than ambient water concentrations (Weiner *et al.* 2003). Mercury trophic transfer begins at the bottom of the food web with sulfur-reducing bacteria (SRB) adsorbing and then methylating dissolved inorganic mercury. The methymercury moves up the food chain to include zooplankton and herbivores consuming the bacteria, small fish feeding on zooplankton, and beyond to large fish feeding on a combination of food from the lower trophic levels. Anoxic environments such as wetland sediments support higher mercury methylation rates and increase the rate of mercury bioaccumulation relative to other ecosystems. This would seem to rule out wetland phytoremediation when mercury is present. Unfortunately, mercury is distributed widely via atmospheric deposition from power plant emissions, so is quite common, even in originally unpolluted regions. Thus, there is a need to determine how to construct a treatment or other type of wetland that will remove other pollutants but not create conditions supporting methylmercury production. Since mercury is both toxic to biota and is not required for biological processes, its permanent immobilization in the environment is desirable. Some recent progress has been made (Mehrotra *et al.* 2003) that suggests increasing the iron in the wetland may reduce mercury methylation by competitive use of the sulfur substrate needed by the SRBs.

Selenium (Se), an element with both metallic and non-metallic properties, can also bioaccumulate rapidly. Depending on its concentration and chemical form, Se serves as either an essential element or a strong toxicant to biota, including humans, livestock, plants, waterfowl, and certain bacteria (Frankenberger and Benson 1994). Bioaccumulation of Se at toxic levels can occur in wetlands as well as lakes and rivers. However, the higher productivity of wetlands compared with other aquatic habitats and the low threshold between sufficiency and toxicity means that damage from Se becomes evident earlier. The well-publicized and dramatic case of Kesterson Reservoir, a marsh located in California's Central Valley, became the first demonstration of widespread Se toxicity in wetlands in the mid 1980s (Ohlendorf and Santolo 1994). Unknown to the farmers, high concentrations of Se were eluted from the soil by irrigation and subsurface agricultural drainage. The water flowed to the marsh, and within five years caused reproductive deformities and death in resident birds. Despite the fact that Se toxicity was tragic for the birds, the Kesterson case is ironic because the wetland trapped almost all of the incoming Se giving considerable protection to the downstream waters of the San Francisco Bay-Delta ecosystem.

It is possible to immobilize Se using treatment wetlands, because under the anoxic conditions characteristic of wetlands, Se is either rapidly immobilized to a red metallic precipitate (Se^0) or captured by organic-Se-H bonds. The metallic precipitate is not bioavailable and remains in the metallic form so long as anoxia persists. Since anoxia is a permanent feature of continually flooded wetlands, they act as traps for Se. If the Se-H bond is attached to dimethylselenide, a volatile compound manufactured by algae or aquatic macrophytes, the Se compound is transported from the wetland to the atmosphere. Since atmospheric Se (as dimethylselelnide or dimethyldiselenide) is a natural component of the global Se cycle and plays a vital role in mammalian biochemistry, the current state of knowledge regarding Se suggests that it is reasonable for small amounts to be released as a byproduct of wetland treatment. If the Se-H bond is attached to dead plant matter it can be immobilized under anoxic conditions. However, when the Se-H bond is attached to living biota, toxicity will occur with more sensitive species, particularly aquatic birds such as the common mallard.

If oxic conditions are re-introduced to the wetland, the metallic Se precipitate or the organically bound Se-H is rapidly converted to oxic forms of Se (SeO_3^{2-}, SeO_4^{2-}), which are soluble and bioavailable. Hence, despite the fact that wetlands are more productive if they have a seasonal drying cycle, for Se immobilization the wetland must be held continuously anoxic. This reduces the role of the treatment wetland as a wildlife habitat and emphasizes the role of aquatic plants for providing the necessary anoxic conditions to immobilize Se. Once again the plants' role is the indirect provision of labile carbon and an anoxic habitat that is the key to Se-phytoremediation. It is advisable in the case of Se, and Hg, to de-emphasize use of the wetland as a wildlife habitat to avoid any possible toxic effects.

Organics removal

Wetlands remove and create many kinds of organic molecules as might be expected if the extensive microbial biofilm is considered as a huge biochemical factory. Inflowing organic compounds can be retained, transformed, and sometimes fully degraded to both simple and complex organic molecules. While total organic carbon (TOC) does not necessarily increase through wetlands (Horne, 2000), there is a tendency for an increase in refractory organic molecules in the effluent. This is evident when considering nitrate removal. When nitrate passes through a wetland most of it is denitrified to nitrogen gas; for example, over 80% of the inflowing 11.5 mg N/L in San Joaquin Marsh. However, the organic-N which made up only 15% (~ 1.7 mg N/L) in the influent river water doubled to represent about half of the lower total-N value in the effluent (Fig. 4). Accounting for the living algal fraction of that effluent, organic-N still accounted for 60 % of the TN in the outflow or about 0.6 mg/L. Since refractory organic compounds are ubiquitous and have beneficial effects, such as binding of toxic heavy metal ions, these changes are not necessarily of concern. However, where the wetlands effluent is to be used for drinking water a rise in any kind of organic that will raise TOC above 2-3 mg/L is of concern since potentially harmful byproducts may occur following disinfection. Once again the selection of the wetland vegetation during design of the wetland, especially the final units can reduce TOC and humic matter.

More usually, the wetland is required to remove, immobilize and if possible destroy complex organic molecules such as pesticides, PAHs and PCBs or exotic compounds such as TNT or other explosives. An exciting new area is the removal of trace quantities of pharmaceutically active compounds such as birth control pills. Most drugs taken by humans or given to livestock pass through the body and soon reach surface waters and thus wetlands. A nice example of removal of several pesticides from agricultural storm runoff was given by Schulz and Peall (2000). Here only a few hours of residence time were required to reduce or eliminate the concentration of azinphos-methyl, chlorpyrophos, and endosulfan pesticides (Table 3). Most of these more modern pesticides are strongly attached to particles so that their removal by the wetland could also have been replicated by a simple storm water detention pond. However, it is likely that more than simple sedimentation was involved. It is in this direction of long-term inert storage or even total degradation that wetlands phytoremediation must progress.

Table 3. Removal of pesticides in storm water by a surface water wetland. Very good removal occurred but it is not clear how much loss was by simple sedimentation and how much by phytoredediation (but see simazine below). Eighteen mm of rain began on December, 1988, at 3 pm. Data modified from Schulz and Peall 2000.

Pesticide (ug/L)	Time				
	3:30 pm	4:30 pm	5:30 pm	6:30 pm	7:30 pm
azinphos-methyl, inlet	0	0.14	0.31	0.85	0.27
azinphos-methyl, outlet	0	0	0.07	0.05	0.05
azinphos-methyl, inlet	0	trace	No data	0.02	0
azinphos-methyl, outlet	0	0	0	0	0
endosulfan, inlet	0	0.06	0	0.2	0
endosulfan, outlet	0	0	0	0	0

Perhaps the best example of ideal wetland treatment of organic contaminants is atrazine and its aquatic form simazine. They are common selective herbicides that target broadleaf and grassy weeds. Atrazine is a small molecule comprising a single mixed ring with three carbon and three nitrogen atoms and two amino side chains. It was widely used for 40 years on farmland and roadside verges before being restricted in 1993. It has been linked with environmental effects on amphibians and can produce human health effects at high levels. Atrazine does not bind to soil very well and is easily leached into surface waters finding a way into groundwater and drinking water wells. Its environmental half-life is quite long (60-100 days) and atrazine is still quite common in low concentrations in many areas.

In a recent study in South Carolina, atrazine flowing into a natural wetland was sampled before and after a storm (Kao *et al.* 2001). The wetland contained swamp areas with cypress, red maple, willow and spruce and marsh regions with cattail and bulrush. Agricultural areas upstream drained into the wetland and contained considerable amounts of nitrate (6-9 mg/L) and ammonia (6-11 mg/L) indicating rapid flushing from the soil. The wetland had a hydraulic residence time of about 10 days and retained all the inflowing atrazine (up to 130 ug/L) during the five wet days (Fig. 7). The outflow concentration of atrazine was >1 ug/L. Unlike many most stormwater studies it is clear from the simple mass balance that the removal of the pesticide was not due to sedimentation since atrazine is soluble in water. Therefore, the wetland outperformed a typical detention basin where only the background degradation would occur (estimated at about 16 ug/L of the measured 130 ug/L drop).

In an excellent addition to the storm monitoring, the authors performed microcosm studies with a wetland inoculum that established that there was a high atrazine removal rate when sucrose or sucrose and nitrogen

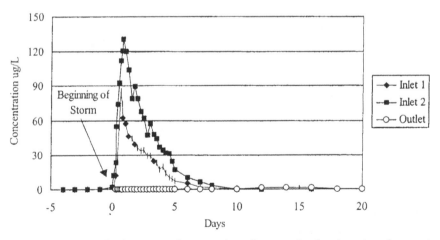

Figure 7. Removal of atrazine by a surface flow wetland. Atrazine does not sediment out in in wetlands or detention basins since it is soluble in water. Microcosms studies carried out with an inoculum from the wetlands showed that atrazine was destroyed when labile carbon was added. Modified from Kao *et al.* 2001.

were added under anoxic conditions. Removals were less in oxic conditions. Naturally, sucrose is a very labile compound but similar small labile carbon fragments are released from decay in wetlands. In wetlands ponds where algae can grow well in the high nutrient agricultural water it is likely that the three-carbon glycollate molecule is common and would further add to atrazine's rapid demise.

Glycol removal in combined surface-flow and subsurface-flow wetlands at Heathrow Airport

Glycols, widely used as anti-freeze in automobile radiators is also used as aeroplant de-icers and anti-icers. Both ethylene and propylene glycols are toxic to many organisms from cats to fish and are thus generally undesirable in surface waters. Unfortunately, glycols are highly soluble in water and not removed by typical sedimentation basins or surface film oil skimmers. A pilot-scale wetland phytoremediation system was developed using a combination of SF and SSF reed beds to remove glycols in runoff from Heathrow Airport, London (Revitt *et al.* 2000). The Heathrow Airport runoff eventually makes its way to the River Thames. Despite their wide use and known detrimental effects on receiving water quality, relatively few airports have recovery systems for glycols (Sabeh and Narasiah 1992). Using the wetland phytoremediation system, average glycol removal efficiencies for a surface water wetland containing was 54%, following shock dosing inputs. As with most systems, the authors suggest that the surface water wetland becomes part of a unit process system where an aeration pond would reduce BOD that is not efficiently degraded in the anoxic conditions of wetlands sediments.

Pathogen Removal

Pathogens in rivers and streams may be one of the most pressing problems for overall watershed health. In recent years there have been an increasing number of beach closures worldwide due to excessive pathogen presence, usually as bacteria. Some idea of the scale of the problem can be seen in Chicago where urban stormwater cannot be disposed of into the nearest large water body, Lake Michigan, since it is also the drinking water supply for that city. Disposal into the nearby upper Mississippi drainage is limited by the size of the canal connection. In the 1970s it was proposed to hold the storm water in an underground chamber and treat and release it over time. The cost at that time was $8 billion due to the size of the excavation needed. Chicago is still wrestling with this problem in 2004. In Southern California frequent closures of Huntington Beach due to high pathogens has been a major cultural shock for the sun-loving local inhabitants. The usual source of the pathogens is storm runoff from the surrounding watershed or wastewater disposal originating from localized activities. Following precipitation events, it is typical to find enormous amounts of bacteria in just about any urban stream or waterway located near livestock operations. Perhaps more sinister is the amount of human protozoan pathogens such as *Giardia* and *Cryprosporidium* found in streams and rivers as a result of non-point source stormwater runoff. Although they are not present in large numbers, these two organisms cause long-term intestinal problems in humans and not a few deaths, especially for imuno-compromised individuals. Many bacteria and viruses have evolved resting stages that enable them to survive harsh conditions such as desiccation and sunlight that would otherwise kill the active phase. It is not surprising that the resting stage of *Cryptosporidium* in particular is very resistant to conventional disinfection treatments such as chlorination, although it is inactivated by ozonation.

How can wetlands phytoremediation assist in the economical removal of pathogens from storm waters? In addition to treatment wetlands accommodating substantial amounts of water, they can also provide more than a passive sedimentation step for pathogen removal. The plants in a treatment wetland can be configured to provide a pathogen removal mechanism just as they can for nitrate removal. As in the case of nitrate or metal removal in wetlands, the role of the plants is an indirect one; to provide the physical and chemical substrates rather than act directly via root-to-shoot uptake. As discussed in the introduction section of this chapter, the attached biofilm community in wetlands is located on submerged plant stems and decaying plant leaves. It contains bacteria, algae, protozoans, rotifers (Fig. 8), and nematodes. Of these, the protozoans and rotifers often feed on small suspended particles in the size range of stormwater pathogens (1-5 µm). Additionally, the biofilm filtering community, primarily sessile or stalked rotifers, has been given little consideration by the wetlands treatment community. However, preliminary studies made using the more easily cultured planktonic rotifers and the human pathogen *Enterococcus* showed that these tiny animals did remove the bacteria (Proakis 2001). Encouraging attached rotifers in wetlands may depend on the plant species, where surface area is most likely the critical factor controlling rotifer density. For this application, the dense stems of aquatic grasses may prove superior to those of cattail or bulrush, and even more so for floating plants. However, floating wetland plants have large surface areas available for biofilm development. Although floating plants cannot provide most of the needed wetlands services (i.e., no anoxic sites) and they need to be physically harvested for

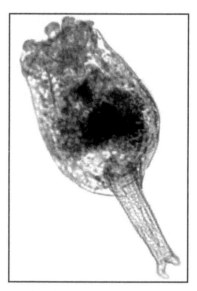

Figure 8. Microphotograph of a planktonic rotifer. Rotifers are microscopic animals that are just visible to the eye and are very common in wetlands. They consume particles that cover the range of many pathogens. The numbers of free-swimming and attached rotifers can presumably be increased by optimizing the plant species they favor. Photo by Dr. Marcie L. Commins.

metal and nutrient removal, the biofilm communities on their roots could provide sites for pathogen consumption. The end product of consumption is primarily carbon dioxide and small inert fragments of cell wall.

The role of rotifers associated with both rooted and free-floating macrophytes in wetland phytoremediation is relatively easy to work on by direct experimentation and could soon provide some very useful results. For practical application of laboratory studies research is also needed on what regulates the rotifers in wetlands. Just as the dominance of cattails, bulrushes, duckweed, and other plants is controlled by initial dominance or planting density, water depth, hydroperiod, and wind exposure, the numbers of stalked rotifers is most likely controlled by grazing pressure. The role of snail or shrimp grazing on the biofilm and the control of grazers by fish or birds is beyond the scope of this review but is essential nonetheless if a real plant-rotifer dominated treatment wetland is to be constructed for pathogen removal.

Plants and their biofilm also provide many natural coagulants such as muco-polysaccarides excreted by the attached bacteria using decaying plants as a carbon source. Some other natural coagulants may be excreted by living plants stems and leaves, but this is not known. The coagulants may assist removal of pathogens of many types but this area has been little studied as yet. The pressing problems of storm water pathogens may stimulate such research.

Removal of pathogens from a dairy wastewater in Arizona

Liquid dairy manure is a major problem in the developed world. It is present in large quantities and contains nutrients, some pharmacologically active growth hormones, and pathogens. The volume of liquid manure, as with many pathogen containing wastes such as storm runoff, is too great for easy disposal via percolation into the ground or for long containment in surface storage reservoirs. Thus phytoremediation's strength in handling large volumes of water can be used for pathogen removal. Phytoremediation of pathogens was tested in a multi-component treatment system for dairy and municipal wastewater in Arizona, USA (Karpiscak et al. 2001). The system consisted of paired solids separators, anaerobic lagoons, aerobic ponds and eight constructed wetland SF cells. Compared with the open ponds, the wetlands were most effective at removing coliphage (95%) and enterococci (74%, Table 3). These wetlands were much less effective than the open ponds for the removal of other pathogens; but, of course, a different wetland design emphasizing the rotifer component of the biofilm (see above) or with a longer residence time could improve performance. Dairy wastewater is very turbid (~ 1,300 NTU compared with 50 NTU for many wastewaters) and may have decreased the overall ability of the treatment system, including the phytoremediation component, to remove some microbial indicators and pathogens. However, since oxygenated

Table 3. Removal efficiencies of pathogens (as percent removal) from dairy waste in a pond-wetland unit process series (Modified from Karpiscak et al. 2001).

Type of Pathogen	Type of unit process treatment system			
	Anoxic ponds (%)	Oxic ponds (%)	Surface flow Wetlands (%)	Cumulative removal (%)[1]
Coliphages	96	44	95	99.9
Enterococci	98	83	74	99.9
Total coliform bacteria	98	92	+20	99.87
Fecal coliform bacteria	99	84	13	99.96
Listeria monocytogenes	98	83	32	99.86
Clostridium perfringens	53	51	20	74.6

[1] Includes a solids separation stage (not shown)

open ponds are part of the unit process system outlined earlier the combination seemed to be effective at about 99.9% (or three log) overall removal. With typical human sewage with ~ 10^6 total coliforms/100 mL, a five log removal is needed so some disinfection may still be needed until the more sophisticated unit designs listed earlier (especially unit # 5, the very shallow pond) are tested and proven.

Should plant uptake of toxic contaminants in wetlands phytoremediation be encouraged or discouraged?

Terrestrial phytoremediation often involves the deliberate planting of plant species which super-accumulate heavy metals. Metal accumulation of over 5% dry weight of zinc and nickel has been shown for some terrestrial plants (Brown *et al.* 1995, Blaylock 2000). Following metals accumulation in plant tissues, the plants are harvested and disposed of. Some wetlands plants are also super-accumulators. Duckweed (*Lemna*), for example uses luxury uptake for heavy metals such as copper, cadmium, and selenium but not nickel or lead (Zayed *et al.* 1998). Large-scale use of such wetland plants is risky. Large amounts of heavy metal in a wetland plant can constitute a "toxic nuisance" if birds or other organisms (e.g., butterfly caterpillars) eat the contaminated vegetation. The super-accumulator strategy is not practical for treatment wetlands because they are extremely productive ecosystems, akin to fertilized agricultural fields as well as the most productive parts of coral reefs and estuaries that possess an energy subsidy from tidal action. More wildlife is likely to be attracted to and potentially killed by super-accumulating wetlands plants than terrestrial counter-parts.

In some terrestrial phytoremediation, the vegetation is not an inherent accumulator, but becomes so after the soil is flooded with EDTA. The strong chelation capacity of EDTA is well known (Schwarzenbach *et al.* 2003) and it acts to pull metals from the soil sites into solution in pore waters. Rather unexpectedly, considering the size of its molecule, the EDTA-metal complex then passes quickly into the plant via the roots. If the plants are harvested relatively quickly, the time of wildlife exposure to toxic vegetation is small. Such an operation is more difficult in the already flooded roots of wetlands.

It is possible that the production of an "attractive or toxic nuisance" is more of a particular plant problem than an overall concern for wetlands phytoremediation. At Kesterson Reservoir (marsh) Se accumulated thousands of times over the water concentrations and found its way into the food web. In particular, the small black seeds of the rooted macrophyte *Ruppia* were found to contain as much as 1,400 ug/L of Se compared with 5-

400 ug/L for other plants and < 1 ug/L for uncontaminated vegetation (Horne and Roth 1989). Since the common name for *Ruppia* is widgeon grass, it is perhaps not surprising that many birds at Kesterson died or produced mutated eggs. Had a plant species other than *Ruppia*, for example bulrush, been predominant, the bird tragedy might have been avoided. *Ruppia* thrives in seasonal wetlands but is not as restricted by deep water as bulrush or cattails. Thus a major part of the bird contamination at Kesterson could have been controlled simply eliminating any water of over 0.5 m depth and thus replacing *Ruppia* with the virtually non-toxic cattails or bulrush.

The future of wetland phytoremediation

There are four areas in which substantial progress is expected in wetlands phytoremediation. The first is increased use of *large-scale phytoremediation* using techniques that are, if not fully mature, are at least well established. The second is the development of wetlands or parts of wetlands for specific pollutant removal using *unit processes*. The third is design for the removal of *dissolved heavy metals* that are hard to remove by other means. The fourth area is the removal of specific pollutants as they arise, for example perchlorate and manure application to land.

Large-scale use of wetlands phytoremediation

Some wetlands phytoremediation processes have been at work for at least a decade or two. These include removal of larger suspended solids including organic ones with a biological oxygen demand (BOD). However, wetlands are not as well suited for these processes as specifically designed solids detention ponds and conventional waste treatment processes such as oxidation ponds, activated sludge, or trickling filters. Thus such wetlands tend to be used for storm runoff from small fields or houses or wastewater from small developments or farms. A major problem for wetlands receiving inorganic silt is that it fills in the marsh converting it to a terrestrial environment. In conventional detention ponds silt is removed regularly by excavation but in wetlands dirt removal also removes the wetlands plants - negating their beneficial action. For BOD, wetlands can easily become overloaded if large amounts of rich waste are added and, unlike other systems there is not an emergency method to increase processing speed.

Converting wetlands to phytoremdiate larger amounts of storm runoff, industrial waste or domestic sewage requires playing to the strength of wetlands. That strength is the processing of large volumes of dilute pollutants. At higher concentrations conventional processes usually are more efficient and reliable (nitrate removal is an exception). However, after the appropriate conventional treatments have been carried out there is often a need to further "polish" the waste before release to surface waters or to supplement drinking water supplies. Here wetlands can play a role and usually are very much cheaper than other methods. As usual it is the length of time that water is processed in wetlands (1-2 weeks) that gives them the edge over conventional processes (few hours). In turn, the longer detention time in wetlands is possible by the other services that properly designed wetlands can give (wildlife habitat, aesthetic beauty).

Large-scale nitrate removal is the most obvious process that can be expanded from research and pilot processes to large scale. There is an urgent need to remove nitrate from many rivers and groundwater. About half of the U.S. population uses drinking water from the ground and over 1.5 million of those drinking groundwater have a source that is over or near the threshold safely standard of 10 mg N/L (Baker 1993). Of those about 44,000 are small infants with no tolerance or protective mechanisms against nitrate poisoning. Nitrate does not poison infants directly but via the formation of methemoglobin in the blood. This chocolate-brown blood does not carry oxygen efficiently and gives rise to the term "blue babies" to describe the result on small children. Adults have an enzyme to resurrect normal hemoglobin but it is lacking in children of up to a few months old. Methemoglobin is formed when nitrite is present in the blood and the presence of nitrate in water or water-derived food such as infant formula is the reason for the disease. Most nitrite is produced in the saliva and passes to the gut and to the bloodstream. In infants additional nitrite is produced in the stomach which has a neutral pH, unlike the very acid stomach of adults. Mid-range pH favors the bacteria that convert nitrate to nitrite giving infants a double dose of toxic nitrite.

Nitrates are the prime cause of eutrophication in almost all estuarine, coastal and ocean waters and the co-cause of eutrophication in most lakes. Eutrophication is the process that causes blooms of often unwanted algae that not only look unsightly but can use up all of the oxygen during overnight respiration. The resulting fish kills and malodors are often dramatic. Fish kills of hundreds of thousands of freshwater fish are becoming more frequent as eutrophication expands along with human developments. Over 7,000 square miles ($\sim 18,000$ km^2) of the Gulf of Mexico regularly suffers a "dead zone" attributed to nitrate-induced eutrophication and has serious effects on the shrimp fishery. Cures using wetlands phytoremediation have been suggested (Horne 2000a) but would require large amounts of land that could only be feasible if they were multipurpose wetlands (nitrate removal, flood control, wildlife preserves, hunting, water drinking storage reservoirs).

In most cases the main source of nitrate is not human sewage but farm runoff, which is more important. The "Green Revolution" that supplies about two-thirds of the planet's inhabitants with food has been rightly attributed to developments in plant species such as short stemmed rice. However, growth of enough of these new plant varieties is depending on fertilizers, in particular nitrate. Nitrogen comprises about 5% of the dry weight of food plants and is needed in much larger quantities than, for example phosphate (0.3% dw) or iron (trace). In addition, unlike particulate-bound phosphate or iron, nitrogen fertilizer soon produces soluble nitrate that passes easily through the soil to the rivers. Farmers can be trained to be more efficient in the use of fertilizers and manure disposal but there is a large irreducible minimum that cannot be contained and still have a productive farm. Lesser application of nitrogen fertilizer is possible but the crop declines. To overcome lower production, more land must then be put into production which destroys more natural land. The net result over the last 50 years is a steady increase in nitrate in rivers and oceans (Fig. 9). The combination of very large volumes of water contaminated with moderate amounts of nitrate create ideal conditions for a wetlands phytoremediation.

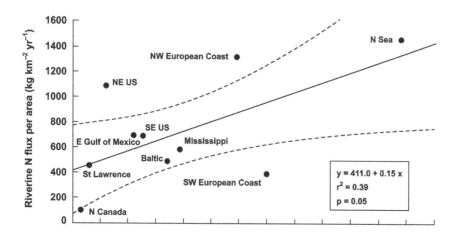

Figure 9. Increases in the nitrate concentration in the North Atlantic Ocean and the amount of nitrogen fertilizer applied to farmland that drains into it.   The correlation is good suggesting that substantial amounts of nitrate are leaking from North America and Europe. Nitrate wetlands could reduce this flow, especially in the warmer regions. Modified from Howarth *et al.* 1996.

Nitrate can be removed, even at high concentrations, using wetlands phytoremediation. An example is the Prado wetlands constructed by the Orange County Water District in 1992. At 200 ha (500 acres) it is perhaps the largest wetland designed to clean up a specific pollutant. A second phase of 400 acres is under construction in 2004 and a third phase is planned. The new system will treat an entire medium sized river (Santa Ana River, 3.5 m^3/s, 120 cfs) to reduce nitrate in the water from ~10 to 1 mg N/L prior to its infiltration into the ground and eventual use as part of a potable water source (Reilly *et al.* 2000). A similar system constructed by Irvine Ranch Water District in 1997 removes nitrate from about one third of a smaller river, San Diego Creek (Fleming-Singer and Horne 2004). In this case the purpose is to reverse eutrophication in Newport Bay, a formerly algae-clogged estuary with high wildlife and recreational use (Horne 2003).

Large-scale phosphorus removal. Phosphate pollution, the other main promoter of eutrophication, is usually caused by discharge of sewage although non-point sources including agricultural fertilizers and manure are important too. It is possible to reduce phosphates in detergents, with some loss in white "brightness". Phosphate can be removed down to low levels with alum flocculation but the process is costly and the sludge is not conveniently recycled. The seriousness of phosphorus pollution can be seen from calculations made in agricultural areas ranging from Northern Ireland to Wisconsin to Florida. These soils now contain sufficient P to last for hundreds of years without any further application of fertilizer!

Phosphorus removal is not a natural fit for wetlands since, unlike nitrate, phosphorus is mostly taken up and used in cellular biochemistry and nucleic acids. In contrast, nitrate is used to provide terminal electron acceptors, denitrified to nitrogen gas, and cycled back to the atmosphere. Over 95% of the P taken up by wetlands plants is recycled back to the water as soluble P. Only if the wetland microbes convert the added P into insoluble calcium phosphate (apatite) is true wetlands removal and long-term storage possible (Fig. 5). However, in the absence of other good methods for the removal of small concentrations of P in large volumes of water, wetlands are being used by default.

Future plans for large nitrate and phosphate removal wetlands include two very large projects; the Mississippi River and the Florida Everglades. Each of these wetlands phytoremediation cleanups will require hundreds of thousands of acres or even millions of acres of land and cost billions of dollars. Nonetheless, for want of any other solutions both solutions have received considerable attention and initial design and some large-scale pilot testing. Some aspects of the solution for the Everglades was given earlier and one estimate for the Mississippi was made by Horne (2000b).

Development wetlands unit processes for phytoremediation

The details of some unit processes were discussed earlier. However, until wetlands phytoremediation can fit into conventional wastewater and drinking water unit process trains, their use will be restricted to rural areas and small scales. Engineers and regulators are not experimental scientists and need some degree of reassurance and reliability before committing themselves.

Phytoremediation of dissolved heavy metals

Dissolved water column metal pollution at moderate levels is a severe problem even in developed countries such as the US and Western Europe and is out of control in many developing countries. For example, mercury pollution associated with "cowboy" gold mining in rivers such as the Orinoco is due to irresponsible practices. However, California's two largest abandoned mercury mines lay for almost a century before even initial attempts were made to control mercury runoff a few years ago. It was estimated that 20,000 km of streams and rivers in the US were polluted with soluble heavy metals (Rosa and Lyon 1997), some so seriously that only a few resistant bacteria and algae can survive. At this time there is simply not enough data to design a reliable phytoremediation marsh for dissolved metals. For example, a typical wastewater treatment plant may have a copper concentration of 10-50 ug/L but the marine and freshwater aquatic habitat standards may be as low as 4-20 ug/L. No data is available to design a surface water wetland to reliably meet this standard, year round for a typical flow of 1.5 m^3/s, the discharge from a city of one million people.

The future role of carbon

Unlike terrestrial phytoremediation, plants in wetlands do their work for us after they die. Work over the last decade has shown that the role of aquatic plants in wetlands phytoremediation is to grow, die, and then act as a source of carbon and provide habitat for bacteria that actually carry out the processes (Hume et al. 2002). In addition, dead plants acts as sorption sites for some organic and inorganic pollutants. This area of absorption research is new and exciting. Both heavy metals and some complex organic pollutants may be routinely removed by physical sorption onto carbon. The kind of carbon produced regulates the kind and amounts of pollutants removed. In turn the kind of carbon produced in a wetland can be selected by the choice of plants. The ratio of lignin to cellulose is one guide for plant choice. However, even more benefit can be gained by further environmental engineering to give specific degradation products and even incorporate the special complex types of carbon provided by insect chitin.

Although the addition of organic matter such as wetland plant leaf litter may remove metals from solution, the effect is likely to be less strong and less long-lived than the formation of sulfide or similar ore-like compounds. Given the sulfide alternative, the use of carbon to remove dissolved metals in wetland phytoremediation should be directed towards the production of food and habitat for sulfur reducing bacteria that produce sulfide. Thus the phytoremediation should concentrate on producing labile organic matter rather than refractory matter. In the unit process this means that dissolved metals would be immobilized in the unit #2 cattail marsh rather than the unit # 4 bulrush-peat marsh. In general, soluble heavy metals can be treated in wetlands but the mechanisms are somewhat uncertain. The eventual fate of many metals is generally to become part of the "insoluble" fraction and much of this fraction seems to be metal sulfides. However, the transport mechanisms, timing and intermediate compounds are not known for certain.

Given the carbon-based nature of wetland phytoremediation, much of the future may depend on various kinds of inorganic sorption rather than the better-known chemical processes of chemical bonding. Although the kind of carbon is important, perhaps the state of decay of natural carbon sources is even more critical. That different kinds of carbon act differently has been know for a long time. In studies of the partitioning of polycyclic aromatic hydrocarbons (PAHs) between pore water and estuarine wetland sediments Maruya et al. (1996) showed that dead leaves washed into the intertidal zone were poor at absorbing PAHs. In contrast, burnt carbon such as soot was able to hold greater quantities of PAHs on a per gram of carbon basis. The reason postulated was the greater amount of open pore and sites for sorption in the sooty material relative to the leaves.

A good recent example for metals is provided for the bio-sorption of cadmium onto chitin (Benguella and Beniassa 2002). These Algerian authors showed that removal occurred following standard chemical equilibriums (Langmuir isotherm gave the best fit) and pseudo-second order kinetics controlled by the usual variables such as the amount of cadmium and chitin particle size. Use of scanning electron microscopy shows that thin chitin sheets become "littered with nodules" of an electron dense material that EDAX-micro-spectroscopy showed to be cadmium. It has long been known that chitin has the potential for bio-sorption (Muzzarelli 1973) but without the details of how and why, it is difficult to design wetlands to amplify the natural process. Once carried out, studies such as those of Benguella and Beniassa (2002) or Tsezos and Mattar (1986) can easily be incorporated into wetlands design. Chitin is shed by many wetland invertebrates, especially insects, each time they molt. And insects as well crustaceans must molt frequently as they grow since the chitin is their inflexible external skeleton. Invertebrates can be encouraged or discouraged in wetlands depending on the design of the ecosystem. When it is considered that the most common larger animals in wetlands are midge larvae and these may be present on the order of thousands/m^2, the possibilities of metal removal are startling.

Perchlorate and manure. Two current examples where wetland phytoremediation could rapidly solve emerging major problems

Over the last year two problems have suddenly risen to prominence in the Southwest of the United States but also occur in the rest of the USA and Europe. They are perchlorate and manure. Perchlorate, a component of rocket fuel, is leaking from abandoned munitions factories. It has leaked into the Colorado River near Las Vegas and is now a ubiquitous contaminant in Southern Nevada and Arizona. The Colorado River is the main water supply of agriculture in Southern California and low levels of perchlorate have been found in cow's milk in California, presumably due to irrigation with Colorado River water. The river is also the raw drinking water supply for the 17 million people who live in Los Angeles; although the perchlorate is removed before it reaches the tap. Perchlorate has been found in several sites in Northern California hundreds of km from the Colorado River (Oakland Tribune, July 2004). It is probably a contaminant elsewhere in the US too since rocket fuel was widely used and there are over 1,000 military bases in the 50 states and various territories of the USA. Perchlorate is very like nitrate in that it is the fully oxidized form and hard to degrade in most surface waters. Like nitrate perchlorate is very soluble in water and travels long distances without degradation. In fact it is so like nitrate that the same denitrification enzyme complex may also degrade perchlorate via accidental co-metabolism. Early studies, however, have shown that simple cattail wetlands that remove nitrate do much less well with perchlorate. Some other plants show promise and the co-metabolism aspect needs attention. For example, do we need to ensure sufficient nitrate to activate the denitrification enzymes?

In 2003, the dairy industry in California was threatened with closure due to overabundance of manure. California surpasses even Wisconsin in the size of its dairy industry; California's is worth over $6 billion annually. The problem was excess nitrate and TDS in the groundwater. The soluble runoff from manure is a problem from Wisconsin to Ireland. The combination of some kind of wetland phytoremediation with nitrate, growth hormones and salt removal is needed. Probably some kind of sustainable energy source should be harnessed for economical TDS removal making use of the wetland as a hydraulic balancing system while still maintaining the phytoremediation capacity for nitrate and organics removal.

SUMMARY

Phytoremediation employs sustainable sunlight energy and photosynthesis to transform harmful substances into harmless forms. Wetlands phytoremediation provides a unique service in two areas: those reactions that require anoxic conditions and those where the pollutant is found in large volumes of contaminated water, although performing best when contaminant concentrations are low or moderate. Wetlands treat dissolved or particulate pollutants. The combination of large volume and lower contamination is most costly to treat using conventional engineering technologies such as flocculation, sorption and coagulation, ion exchange, membrane filtration, and reverse osmosis. This review mostly concerns free surface wetlands where plants play the major role in both treatment and provision of wildlife habitat and aesthetic appeal. Free surface wetlands contain various kinds of emergent and submergent plants and algae growing in 25-50 cm of water.

Natural wetlands have a good reputation as "nature's filters" but their performance usually pales in comparison with well-designed constructed wetlands. The removal of particulates (and attached metals and organics), BOD, and nitrate is quite well quantified, if not as well understood. Wetlands phytoremediation remains partially in the "black box" stage. In particular, the uptake of dissolved contaminants has been neglected. Thus although wetlands are widely used to clean up water, there is much room for improvement and increases in efficiency, especially in urban and arid areas where both water and space are costly. Passive sorption of contaminants onto dead plant carbon and increasing the roles of pathogen-devouring protozoa and rotifers by selecting favorable plant species are currently almost blank slates in wetlands research. The importance of different plant species is not fully recognized even though they provide all wetland substrates. In turn, different substrates control much of the contaminant removal in wetlands. For example, humic substances that absorb and hold organics such as pesticides can be increased by using plants with high lignin content (bulrushes), while those needing bacterial action to destroy or metabolize pollutants require more labile carbon sources (cellulose in cattails or duckweed). Substrates that selectively absorb metals (cadmium on chitin) require development of larger larval and adult insect populations in the wetland. This aspect of ecological engineering is an exciting part of the new era of wetlands phytoremediation.

Major problems for the developed and developing societies alike are the control of large volumes of polluted irrigation return water, partially treated sewage, urban and rural storm waters. Contaminants in such waters range from simple nutrients such as nitrate and phosphate to heavy metals such as lead or copper to complex organic moieties such as PAHs and pesticides. Added to such problems are pathogens such as bacteria, protozoans, and viruses. Some phytoremediation will require very large wetlands but since these can also serve many other functions such as flood control, wildlife habitat, hunting, and public education, they are less costly than the alternatives. The new field of myco-remediation using designed fungal arrays is an exciting partner to wetland phytoremediation since there is a natural symbiosis between the two (wetlands provide the carbon source; fungal peroxidases break down ring compounds that bacteria in wetlands usually cannot). New research in the area of reliable removal of small to moderate concentrations of dissolved metals by wetlands is needed. Also desirable is more knowledge about the role of plant carbon and its microbial biofilm in absorption of pesticides and pharmaceutically-active compounds such as birth control pill residues, and perhaps heavy metals. Finally, there is a need to be able to design phytoremediation wetlands to rapidly respond to emergencies such as the recent perchlorate scare and the growing problem of runoff from manure applications in the Western World.

REFERENCES

Alvord, H.H. and R.H. Kadlec. 1996. Atrazine fate and transport in the Des Plaines Wetlands. *Ecol. Model.* 90: 97-107

Bachand, P.A.M., and A.J. Horne. 2000a. Denitrification in constructed free-water surface wetlands: I. Very high nitrate removal rates in a macrocosm study. *Ecol. Engineer.* 14: 9-15.

Bachand, P.A.M., and A.J. Horne. 2000b. Denitrification in constructed free-water surface wetlands: II. Effects of vegetation and temperature. *Ecol. Engineer.* 14: 17-32.

Baker, L.A. 1993. Introduction to non-point source pollution and wetland mitigation. Pages 7-41 in *Created and Natural Wetlands for Controlling Non-point Source Pollution,* R.K. Olsen, ed., USEPA, Washington DC. Smoley/CRC Press, Boca Raton, Florida.

Barrett, E.C., M.D. Sobsey, C.H. House, and K.D. White. 2001. Microbial indicator removal in onsite constructed wetlands for wastewater treatment in the southeastern U.S. *Water Sci. Technol.* 44: 177-182.

Benguella, B., and H. Beniassa. 2002. Cadmium removal from aqueous solution by chitin: kinetic and equilibrium studies. *Water Res.* 36: 2475-2482.

Blaylock, M.J. 2000. Field demonstration of phytoremediation of lead-contaminated soils. Pages 1-12 in *Phytoremediation of Contaminated Soil and Water*, N. Terry and G. Banuelos, eds., Lewis, Boca Raton, Florida.

Brown, S.L., R.L. Chaney, J.S. Angle, and A.J.M. Baker. 1995. Zinc and cadmium uptake by hyper-accumulator *Thalspi caerulenscens* grown in nutrient solution. *Soil Sci. Soc. Am. J.* 59: 125-133.

Cerezo, R.G., M.L. Suarez., and M.R. Vidal-Abarca. 2001. The performance of a multi-stage system of constructed wetlands for urban wastewater treatment in a semiarid region of SE Spain. *Ecol. Engineer.* 16: 501-517.

CH2M-Hill. 2001. Periphyton-based stormwater treatment area (PSTA) research and demonstration project. Phase 2 interim report (April-December 2000). CH2M-Hill Consulting Engineers, Florida.

Coates, J.D., J. Woodward, J. Allen, P. Philip, and D.R. Lovely. 1997. Anaerobic degradation of polycyclic aromatic hydrocarbons and alkenes in petroleum-contaminated marine harbor sediments. *Appl. Envir. Microbiol.* 63: 3589-3593.

Cronk, J.K., and M. Siobhan-Fennessy 2001. *Wetland Plants: Biology and Ecology*. CRC Press, Boca Raton, Florida.

Brix, H. 1997. Do macrophytes play a role in constructed treatment wetlands? *Water Sci. Technol.* 35: 11-17.

Douglas, M.S. 1988. *The Everglades: River of Grass*, 2nd ed. Pineapple Press, Sarasota, Florida.

Eger, P., and K. Lapakko. 1988. Nickel and copper removal from mine drainage by a natural wetland. Pages 301-309 in Mine Drainage and Surface Mine Reclamation Conference. USBM Pittsburg.

Elder, J.F. 1985. Nitrogen and phosphorus speculation and flux in a large Florida river-wetland system. *Water Resources Res.* 21: 724-732.

Esteves da Silva, J.C.G., and C.J.S. Oliveira. 2002. Metal ion complexation properties of fulvic acids extracted from composted sewage sludge as compared to a soil fulvic acid. *Water Res.* 36: 3404-3409.

Fleming-Singer, M.S. 2002. Optimization of Nitrate Removal in Treatment Wetlands Using an Episediment Layer for Increased Denitrification Potential. Dept. Civil and Environmental Engineering. Berkeley, University of California: 242.

Fleming-Singer, M., and A.J. Horne 2004. Balancing nitrate removal and bird populations in a constructed wetland. Submitted for publication.

Frankenberger, W.T., Jr., and S. Benson. 1994. Kesterson Reservoir - Past, Present, and Future: An Ecological Risk Assessment. In *Selenium in the Environment*, W.T. Frankenberger, Jr and S. Benson eds., Marcel Dekker, Inc., New York, New York.

Hauri, J.F. 2001. Measurement and manipulation of copper speciation and toxicity in urban runoff, acid mine drainage, and contaminated discharged groundwater. Ph.D. dissertation, Dept. Civil and Environmental Engineering. University of California, Berkeley.

Horne, A.J., and J.C. Roth. 1989. Selenium detoxification studies at Kesterson Reservoir wetlands: Depuration and biological population dynamics measured using an experimental mesocosm and pond 5 under permanently flooded conditions. Univ. Calif., Berkeley. Envir. Eng. Health Sci. Lab. Rept. No. 89-4. 107 p. + Appendix (89 p.).

Horne, A.J., 1995. Nitrogen removal from waste treatment pond or activated sludge plant effluents with free-surface wetlands. *Water Sci. Technol.* 33: 341-351.

Horne, A.J., M.S. Fleming, and N.P. Hume. 1999. Irvine Ranch Water District, Wetlands Water Supply Project, Pond System Organic Carbon Evaluation 1998-1999. Final Report. Irvine, CA, Irvine Ranch Water District: 101.

Horne, A.J. 2000a. Phytoremediation by Constructed Wetlands. Pages 13-60 in *Phytoremediation of Contaminated Soil and Groundwater*, N. Terry, ed., CRC Press. Boca Raton, Florida.

Horne, A.J. 2000b. Potential value of constructed wetlands for nitrate removal along some large and small rivers. *Ver. Int. Ver. Limnol.* 27: 4057-4062.

Horne, A.J. 2003a. Eutrophication in the Newport Bay-Estuary in 2002: trends in the abundance of nuisance macroalgae (seaweed) in 1996-2002. Report to Orange County Public Facilities & Resources Department, Santa Ana, California. May 2003. Alex Horne Associates. 18 p. + Appendices.

Horne, A.J. 2003b. *Wetlands & Rivers: Ecology & Management.* Pages. 181-184 in Reader for CEE 118 Dept. Civil and Environmental Engineering, University of California Berkeley.

Horne, A.J. 2000c. Design of a winery treatment wetland for a sloping site on the Central California Coast. Integrated Structures, Berkeley CA.

Horne, A.J. 2002. Control of mosquitoes in California constructed wetlands. Talk given at Calif. Vector Control Meeting, July 2002.

Horne, A.J. and M. Fleming-Singer. 2003. Evaluation of nitrate removal in San Joaquin Marsh 1997-2003. Alex Horne Associates report to Irvine Ranch Water District, Irvine CA. Aug. 2003.

Horne, A.J., and M.S. Fleming-Singer. 2004. Evaluation of nitrate removal in San Joaquin Marsh 1997-2002. Report to the Irvine Ranch Water District. Alex Horne Associates, January 25, 2004. 52 p. + Appendices.

Howarth, R.W., G. Billen, D. Swany, A. Townsend, N. Jaworski, K. Lajtha, J.A. Downing, R. Elmgren, N. Caraco, T. Jordan, F. Berendse, J. Freney, V. Kudeyarov, P. Murdoch, and Z. Zhao-Liang. 1996. Regional nitrogen budgets and riverine N & P fluxes for the drainages to the North Atlantic Ocean: Natural and human influences. *Biochemistry* 35: 75-139.

Hume, N. P. 2000. Effects of plant carbon quality on microbial nitrate reduction in wetlands. Ph. D. Diss. Dept. Civil and Environmental Engineering. Univ. California, Berkeley.

Hume, N.P., M. Singer-Fleming, and A.J. Horne. 2002. Plant carbohydrate limitation on nitrate reduction in wetland microcosms. *Water Res.* 36: 577-584.

IRWD, 2003. EIR for Natural Treatment Systems. Irvine Ranch Water District, Irvine CA. irwd.com.

Gersberg, R.M., R.A. Gearheart, and M. Ives. 1989. Pathogen removal in constructed wetlands. Pages 431-444 in *Constructed Wetlands for Wastewater Treatment*, D.A. Hammer ed., Lewis Publishers, Chelsea, Michigan.

Goulet, R.R., F.R. Pick, and R.L. Droste. 2001. Test of the first-order model for metal retention in a young constructed wetland. *Ecol. Engineer.* 17: 357-371.

Kadlec, R.H., and Knight, R.L. 1996. *Treatment Wetlands.* Lewis Publishers, Boca Raton, Florida.

Kao, C.M., J.Y. Wang, and M.J. Wu. 2001. Evaluation of atrazine removal processes in a wetland. *Water Sci. Technol.* 44: 539-544.

Kerndorff, H., and M. Schnitzer. 1980. Sorption of metals on humic acid. *Geochim. Cosmochim. Acta* 44: 1707-1708.

Knight, R.L., B. Gu, R.A. Clarke, and J.M. Newman 2003. Long-term phosphorus removal in Florida aquatic systems dominated by submerged aquatic vegetation. *Ecol. Engineer.* 20: 45-64.

Karpiscak, M.M., L.R. Sanchez, R.J. Freitas and, C.P. Gerba. 2001. Removal of bacterial indicators and pathogens from dairy wastewater by a multi-component treatment system *Water Sci. Technol.* 44: 183-190.

Karpiscak, M.M., C.P. Gerba, P.M. Watt, K.E. Foster, and J.A. Falabi. 1996. Multi-species plant systems for wastewater quality improvements and habitat enhancement. *Water Sci. Technol.* 33: 231-236.

Lyon, J. C. 1993. *Practical Handbook for Wetland Identification and Delineation.* Lewis Publishers, Boca Raton, Florida.

Maruya, K., R.W. Risebrough, and A.J. Horne. 1996. Partitioning of polynuclear aromatic hydrocarbons between sediments from San Francisco Bay and their porewaters. *Chemosphere* 30: 2945-2947.

Mays, P.A., and G.S. Edwards. 2001. Comparison of heavy metal accumulation in a natural wetland and constructed wetlands receiving acid mine drainage. *Ecol. Engineer.* 16: 487-500.

McCutcheon, S.C. and J.L. Schnoor. 2003. *Phytoremediation : Transformation and Control of Contaminants.* Wiley, New York.

Meade, R.H., and R.S. Parker 1985. Sediments in rivers of the United States. National Water Summary, 1984. Water Supply Paper 2275. U. S. Geological Survey, Reston Virginia.

Mehrotra, A., A.J. Horne, and D. Sedlak. 2003. Mercury control in wetland using iron additions. Submitted for publication.

Metcalf and Eddy, Inc. 1991. *Wastewater Engineering, Treatment Disposal and Reuse.* 3rd Ed., Revised by G. Tchobanoglous and F.L. Burton, McGraw-Hill, New York.

Moore, M.T., J.H. Rodgers, C.M. Coopes and S. Smith. 2000. constructed wetlands for mitigation of atrazine associated agricultural runoff. *Environ. Pollut.* 110 : 393-399.

Muzzarelli, R. 1973. *Natural Chelating Polymers.* Pergamon Press, Oxford.

Mitch, W.J., and K.M. Wise. 1998. Water quality, fate of metals, and predictive model validation of a constructed wetland treating acid mine drainage. *Water Res.* 32: 1888-1900.

Mitsch, W.J., and J.G. Gosselink. 2000. *Wetlands,* 3rd ed., John Wiley, New York.

Rosa, C.D., and J.S. Lyon. 1997. *Golden Dreams, Poisoned Streams.* Mineral Policy Center. Washington, DC.

Oakland Tribune. 2004. Perchlorate threat needs to be clarified, addressed. Opinion page Local 7. @ July 2004.

Ohlendorf, H.M., and G.M. Santolo. 1994. Kesterson Reservoir - Past, Present, and Future: An Ecological Risk Assessment. In *Selenium in the Environment,* W.T. Frankenberger and S. Benson, eds., Marcel Dekker, New York.

Philips, R.G., and W.G. Crumpton. 1994. Factors affecting nitrogen loss in experimental wetlands with different hydrologic loads. *Ecol. Engineer.* 3: 399-408.

Proakis, E. 2001. Removal of Enterococcus bacteria from water using rotifers. Master's thesis, Dept. Civil & Environmental Engineering. University of California, Berkeley.

Quiñónez, M.J., M. Karpiscak, and C.P. Gerba, 1997. Constructed wetlands for wastewater quality improvement. In: *Proceedings of the 97th General Meeting of the American Society for Microbiology,* May 4-8, 1997, Miami Beach, Florida.

Reed, S.C., D.I. Siegel, and E.J. Middlebrooks. 1995. *Natural Systems for Waste Management and Treatment,* 2nd ed., McGraw-Hill, New York.

Reilly, J.F., A.J. Horne, and C.D. Miller. 2000. Nitrogen removal in large-scale free-surface constructed wetlands used for pre-treatment to artificial recharge of groundwater. *Ecol. Engineer.* 14: 33-47.

Revitt, D.M., P. Worrall, and D. Brewer. 2000. The integration of constructed wetlands into a treatment system for airport runoff. *Water Sci. Technol.* 44: 469-476.

Richardson, C.J., S. Qian, C.B. Craft and R.G. Qualls. 1997. Predictive models for phosphorus retention in wetlands. *Wetlands Ecol. Management* 4: 159-175.

Rodgers, J.H. and A. Dunn. 1992. Developing design guidelines for constructed wetlands to remove pesticides from agricultural runoff. *Ecol. Engineer.* 1: 83-95.

Russell, R.C. 1999. Constructed wetlands and mosquitoes: Health hazards and management options - an Australian perspective. *Ecol. Engineer.* 12: 107-124.

Sabeh, Y., and K.S. Narasiah. 1992. Degradation rate of aircraft de-icing fluid in a sequential biological reactor. *Water Sci. Technol.* 26: 2061-2064.

Schulz, R. and S.K.C. Peall. 2001. Effectiveness of a constructed wetland for retention of non-point source pesticide pollution in the Lourens River catchment, South Africa. *Environ. Sci. Technol.* 35: 422-426

Schwarzenbach, R.P., P.M. Gschwend, and D.M. Imboden 2003. *Environmental Organic Chemistry*, 2nd Ed. John Wiley and Sons, Hoboken, New Jersey.

SFWMD. 2002. Demonstration of submerged aquatic vegetation/lime rock treatment technology for phosphorus removal from Everglades Agricultural Area waters. South Florida Water Management District & Florida Dept. Environmental Protection, West Palm Beach. Report Prepared by BD Environmental Rockledge, Florida. Final Draft Rept., March 2002.

Shutes, R.B.E., D.M. Revitt, L.N.L. Scholes, M. Forshaw, and B. Winter. 2001. An experimental constructed wetland system for the treatment of highway runoff in the UK. *Water Sci. Technol.* 44: 571-578.

Sposito, G. 1989. *The Chemistry of Soils.* Oxford Univ. Press, New York.

Sposito, G. (1986). Sorption of trace metals by humic materials in soils and natural waters. *Critical Reviews in Environmental Control.* 16: 193-229.

Srebotnik, E., K.A. Jensen, and K.E. Hammel. 1994. Fungal degradation of recalcitrant nonphenolic lignin structures without lignin peroxidase. *Proc. Natl. Acad. Sci. USA* 91: 12794-12797.

Sundaravadivel, M., and S. Vigneswaran. 2001. Constructed wetlands for wastewater treatment. *Crit. Rev. Environ. Sci. Technol.* 34: 351-409.

Tchobanoglous, G. and E.D. Schroeder. 1985. *Water Quality.* Addison-Wesley, Reading, Massachusetts.

Tsezos, M., and S. Mattar. 1986. A further insight into the mechanism of biosorption of metals by examining EPR spectra. *Talanta* 33: 225-232.

USEPA 1998. A citizen's guide to phytoremediation. Report No. EPA 542-F-98-011. Washington, D.C. Office of Solid Waste and Emergency Response, 6 pp.

USEPA. 1999. *Manual of Constructed Wetlands Treatment of Municipal Wastewaters.* USEPA Washington, DC.

Walker, D.J., and S. Hurl. 2002. The reduction of heavy metals in a stormwater wetland. *Ecol. Engineer.* 18: 407-414.

Weiner, J.G., C.C. Gilmour, and D.P. Krabbenhoft. 2003. Mercury Strategy for the Bay-Delta Ecosystem: A Unifying Framework for Science, Adaptive Management, and Ecological Restoration. Final Report to the California Bay Delta Authority, December 31, 2003, 67 p. (http:// science.calwater.ca.gov/pdf/MercuryStrategyFinalReport.pdf).

Zayed, A., E. Lytle, and N. Terry. 1998. Phyto-accumulation of trace elements by wetland plants: 1. Duckweed. *Environ. Q.* 27: 715.

Zoh, K.-D., and A.J. Horne. 1999. The removal of TNT using plants in constructed wetlands. Pages 357-364 in *Wetlands & Remediation*. J.L. Means and R.E. Hinchee, eds., Battelle Press, Columbus, Ohio.

Engineering of Bioremediation Processes: A Critical Review

*Lisa C. Strong and Lawrence P. Wackett**

Department of Biochemistry, Molecular Biology and Biophysics and
Biotechnology Institute, University of Minnesota, St. Paul, MN 55108
* Author to whom correspondence should be addressed:
Phone: 612-625-3785; Fax: 612-625-5780
email : wackett@biosci.cbs.umn.edu

Introduction

Xenobiotic compounds end up in the local environment because spills or leaks are not properly handled, equipment may leak during loading, and rinsewater and wastewater are sometimes not treated. These chemicals move via runoff into surface waters, ground waters and aquatic sediments. The chemical properties of a compound govern its transport, persistence, and bioavailability in the environment. These, in turn, determine whether a compound binds irreversibly to the soil matrix and causes no adverse effects, as has been demonstrated for anthracene (Richnow *et al.* 1998) or certain heavy metals (Brown *et al.* 2003), or becomes widely distributed and of general concern (Alexander 1999). Traditional cleanup methods include containment using barrier wells or caps, incineration, soil washing, solidification and stabilization, or simple storage. These methods involve physically moving large volumes of material and can be quite expensive (Cookson 1995). If left undisturbed, the fate of most chemicals introduced into the environment is dominated by biological degradation (Wiedemeier 1998).

Engineered bioremediation systems endeavor to channel natural processes for enhancing biodegradation of organic constituents dissolved in groundwater and adsorbed onto the soil or aquifer matrix. This approach minimizes site disturbance, and is reported to average 10% to 50% less expensive than traditional cleanup methods (http://www.ensr.com/services/waste/brownfields.htm). So far, intentionally using microbes to transform organic contaminants to non-toxic end products has had mixed success, with much effort directed towards identifying systems that can be treated biologically and *in situ* (Hapeman *et al.* 2003, Parales *et al.* 2002, Dua *et al.* 2002, Samanta *et al.* 2002). A wide range of bioremediation strategies have been developed for the treatment of contaminated soils using natural and modified microorganisms (Pieper and Reineke 2000, Strong *et al.* 2000). Summaries of field studies available from the US government are categorized by contaminant or by the technology applied, and a searchable database of 274 remediation case studies performed in the United States is available (Federal Remediation Technologies Roundtable 2001, http://www.frtr.gov/).

Factors that limit biological treatment include the poor bioavailability of the pollutant(s), toxicity of pollutants or metabolites, lack of one or more trace nutrients, and the lack of microbes with the appropriate biocatalytic potential in the environment to be treated (Watanabe *et al.* 2002, Widada *et al.* 2002). These potential deficiencies thus become the foci of *in situ* biostimulation engineering methods, which concentrate on adding detergents, nutrients, electron donors/acceptors into the target area. Biostimulation can be used if indigenous bacteria present on site are capable of metabolically utilizing the chemical contaminant.

Bioaugmentation, adding bacteria possessing targeted catalytic pathways, can be effective in areas where native organisms are unable to degrade specific contaminants. Advances in bioinformatics may help maximize bioremediation success rates by helping match appropriate bioremediation strategies with the contaminating compound and the environment where it is present (Ellis *et al.* 2003, Hou *et al.* 2003, Wackett and Ellis 1999). In some cases, *ex situ* engineering solutions involving bioreactors may be required. The delivery of electron donors or acceptors has been a serious challenge for *in situ* and enhanced bioremediation.

Redox processes important for *in situ* bioremediation

Microbial degradation of xenobiotic compounds often involves redox chemistry, or reactions involving a transfer of electrons. Microbes harness redox reactions to gain energy by trapping some of the energy evolved as the electrons transfer between the electron acceptor and the electron donor. They may also capture carbon, nitrogen, or trace elements from xenobiotic compounds to build their cellular structures. When an organic compound undergoes a redox reaction, the compound is transformed into a new organic compound. When microbial degradation results in complete conversion of a chemical contaminant to its inorganic constituents such as carbon dioxide, ammonia, phosphate, and chloride anion, the process is called mineralization.

To illustrate these concepts, Figure 1 shows the microbial degradation pathway for trichloroethene, a widely used industrial solvent that is also a major ground water pollutant. Three separate 2-electron reduction reactions transform trichlorethene into ethene. The reactions occur in an oxygen-limited, or anaerobic, environment. These three reactions are known to be mediated by a single organism, *Dehalococcoides ethenogenes* strain 195, which was shown to use H_2, ethanol, or butyrate as electron donors (Maymo-Gatell *et al.* 1997, 1999). Note that the cells use the trichloroethene in the manner that aerobic organisms use oxygen, with the cells gaining energy by linking the oxidation of other compounds to the reduction of a final electron acceptor, in this case the pollutant trichloroethene.

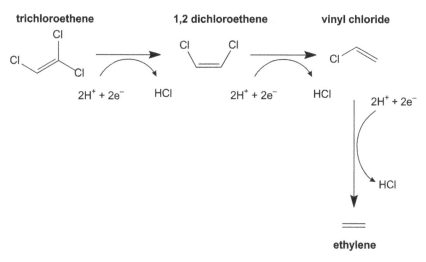

Figure 1. Reduction of 1,1,2-trichloroethene yields lesser chlorinated ethenes.

The redox potential (E, volts), measures the availability of electrons for transfer between molecules, and is compared with hydrogen which has E = 0 by definition. The redox potential is used to determine which type of metabolism dominates in any particular environment. A more positive E means oxidation is favorable, while negative E favors reduction. Environments in contact with atmospheric oxygen typically have E values of around +0.8 V, and microbes that use oxygen as an electron acceptor dominate. As oxygen becomes depleted, the redox potential becomes more negative and microbial populations shift to those capable of metabolism utilizing other electron acceptors such as nitrate (NO_3^-, +0.74 V), manganese (Mn, +0.52 V), iron (III) (Fe^{+3}, -0.05 V), sulfate (SO_4^{-2}, -0.22 V), or bicarbonate (CO_3H^-, -0.24 V).

Aerobic in situ bioremediation

In oxygen-rich environments, aerobic oxidation of compounds is the most efficient pathway for degradation. The organic compound serves as an electron donor, and oxygen serves as an electron acceptor. Aerobic remediation has been very successful in reducing levels of aliphatic and aromatic petroleum hydrocarbons; benzene, toluene, ethylbenzene and total xylenes (BTEX compounds), methyl t-butyl ether (MTBE), pentachlorophenol (PCP), polychlorinated biphenyls (PCBs) and nitroaromatics. These compounds are fed directly into central metabolism as acetyl CoA, although it should be noted that aromatic or aliphatic hydrocarbons substituted with halogens or functional groups at the ß-carbon are resistant to this degradation pathway.

To establish aerobic conditions, solid peroxygens, such as magnesium peroxide or calcium peroxide, can be injected into the subsurface where they solidify into reactive plugs. Oxygen is released over several months as the plugs react with water. However, the relatively small surface area of the solid plugs limits total oxygen release. Directly oxygenating groundwater by sparging air through an injection well, or injecting oxygen-saturated water, is very effective, but has the disadvantage of being a continuous process that is relatively expensive. The most promising technologies for establishing aquatic aerobic zones involve liquid sources of oxygen.

Liquid sources of oxygen include hydrogen peroxide, and potassium, sodium, and calcium permanganates that release free oxygen when mixed with water. These chemicals provide oxygen for short periods, on the order of minutes for hydrogen peroxide to several hours for the permanganates. Recently, this limitation has been addressed by oxygen-releasing compound (ORC). ORC is a proprietary formulation of phosphate-intercalated magnesium peroxide that releases oxygen slowly when hydrated:

$$MgO_2 + H_2O \rightarrow 1/2O_2 + Mg(OH)_2.$$

ORC is typically directly injected into the aquifer, where ORC particles slowly release oxygen for periods of up to 1 year. ORC was added to a Naval Air Station site contaminated with chlorinated benzenes that were undergoing biodegradation naturally. However, after the addition of ORC, concentrations decreased 97%, from 4060 μg/L to 98.5 μg/L in 4 weeks, with a cost-savings of 75% over that estimated for pump-and-treat technology   (http://toxics.usgs.gov/topics/rem_act/benzene_plume.html).

Anaerobic in situ bioremediation

Anaerobic microbial degradation of xenobiotic compounds has generally been less-studied than aerobic metabolism, although our understanding has increased dramatically in the last two decades. The anaerobic conversion of six different groups of aromatic xenobiotics--surfactants, phthalate esters, polycyclic aromatic hydrocarbons (PAHs), PCBs, halogenated phenols, and pesticides has been comprehensively reviewed recently (Mogensen et al. 2003). It has been recently reported that the monoaromatic hydrocarbon compounds benzene, toluene, ethylbenzene, and all three xylene isomers serve as carbon and energy sources for bacteria, with nitrate, manganese, ferric iron, sulfate, or carbon dioxide as the sole electron acceptor (Chakraborty and Coates 2004).

Some chlorinated solvents are directly oxidized anaerobically, which means they serve as electron donors in microbial reactions. Vinyl chloride, dichloroethene, and carbon tetrachloride degrade in this way, with naturally occurring humic acids used as terminal electron acceptors (Bradley et al. 1998, Cervantes et al. 2004). A large variety of other contaminants have been reported to function as electron donors, including methane, ethane, ethene, propane, butane, aromatic hydrocarbons such as toluene and phenol, and ammonia.

During the last decade, it has been established that oxygen-poor environments contain bacteria that utilize some pollutants as electron acceptors, in place of oxygen. For example, anaerobic reductive dechlorination occurs when chlorinated solvents are used as electron acceptors instead of oxygen; bacteria gain energy and grow as one or more chlorine atoms on a chlorinated hydrocarbon are replaced with hydrogen. Reductive dechlorination has been shown to be an active mechanism in the degradation of chlorinated ethenes and ethanes (perchloroethene, trichloroethene, dichloroethenes, 1,1,1-trichloroethane, dichloroethanes, and vinyl chloride). These systems are comprehensively reviewed in Lee et al. (1998).

Anaerobic recalcitrance is associated with hydrocarbons lacking functional groups, branched molecules (gasoline oxygenates), aromatic amines and aromatic sulfonates. However, recently anaerobic microorganisms have been discovered to degrade compounds previously considered to be recalcitrant (Alexander 1999), and it is likely that more pathways will be discovered as our knowledge expands.

Creating stable anaerobic zones, areas where E < -0.2V, inhibits aerobic metabolic pathways. This is important when aerobic microbial decomposition of chemicals either does not proceed at all, or results in accumulations of toxic and carcinogenic intermediates. An example of this is the aerobic degradation of PCB's; some monohydroxylated PCBs are potent endocrine disrupters, and some PCB metabolites with a hydroxy group in the meta or para position have been reported to be involved in developmental neurotoxicity (Maltseva *et al.* 1999).

It is not easy to change the redox status of *in situ* systems, but it has been accomplished by adding a large amount of a substrate that rapidly degrades aerobically (Roberts *et al.* 1993). This consumes all of the ambient oxygen producing anaerobic conditions. Anaerobic conditions sufficient to support mineralization of the pesticide dinoseb were established in aerobic soils by adding potato processing by-product from a local factory, then flooding the area with phosphate buffer (Kaake *et al.* 1992). In this case, two contamination scenarios were studied. The first was chronic exposure to low dinoseb concentrations as the result of several decades of rinsing crop-dusters; the second was a highly concentrated dinoseb spill from storage barrels in transit. In the case of the acute spill, remediation was improved by addition of bacterial dinoseb degraders, which did not affect the degradation rate in the chronically exposed soil.

Recently, advances have been made with substances that release hydrogen in a controlled, time-released manner. Hydrogen-releasing compound (HRC) is a proprietary formulation of polylactate ester that slowly releases lactic acid when contacted with water. HRC was used successfully to create stable anaerobic subsurface zones to promote tetrachlorethene and trichloroethene mineralization (He *et al.* 2003).

In situ treatment: Bioaugmentation

Environmental remediation by bioaugmentation means adding supplemental microbes to an environment where native organisms are unable to degrade the contaminants present at the site. The supplemental microbial culture is selected for the ability to exploit organic wastes as carbon or nitrogen sources. Inoculation of soils has been applied for decades, but it has yielded variable results. Bacteria are sensitive to pH, nutrient starvation, osmotic stress, temperature, redox conditions, and soil characteristics (Poolman and Glaasker 1998). A review of the biotic and abiotic interactions in soil, soil properties, and the physiological status of the inoculant cells is found in van Veen *et al.* (1997). The application of bioaugmentation technology is site-specific and highly dependent on the microbial ecology and physiology of the environment.

Usually, it is necessary to grow the bacterial inoculum in a fermentor, transport the cells to the site, and then distribute them into the area to be remediated. Often, growing microbial cultures for quick production of cell mass is a prescription for inducing culture death during transport to the site, or shortly after introduction into the larger environment. During growth, transport, and distribution, there are many places that bacterial cells can lose activity or even viability.

While the biotechnology industries have effectively harnessed several microbial workhorses such as *Escherichia coli, Bacillus* spp., *Streptomyces* spp. and various yeast strains, these bacteria do not typically tolerate and metabolize high levels of pollutant compounds. Some of the biodegrading bacteria, strains of *Pseudomonas, Rhodococcus* and *Arthrobacter*, are significantly different in growth habit and nutritional requirements from industrial fermentation strains. Thus, fermentation methods to produce highly active and robust strains for bioremediation are likely to require significant modification from methods used with industrial strains. In particular, more work is needed for the development of reproducible methods to formulate microorganisms for deployment after extended periods of storage.

Formulation, storage, and delivery of bacteria for bioremediation

Formulation of bacteria with the intent of storing them until needed for an environmental remediation is currently an active area, and there is limited information on this currently. Freeze-drying, and vacuum- or spray-drying are common techniques used to preserve biological cultures. However, cells can be irreversibly damaged during dehydration treatments (Potts 1994, 2001). Bacteria produce a number of small molecules to protect themselves against adverse environmental conditions, and it may be that the best strategy involves growing and harvesting cells such that the culture self-produces those things it needs to survive.

There are three main parameters that are easily manipulated to induce bacteria to produce small protective molecules: temperature, osmotic pressure, and food supply (Denich *et al.* 2003, Lowder *et al.* 2000, Malwane and Deutch 1999, Zevenhuizen 1992). An example of a small protective molecule produced by bacterial cells is the non-reducing sugar trehalose. An excellent review of its properties is available (Crowe *et al.* 2001). There is evidence that other compounds may provide superior cellular protection than trehalose (Manzanera *et al.* 2002).

Freeze-drying is commonly used to preserve and store microbial cultures, allowing long-term maintenance and easy distribution (Billi *et al.* 2000). Freeze-dried bacteria have survived for at least 35 years (Kirsop and Snell 1984). However, this storage method was largely developed with *Escherichia coli* strains, environmental strains often prove difficult to revive following freeze-drying (Parthuisot *et al.* 2003). Some critical issues involved in the lack of survivability include the denaturation of sensitive proteins (Arguelles 2000, Reich and Morien 1982).

Spray drying and vacuum drying have less of a history in microbial engineering. Spray drying is a well established method of food preservation, and could be a suitable method for preserving microorganisms for bioremediation, as it allows large-scale culture dehydration at low cost. In spray drying, various bulking substances are used to carry and support the bacteria during drying and storage. Common bulking agents in the food industry are salt solutions ($MgSO_4$, K_2SO_4. and Na_2CO_3) or dairy products (nonfat skimmed milk). Outlet temperatures from the drying equipment that are too high are thought to limit bacterial viability during the drying process (Costa et al. 2002), and lignin-based spray-dried formulations of bacteria are reported with a shelf-life of up to 3 months at 30°C and up to 30 months at 4°C (Behle et al. 2003). However, it is likely that improvements can be obtained from optimization of the equipment for application in bioremediation. Two main questions need to be addressed: (1) how to quickly determine cell viability when a stored cell culture is sent to a site for remedial inoculation, and (2) how to maintain metabolic activity during storage and in the remediation environment.

Flow cytometry can be used to determine the ratio of live to dead cells in a bacterial culture. Moreover, cells can be dyed to discern more about their physiological state. Dyes can be used to detect total cells based on the detection of nucleic acids, others can differentiate between live from dead cells, and a third class detect specific metabolic activity of bacterial cells (Porter and Pickup 2000, Nebe-von-Caron et al. 2000, and Parthuisot et al. 2003). These methods require specialized equipment, and will thus not help in assessing viability or efficacy of an inoculant that has arrived on a remediation site.

While flow cytometry can be useful in determining the microbial health of cell cultures and aquatic samples, the question of viability of cultures after they have been introduced into a remediation environment is more difficult, as it involves the simultaneous identification of the specific strain that was introduced and its metabolic state. Recent advances into these questions have been successfully addressed by using recombinant strains that metabolically produce green fluorescent protein, which enables the cells to be monitored in situ (Lowder et al. 2000, Unge et al. 1999).

Release of genetically-engineered strains for bioaugmentation

There are many issues associated with the release of genetically-engineered, or recombinant, microbial species into an open environment.

Two such cases for the bioremediation of contaminated soil are described below. However, difficulties in obtaining permission to use genetically-engineered microorganism from government regulatory agencies has made companies reluctant to develop such strategies. Thus, most bioaugmentation situations have used naturally-occurring bacteria for which obtaining regulatory approval is relatively easy.

Tracking field-released bacteria using the *lux* gene casette

In 1996, the United States Environmental Protection Agency approved the first controlled field-release of a recombinant bacterium, *Pseudomonas fluorescens* HK44. *P. fluorescens* HK44 contained a *lux* gene casette that expresses proteins which emitted light upon exposure to naphthalene, salicylate, and other substituted naphthalene analogs. The bacterium was inoculated into intermediate-scale field lysimeters and population dynamics were monitored over two years. The study found that the standard selective plating technique overestimates the population of a single species within the larger community, and that bioluminescence is an accurate tool for monitoring populations of *lux*-containing microorganisms (Ripp *et al.* 2000).

This strain was originally developed as a real-time reporter for measuring salicylate and naphthalene catabolism, as it emits light in proportion to the reaction rate of these species. However, the light emission of the strain was found to vary over four orders of magnitude during a period of constant naphthalene degradation. It has been documented that a temperature change of 1°C or a pH change of 0.2 significantly changes the emission of light, and that the change can be calibrated, providing hope that a predictable luminescent response can be obtained (Dorn *et al.* 2003).

In situ bioaugmentation using non-viable cells

To our knowledge, there is only one record of recombinant bacteria being used in an open, field-scale remediation project in the United States. A spill of the herbicide atrazine occurred in a confined area, producing soil concentrations up to 29,000 ppm. This site was cleaned using chemically-crosslinked, non-viable, recombinant organisms engineered to over-produce atrazine chlorohyrolase, AtzA. AtzA catalyzes the dechlorination of atrazine, producing non-toxic hydroxyatrazine.

Initially, this field study compared the performance of biostimulation or bioaugmentation for atrazine removal from the contaminated soil. Control plots contained moistened soil; biostimulation plots received 300ppm phosphate; bioaugmentation plots received recombinant *E. coli* cells encapsulating AtzA; and combination plots received phosphate plus the enzyme-containing cells. After 8 weeks, atrazine levels declined 52% in plots containing recombinant *E. coli* cells, and 77% in combination plots. In contrast, atrazine levels in control and biostimulation plots did not decline significantly. These data indicate that genetically engineered bacteria overexpressing catabolic genes significantly increased degradation in soil heavily contaminated with atrazine (Strong *et al.* 2000).

Ex situ bioremediation: Pump and treat

Pump and treat is the most common treatment for water quality restoration and plume containment since the 1980's, and is currently used at three quarters of the Superfund sites treating ground water (National Research Council 1994). It usually involves pumping contaminated water to the surface for physical treatment, followed by release of the water into sewer treatment systems or re-injecting back into the water table. The effectiveness of pump-and-treat systems is influenced by site geology, and practical guidelines for evaluating a site for this treatment are available from the U.S. EPA (Environmental Protection Agency 1996).

Complications observed at sites using pump and treat technology are tailing and rebound. With tailing, contaminant concentrations initially decrease rapidly, but then degradation rates slow down. Eventually contaminant concentration can stabilize above the cleanup standard. Rebound is the increase in contaminant concentration that can occur after pumping has been discontinued. The reasons for tailing and rebound often involve chemical partitioning at the soil/water interfaces, and variability in ground water velocity (Berglund and Cvetkovic 1995).

Pump and treat with aerobic/anaerobic cycling

Recent research has indicated that some compounds, for example multiply-substituted aromatic rings, cannot be mineralized by single micro-organisms. In some of those cases, biodegradation can best be achieved by cycling through sequential anaerobic and aerobic treatment stages. Azo dyes provide an illustration of this point.

Azo dyes typically contain the azo group bridging two aromatic rings. They are often colored species with high extinction coefficients, and thus constitute an important class of dyes. They degrade very slowly in some environments. Anaerobic bacterial metabolism of azo dyes has been demonstrated to occur via reductive cleavage of the azo bond, which results in the formation of colorless amines (Fig. 2). Azo dyes are largely non-toxic, but some of the aromatic amines generated by anaerobic metabolism are carcinogenic. These amines typically resist anaerobic degradation, but readily biodegrade aerobically. Therefore, anaerobic/aerobic reactor processes for the treatment of wastewaters may speed azo dye mineralization substantially.

A representative azo reaction is shown in Figure 2. In a system where the anaerobic and aerobic cycles are temporally separated, the reactors are started anaerobically, with any of a wide variety of bacterial cultures and a primary electron donor. The exact species of the bacteria and the electron donor are not critical, many different bacterial species and donors are capable of mediating this reaction (Stolz 2001). *Shigella dysenteriae* (Ghosh *et al.* 1992) *Escherichia coli* (Ghosh *et al.* 1993) and *Lactobacillus* (Suzuki *et al.* 2001) strains have all been shown to catalyze reduction of azo dyes. The redox potential for azo group reduction of the dyes is approximately 100mV allowing for potential reduction by many redox mediators in bacterial cells (Spain 1995).

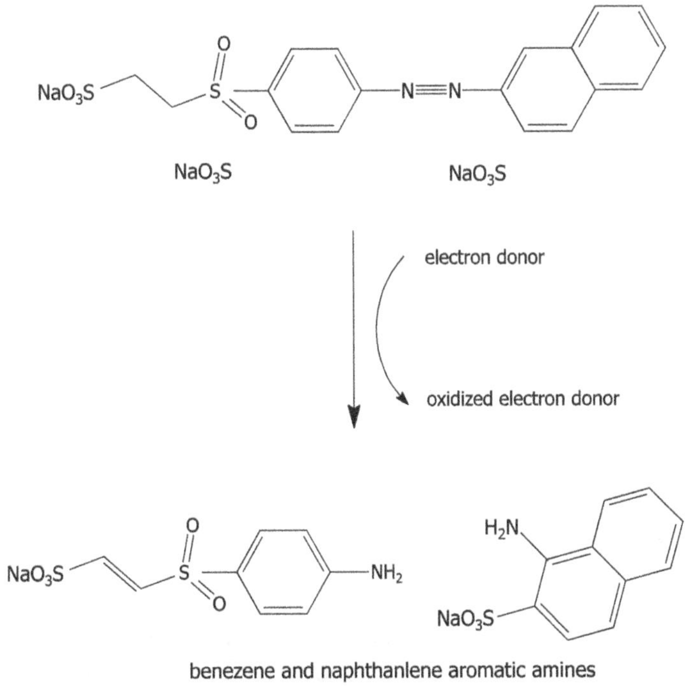

benezene and naphthanlene aromatic amines

Figure 2. Simplified diagram showing metabolism of azo dyes. The reaction shown typically occurs in anaerobic environments. The aromatic amines thus generated can be metabolized by oxygenase-catalyzed reactions carried out by aerobic bacteria.

In one practical bioremediation scenario, decolorization of the dyes can be followed to follow anaerobic biodegradation. When the solution loses 70-100% of its color, typically after 12-24 hours, the azo dyes are known to have been converted into aromatic amines. Aromatic amines are typically resistant to degradation under anaerobic conditions, so aeration is begun. Aerobic conditions are established and enrichment bacterial cultures capable of degrading the appropriate amines are inoculated into the reactor (Tan *et al.* 2000). This stage of the reaction generally goes more quickly than the anaerobic cycle, and is complete in approximately 8 hours. Then the reactor can be emptied and the cycle can begin again.

Pump and treat with hollow fiber biofilm reactor

An issue with pump and treat technology is that it is often unreliable for removing oxidized contaminants from water systems. In the case of perchlorate, (ClO_4^-), pump and treat technology has been coupled with biological treatments to enhance its efficacy. A hollow-fiber membrane-biofilm reactor (MBfR) system utilizes bacterially catalyzed reduction to decrease chemical concentrations (Lee and Rittmann 2000). This is a departure from conventional use of membrane technology for physical separations (Rittmann and McCarty 2001).

This MBfR uses hydrogen gas (H_2) as an electron donor, and the groundwater contaminant acts as the electron acceptor. Ground water is pumped from the ground and over bundles of composite, hydrophobic hollow-fiber membrane tubes filled with H_2 gas (Fig. 3). The H_2 gas diffuses out through the walls of the membrane where it meets the contaminant in the water. A biofilm of bacteria forms on the waterside of the membrane, where they oxidize H_2 and reduce the oxidized contaminants to gain energy for growth (Nerenberg *et al.* 2002).

The MBfR avoids both contamination from overdose and ineffective bioremediation that result if the electron donor is provided at a non-optimal concentration. The biofilm is self-regulating: when contaminant load in the water increases, the bacteria utilize more hydrogen, and diffusion from the inside of the tube increases. The opposite also occurs, automatically balancing hydrogen-contaminant stoichiometry in a very safe manner. The reactor size is relatively small, because the biofilm on the hollow tube surface has a large specific area. Finally, the bacteria do not pose a fouling problem, because water does not pass the membrane surface.

However, this reactor has trouble maintaining an adequate biofilm if perchlorate is the sole groundwater contaminant. This conclusion is drawn from examining the equation for biomass generation (Rittmann and McCarty 2001) :

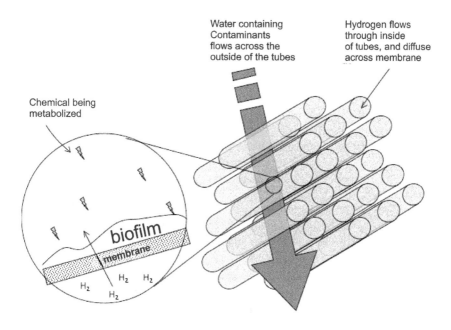

Figure 3. Configuration of a hollow-fiber membrane–biofilm reactor (MBfR).

$$\frac{dX}{dt} = q_{max} \frac{S}{S + K} YX - bX$$

X   = biomass concentration = [g cells/L³]

S   = rate -limiting substrate concentration = [g substrate/L³],

q_{max} = maximum specific substrate utilization rate = [g substrate/g cells/minute]

K   = half-maximum-substrate-utilization constant = [g substrate/L³]

Y   = biomass true yield = [g cells/g substrate], and

b   = endogenous decay rate = [1/minute].

When S is smaller than the positive term on the right side of the equation may become smaller than the negative term, and biomass declines. The minimum concentration that can support steady-state biomass for a continuous suspended or biofilm system S_{min} is (Rittmann and McCarty 2001):

$$S_{min} = \frac{Kb}{Yq_{max} - b}$$

Fortunately, there is often co-contamination, and other molecules that are primary electron acceptors contribute to biomass generation. Simultaneous perchlorate and nitrate reduction has worked very well in a pilot-scale system operating in California since 2002. In this system, a 0.6 L/min MBfR with a 55 -minute retention time reduced 55 mg/L perchlorate to less than 4 mg/L, which is the California Action Level. Nitrate levels were concurrently reduced from 5.5 mg/L to less than 0.02 mg N/L (Nerenberg *et al.* 2004).

The MBfR has successfully been demonstrated to remove nitrate (NO_3^-) and nitrite (NO_2^-), chlorinated solvents like trichloroethene (TCE) and dichloromethane (DCM), explosives like trinitrotoluene (TNT), and perchlorate (ClO_4^-). Perchlorate is used in the manufacture of rocket propellants and has been linked to thyroid gland dysfunction at μg/L levels. The MbfR has also been effective to detoxify bromate generated from ozone disinfection of water, and selenate leached from some agricultural soils (Rittmann and Evans 2003).

REFERENCES

Alexander, M. 1999. *Biodegradation and Bioremediation*, Academic Press, San Diego.

Arguelles, J.C. 2000. Physiological roles of trehalose in bacteria and yeasts: a comparative analysis. *Arch. Microbiol.* 174: 217-224.

Behle, R.W., P. Tamez-Guerra, and M.R. McGuire. 2003. Field activity and storage stability of *Anagrapha falcifera* nucleopolyhedrovirus (AfMNPV) in spray-dried lignin-based formulations. *J. Econ. Entomol.* 96: 1066-1075.

Berglund, S., and V. Cvetkovic. 1995. Pump and treat remediation of heterogeneous aquifers: Effects of rate-limited mass transfer. *Ground Water* 33: 675-685.

Billi, D., D.J. Wright, R.F. Helm, T. Prickett, M. Potts, and J.H. Crowe. 2000. Engineering desiccation tolerance in *Escherichia coli*. *Appl. Environ. Microbiol.* 66: 1680-1684.

Bradley, P.M., F.H. Chapelle, and D.R. Lovley. 1998. Humic acids as electron acceptors for anaerobic microbial oxidation of vinyl chloride and dichloroethene. *Appl. Environ. Microbiol.* 64: 3102-3105.

Brown, S., R.L. Chaney, J.G. Hallfrisch, and Xue, Q. 2003. Effects of biosolids processing on lead bioavailability in an urban soil. *J. Environ. Qual.* 32: 100-108.

Cervantes, F.J., L. Vu-Thi-Thu, G. Lettinga, and J.A. Field. 2004. Quinone-respiration improves dechlorination of carbon tetrachloride by anaerobic sludge. *Appl. Microbiol. Biotechnol.* 64: 702-711.

Chakraborty, R., and J.D. Coates, 2004. Anaerobic degradation of monoaromatic hydrocarbons. *Appl. Microbiol. Biotechnol.* 64: 437-446.

Cookson, J.T. 1995. *Bioremediation Engineering : Design and Application,* McGraw-Hill, New York.

Costa, E., N. Teixido, J. Usall, E. Fons, V. Gimeno, J. Delgado, and I. Vinas. 2002. Survival of *Pantoea agglomerans strain* CPA-2 in a spray-drying process. *J. Food Prot.* 65: 185-191.

Crowe, J.H., L.M. Crowe, A.E. Oliver, N. Tsvetkova, W. Wolkers, and F. Tablin. 2001. The trehalose myth revisited: introduction to a symposium on stabilization of cells in the dry state. *Cryobiology* 43: 89-105.

Denich, T.J., L.A. Beaudette, H. Lee, and J.T. Trevors. 2003. Effect of selected environmental and physico-chemical factors on bacterial cytoplasmic membranes. *J. Microbiol. Methods* 52: 149-182.

Dorn, J.G., R.J. Frye, and R.M. Maier. 2003. Effect of temperature, pH, and initial cell number on *luxCDABE* and *nah* gene expression during naphthalene and salicylate catabolism in the bioreporter organism *Pseudomonas putida* RB1353. *Appl. Environ. Microbiol.* 69: 2209-2216.

Dua, M., A. Singh, N. Sethunathan, and A.K. Johri. 2002. Biotechnology and bioremediation: successes and limitations. *Appl. Microbiol. Biotechnol.* 59: 143-152.

Ellis, L.B., B.K. Hou, W. Kang, and L.P. Wackett. 2003. The University of Minnesota Biocatalysis/Biodegradation Database: post-genomic data mining. *Nucleic Acids Res.* 31: 262-265.

Environmental Protection Agency. 1996. Pump and treat groundwater remediation: A guide for decision makers and practitioners.

Ghosh, D.K., S. Ghosh, P. Sadhukhan, A. Mandal, and J. Chaudhuri. 1993. Purification of two azoreductases from *Escherichia coli* K12. *Indian J. Exp. Biol.* 31: 951-954.

Ghosh, D.K., A. Mandal, and J. Chaudhuri. 1992. Purification and partial characterization of two azoreductases from *Shigella dysenteriae* type 1. *FEMS Microbiol. Lett.* 77: 229-233.

Hapeman, C.J., L.L. McConnell, C.P. Rice, A.M. Sadeghi, W.F. Schmidt, G.W. McCarty, J.L. Starr, P.J. Rice, J.T. Angier, and J.A. Harman-Fetcho. 2003. Current United States Department of Agriculture-Agricultural Research Service research on understanding agrochemical fate and transport to prevent and mitigate adverse environmental impacts. *Pest. Manag. Sci.* 59: 681-690.

He, J., K.M. Ritalahti, K.L. Yang, S.S. Koenigsberg, and F.E. Loffler. 2003. Detoxification of vinyl chloride to ethene coupled to growth of an anaerobic bacterium. *Nature* 424: 62-65.

Hou, B.K., L.P. Wackett, and L.B.M. Ellis. 2003. Microbial pathway prediction: a functional group approach. *J. Chem. Inf. Comput. Sci.* 43: 1051-1057.

Kaake, R.H., D.J. Roberts, T.O. Stevens, R.L. Crawford, and D.L. Crawford. 1992. Bioremediation of soils contaminated with the herbicide 2-*sec*-butyl-4,6-dinitrophenol (dinoseb). *Appl. Environ. Microbiol.* 58: 1683-1689.

Kirsop, B.E., and J.J.S. Snell. 1984. *Maintenance of Microorganisms: A Manual of Laboratory Methods*, Academic Press, London; Orlando.

Lee, K.C., and B.E. Rittmann. 2000. A novel hollow-fiber membrane biofilm reactor for autohydrogenotrophic denitrification of drinking water. *Water Sci. Technol.* 41: 219-226.

Lee, M.D., J.M. Odom, and R.J. Buchanan, Jr. 1998. New perspectives on microbial dehalogenation of chlorinated solvents: insights from the field. *Annu. Rev. Microbiol.* 52: 423-452.

Lowder, M., A. Unge, N. Maraha, J.K. Jansson, J. Swiggett, and J.D. Oliver. 2000. Effect of starvation and the viable-but-nonculturable state on green fluorescent protein (GFP) fluorescence in GFP-tagged *Pseudomonas fluorescens* A506. *Appl. Environ. Microbiol.* 66: 3160-3165.

Maltseva, O.V., T.V. Tsoi, J.F. Quensen, 3rd, M. Fukuda, and J.M. Tiedje. 1999. Degradation of anaerobic reductive dechlorination products of Aroclor 1242 by four aerobic bacteria. *Biodegradation* 10: 363-371.

Malwane, S., and C.E. Deutch. 1999. Adaptive characteristics of salt-induced myceloids of *Arthrobacter globiformis*. *Antonie Van Leeuwenhoek* 75: 335-344.

Manzanera, M., A. Garcia de Castro, A. Tondervik, M. Rayner-Brandes, A.R. Strom, and A. Tunnacliffe. 2002. Hydroxyectoine is superior to trehalose for anhydrobiotic engineering of *Pseudomonas putida* KT2440. *Appl. Environ. Microbiol.* 68: 4328-4333.

Maymo-Gatell, X., T. Anguish, and S.H. Zinder. 1999. Reductive dechlorination of chlorinated ethenes and 1, 2-dichloroethane by *Dehalococcoides ethenogenes* 195. *Appl. Environ. Microbiol.* 65: 3108-3113.

Maymo-Gatell, X., Y. Chien, J.M. Gossett, and S.H. Zinder. 1997. Isolation of a bacterium that reductively dechlorinates tetrachloroethene to ethene. *Science* 276: 1568-1571.

Mogensen, A.S., J. Dolfing, F. Haagensen, and B.K. Ahring. 2003. Potential for anaerobic conversion of xenobiotics. *Adv. Biochem. Eng. Biotechnol.* 82: 69-134.

National Research Council. Committee on Ground Water Cleanup Alternatives. 1994. Alternatives for ground water cleanup. National Academy Press, Washington, DC, 315 pp.

Nebe-von-Caron, G., P.J. Stephens, C.J. Hewitt, J.R. Powell, and R.A. Badley. 2000. Analysis of bacterial function by multi-colour fluorescence flow cytometry and single cell sorting. *J. Microbiol. Methods* 42: 97-114.

Nerenberg, R., B.E. Rittmann, T.E. Gillogly, G.E. Lehman, and S.S. Adham. 2004. Perchlorate reduction using a hollow-fiber membrane biofilm reactor: kinetics, microbial ecology, and pilot-scale studies. *Water Sci. Technol.* (in press).

Nerenberg, R., B.E. Rittmann, and I. Najm. 2002. Perchlorate reduction in a hydrogen-based membrane-biofilm reactor. *Amer. Waterworks Assoc.* 94: 103-114.

Parales, R.E., N.C. Bruce, A. Schmid, and L.P. Wackett. 2002. Biodegradation, biotransformation, and biocatalysis (b3). *Appl. Environ. Microbiol.* 68: 4699-4709.

Parthuisot, N., P. Catala, P. Lebaron, D. Clermont, and C. Bizet. 2003. A sensitive and rapid method to determine the viability of freeze-dried bacterial cells. *Lett. Appl. Microbiol.* 36: 412-417.

Pieper, D.H., and W. Reineke. 2000. Engineering bacteria for bioremediation. *Curr. Opin. Biotechnol.* 11: 262-270.

Poolman, B., and E. Glaasker. 1998. Regulation of compatible solute accumulation in bacteria. *Mol. Microbiol.* 29: 397-407.

Porter, J., and R.W. Pickup. 2000. Nucleic acid-based fluorescent probes in microbial ecology: application of flow cytometry. *J. Microbiol. Methods* 42: 75-79.

Potts, M. 2001. Desiccation tolerance: a simple process? *Trends Microbiol.* 9: 553-559.

Potts, M. 1994. Desiccation tolerance of prokaryotes. *Microbiol. Rev.* 58: 755-805.

Reich, R.R., and L.L. Morien. 1982. Influence of environmental storage relative humidity on biological indicator resistance, viability, and moisture content. *Appl. Environ. Microbiol.* 43: 609-614.

Richnow, H.H., A. Eschenbach, B. Mahro, R. Seifert, P. Wehrung, P. Albrecht, and W. Michaelis. 1998. The use of ^{13}C-labelled polycyclic aromatic hydrocarbons for the analysis of their transformation in soil. *Chemosphere* 36: 2211-2224.

Ripp, S., D.E. Nivens, C. Werner, and G.S. Sayler. 2000. Bioluminescent most-probable-number monitoring of a genetically engineered bacterium during a long-term contained field release. *Appl. Microbiol. Biotechnol.* 53: 736-741.

Rittmann, B., and J. Evans. 2003. Hydrogen-based membrane-biofilm reactor solves thorny problems of oxidized contaminants. Worldwide web. URL = http://www.aptwater.com/_news/article03_6.html.

Rittmann, B.E., and P.L. McCarty. 2001. *Environmental Biotechnology: Principles and Applications*, McGraw-Hill, Boston.

Roberts, D.J., R.H. Kaake, S.B. Funk, D.L. Crawford, and R.L. Crawford. 1993. Anaerobic remediation of dinoseb from contaminated soil. An on-site demonstration. *Appl. Biochem. Biotechnol.* 39-40: 781-789.

Samanta, S.K., O.V. Singh, and R.K. Jain. 2002. Polycyclic aromatic hydrocarbons: environmental pollution and bioremediation. *Trends Biotechnol.* 20: 243-248.

Spain, J.C. 1995. *Biodegradation of Nitroaromatic Compounds*, Plenum Press, New York.

Stolz, A. 2001. Basic and applied aspects in the microbial degradation of azo dyes. *Appl. Microbiol. Biotechnol.* 56: 69-80.

Strong, L.C., H. McTavish, M.J. Sadowsky, and L.P. Wackett. 2000. Field-scale remediation of atrazine-contaminated soil using recombinant *Escherichia coli* expressing atrazine chlorohydrolase. *Environ. Microbiol.* 2: 91-98.

Suzuki, Y., T. Yoda, A. Ruhul, and W. Sugiura. 2001. Molecular cloning and characterization of the gene coding for azoreductase from *Bacillus sp.* OY1-2 isolated from soil. *J. Biol. Chem.* 276: 9059-9065.

Tan, N.C. A. Borger, P. Slenders, A. Svitelskaya, G. Lettinga, and J.A. Field. 2000. Degradation of azo dye Mordant Yellow 10 in a sequential anaerobic and bioaugmented aerobic biore. *Water Sci. Technol.* 42: 337-344.

Unge, A., R. Tombolini, L. Molbak, and J.K. Jansson. 1999. Simultaneous monitoring of cell number and metabolic activity of specific bacterial populations with a dual gfp-*luxAB* marker system. *Appl. Environ. Microbiol.* 65: 813-821.

Van Veen, J.A., L.S. Van Overbeek, and J.D. Van Elsas. 1997. Fate and activity of microorganisms introduced into soil. *Microbiol. Mol. Biol. Rev.* 61: 121-135.

Wackett, L.P., and L.B.M. Ellis. 1999. Predicting biodegradation. *Environ. Microbiol.* 1: 119-124.

Watanabe, K., H. Futamata, and S. Harayama. 2002. Understanding the diversity in catabolic potential of microorganisms for the development of bioremediation strategies. *Antonie Van Leeuwenhoek.* 81: 655-663.

Widada, J., H. Nojiri, and T. Omori. 2002. Recent developments in molecular techniques for identification and monitoring of xenobiotic-degrading bacteria and their catabolic genes in bioremediation. *Appl. Microbiol. Biotechnol.* 60: 45-59.

Wiedemeier, T.H. 1998. Technical protocol for evaluating natural attenuation of chlorinated solvents in ground water. National Risk Management Research Laboratory, Office of Research and Development, U.S. Environmental Protection Agency, Cincinnati, Ohio. World wide web, URL = http://www.epa.gov.superfund/resources/gwdocs/protocol.htm.

Zevenhuizen, L.P. 1992. Levels of trehalose and glycogen in *Arthrobacter globiformis* under conditions of nutrient starvation and osmotic stress. *Antonie Van Leeuwenhoek.* 61: 61-68.

Index